Introduction to Statistical Decision Theory

Utility Theory and Causal Analysis

Introduction to Statistical Decision Theory

Utility Theory and Causal Analysis

Silvia Bacci
Bruno Chiandotto

CRC Press
Taylor & Francis Group
Boca Raton London New York

CRC Press is an imprint of the
Taylor & Francis Group, an **informa** business

A CHAPMAN & HALL BOOK

CRC Press
Taylor & Francis Group
6000 Broken Sound Parkway NW, Suite 300
Boca Raton, FL 33487-2742

First issued in paperback 2021

© 2020 by Taylor & Francis Group, LLC
CRC Press is an imprint of Taylor & Francis Group, an Informa business

No claim to original U.S. Government works

ISBN-13: 978-1-03-209175-4 (pbk)
ISBN-13: 978-1-138-08356-1 (hbk)

Visit the Taylor & Francis Web site at
http://www.taylorandfrancis.com

and the CRC Press Web site at
http://www.crcpress.com

To my parents, Mauro and Giovanna.
And to my decisions without uncertainty:
Massimo, Lorenzo, Serena.
S. Bacci

In memory of my first teacher
of probability and statistics: Giuseppe Pompilj.
B. Chiandotto

Contents

Authors

Silvia Bacci is Assistant Professor of Statistics at the Department of Statistics, Computer Science and Applications "G. Parenti", University of Florence (Italy). Her research interests are addressed to statistical decision theory, with focus on utility theory, and latent variable models, with focus on item response theory models, latent class models, and models for longitudinal and multilevel data.

Bruno Chiandotto is adjunct Full Professor of Statistics at the Department of Statistics, Computer Science and Applications "G. Parenti", University of Florence (Italy). He is mainly interested in the definition and estimation of linear and nonlinear statistical models, multivariate data analysis, customer satisfaction, causal analysis, statistical decision theory and utility theory. A large part of his research activity has been carried out under projects funded by international, national and local institutions.

Preface

"The adoption of the position outlined in this paper would result in a widening of the statistician's remit to include decision-making, as well as data collection, model construction and inference. Yet it also involves a restriction in their activity that has not been adequately recognized. Statisticians are not masters in their own house. Their task is to help the client to handle the uncertainty that they encounter. The 'you' of the analysis is the client, not the statistician."

D. V. Lindley, *The Philosophy of Statistics*

Statistics is the discipline that deals with collection and scientific treatment of data. Statistics transforms data into information and, subsequently, into knowledge. Whenever the knowledge is used for decision-making purposes, the *decision theory* frame is involved. *Statistical decision theory* is the new scientific discipline that comes from merging statistics and decision theory. Many authors [12, 45, 199] state that statistical inference and statistical decision theory must be considered as distinct disciplines, but other authors (see, for instance, [40, 144, 145, 147, 190, 217]) consider statistical decision theory as a natural and necessary generalization of statistical inference. Furthermore, the decision-making approach, through the combination of various theories of statistical inference, avoids the dogmatisms that can lead to paradoxical situations, is free from logical errors, is more effective in practical contexts, and can successfully address a wider range of problems than the traditional approaches.

The aim of this book is to provide the theoretical background to approach decision theory from a statistical perspective, taking into account both the traditional approaches in terms of value theory and expected utility theory, other than the generalized utility theories, and the recent developments in terms of causal inference. The integration among decision theory, classical and Bayesian statistical inference, and causal inference is illustrated step-by-step, showing how sample evidence, prior information on the states of nature, and effects of actions of the decision-maker affect the formalization and the solution of a (statistical) decisional problem. Furthermore, some examples and case studies that rely on actual decisional problems are illustrated throughout the book.

The book is specifically designed to appeal to students who intend to acquire a broader knowledge of statistics with respect to what is usually taught in the upper level university curricula. It provides a rethink of statistical science from the decisional point of view. The book is also devoted to practitioners worldwide involved in decision-making tasks within several fields of knowledge. For such target populations, the book will be a reference work that will provide both the theoretical framework behind decision-making and, through illustrative examples and case studies, a practical guide to carry out decisional processes supported by a quantitative approach. Case studies aim at showing the potentialities of the statistical approach in some unconventional decisional contexts (e.g., banking sector, management of a public service) and some of them rely on actual work and study experiences of the two authors of this book.

The prerequisites to properly understand the content of this book are at the level of a conventional full-year mathematics course, whereas no previous course in probability and statistics is required, as a synthetic introduction to probability and statistical inference is provided in Chapter 2. However, the reader who is totally unfamiliar with statistics might find it helpful to consult other introductory statistical books, such as volumes by Mood, Graybill and Boes [158], Casella and Berger [32], and Olive [161].

The book is organized into six chapters. Chapter 1 addresses preliminary aspects related to decision theory and key concepts that will be employed in the subsequent chapters. After a brief illustration of the historical origins of decision theory, the fundamentals of normative theory, descriptive theory, and prescriptive theory are introduced, together with basic elements of modern decision theory, that is, concepts of value function and utility function. Next, the distinction of decisions under certainty, risk, and uncertainty is illustrated, in accordance with the informational background available to the decision maker. Particular attention is then devoted to decisional criteria under uncertainty (i.e., Wald's criterion, max-max criterion, Hurwicz's criterion, Savage's criterion, Laplace's criterion). The remainder of the chapter deals with the relationship between decision theory and (descriptive and, mainly,) inferential statistics, which in turn distinguishes in classical and Bayesian statistical inference. To conclude, a clarification is provided to discern among classical decision theory, Bayesian decision theory, classical statistical decision theory, and Bayesian statistical decision theory.

Chapter 2 examines the basic concepts of probability theory and statistical inference and it is designed mainly for readers lacking a solid statistical background; it can be skipped by other readers. The first part of the chapter focuses on Kolmogorov's axioms and some relevant theorems with special attention to Bayes' rule. Attention is also devoted to the concept of random variable and to the detailed description of some relevant families of random variables. In the second part of the chapter, we deal with sample distributions and the concept of likelihood function. Then, we focus on the typical inferential problems of point and interval estimation and hypothesis testing, which are investigated both from the frequentist and the Bayesian perspectives. The last part of the chapter carries out a thorough investigation into regression and structural equation modeling.

Chapter 3 is aimed at in-depth illustration of normative decision theories that concern how decisions should be taken by ideally rational individuals. We first introduce the problem of *decisions under certainty* (i.e., when consequences of actions are known), describing a set of axioms of rational behavior and the concept of value function. Then, decisions under risk (i.e., when consequences of actions are not known, but prior information is available) are investigated. In detail, we speak about *decisions under risk* when prior probabilities of states of nature are given (objectively or subjectively defined). In the former case, we refer to the expected utility theory formulated by von Neumann and Morgenstern, while in the latter case we explicitly refer to the subjective expected utility theory formulated by Savage. In both cases, we illustrate the rational behavior axioms and demonstrate the existence and uniqueness (upto a positive linear transformation) of the utility function, which is a function defined along a quantitative scale that allows representing the decision maker's preferences. The remainder of the chapter is devoted to describing several empirical failures of the basic axioms of expected utility theory. Then, alternative utility theories are illustrated that are based on weak formulations of some of the basic axioms, with a special focus on the rank-dependent utility theory and the prospect and cumulative prospect theories.

Chapter 4 focuses on the practical issue concerning the elicitation of the utility function. First, the different attitudes towards risk are introduced, distinguishing among risk aversion, risk seeking, and risk neutrality, and the relationship with the shape (i.e., convex, concave, S-shaped, inverted S-shaped) of the utility function is investigated. Next, the main approaches to assess the utility function are described. The classical paradigm is first illustrated, which is based on using complete information for the utility elicitation. In such a setting, two main types of approaches are distinguished: the standard gamble methods (e.g., lottery equivalent method, certainty equivalent method, probability equivalent method) and the paired gamble methods (e.g., preference comparisons method, probability equivalent-paired method, tradeoff method). Then, alternative elicitation procedures are described that rely on partial preference information, such as approaches based on the concepts of Bayesian elicitation and double expected utility, as well as on clustering of utility functions on the basis of some easily observable characteristics of decision makers. Hints are also provided for the problem of elicitation when the decision maker is a collective body consisting of a multitude of individuals.

Chapter 5 reviews the decisional setting of previous chapters from a genuine statistical perspective. The chapter starts retrieving the main concepts about decision-making criteria when no information (*classical decision theory*) and only prior information (*Bayesian decision theory*) on the probability distribution of the states of nature are available. It proceeds with the fundamentals of *classical statistical decision theory*, which distinguishes from classical decision theory for the presence of sample information to improve the knowledge of the decision maker about the states of nature. In such a context, the relevant concepts of decision function and risk (expected loss) are introduced. Then, attention turns towards the *Bayesian statistical decision theory*, where information coming from sample surveys and information originating from prior knowledge are available and come together in the decisional process. In such a context, the concepts of expected risk and posterior probabilities are introduced and the equivalence between decisional analysis in normal form and in extensive form is illustrated. The last part of the chapter is devoted to the costs involved in the acquisition of sample information. We introduce the concepts of expected value of perfect information, expected value of sample information, and net gain associated with a sample survey and we formalize them in the decisional process.

Chapter 6 aims at providing an overview of the links between causality, statistics, and decision theory. After having recalled the historical and philosophical origins of causality, we focus on the debate concerning the difference between correlation and causality and between statistical inference and causal inference. In such a setting, we outline the relevance of the concept of manipulation of a variable (or intervention on a variable) as opposite to the mere conditioning as a distinctive element to interpret in a causal sense the results of a statistical analysis. Then, the idea that manipulation of variables can modify the probability distribution of the states of nature makes it necessary to revise the causal analysis from a decision-making point of view. The Bayesian statistical decisional setting treated in Chapter 5 and based on prior information and sample data is therefore extended to the *causal (Bayesian) statistical decisional setting* by allowing that posterior probabilities of the states of nature change in accordance with actions chosen by the decision maker. The central role of structural equation modeling and path analysis is well outlined throughout the chapter.

Acknowledgments: The authors thank students who during a period of 25 years (from 1985 to 2010), contributed to improving the lecture notes that B. Chiandotto adopted in the graduate course of Statistical Decision Theory and that have been used as a starting point for this book. The authors are also grateful to all the colleagues in the Department of Statistics, Computer Science, and Applications "G. Parenti" of the University of Florence (IT) and to the anonymous referee for their valuable suggestions. We also wish to thank the RAI - Radiotelevisione Italiana for the permission to use an adaptation of the Reports written on behalf of the RAI by one of the authors of this book. Finally, we address a special thanks to the Editorial Team of the CRC Press for its valuable support in the production of this book.

1

Statistics and decisions

CONTENTS

1.1 Introduction

This chapter provides a general overview of topics related to decision theory and statistics that will be in-depth analyzed throughout the subsequent chapters.

We first introduce some basic concepts in the field of decision theory, distinguishing between the normative and the descriptive approaches. We also outline how a prescriptive approach may represent a satisfactory compromise to compound theoretical rules of rational behavior, which are fundamentals of normative decision theory, with empirical evidence that characterizes the descriptive decision theory. We also point out the focus of this book for individual and rational decisions as opposed to group and moral decisions.

Next, attention focuses on the fundamentals of modern decision theory, also known as value theory or utility theory. The elements that are involved in any decisional setting are defined, that is, actions, states of nature, and consequences or outcomes that depend on the interaction between actions and states of nature. In addition, concepts of value function and utility function are introduced to represent the preferences of a decision-maker for alternative actions. Finally, the decision table and the decision tree are introduced as instruments to describe a decisional problem.

The relation between decisions and information is then investigated, distinguishing the decisional criteria according to the information that the decision maker possesses about the states of nature. More precisely, we introduce the distinction among decisions under certainty, when the state of nature is known; decisions under risk, when a probability distribution about the states of nature is given or may be subjectively elicited by the decision maker; decisions under uncertainty, when no information about the plausibility of the states of nature is given. Special attention is devoted to decisional criteria under uncertainty, that is, the Wald's criterion, the max-max criterion, the Hurwicz's criterion, the Savage's criterion, and the Laplace's criterion.

The remainder of the chapter discusses the relations between statistics and decision theory. First, an overview of the statistics discipline is provided, distinguishing between descriptive and inferential statistical methods and between classical and Bayesian approaches. Second, a classification of decision theory is provided according to the information available about the states of nature (i.e., no information, prior information, sample information, and posterior information). We distinguish among classical decision theory, when no information is given, Bayesian decision theory when only prior information is available, classical statistical decision theory when only sample evidence is available, and Bayesian statistical decision theory when both sample evidence and prior information are explicitly taken into account. In conclusion, the opportunity of a decision-making approach to statistics is outlined to favor the transformation of raw data into useful information finalized to take decisions.

1.2   Decision theory

Decision theory is concerned with processes of decision-making. Through the analysis of the behavior of actors (individuals or groups) involved in a decisional process, it is possible to examine and study how decision makers take or should take decisions.

Making decisions is an everyday activity in many professions and sciences that involves challenging aspects, and this process is of keen interest to researchers from different scientific fields: philosophy and logic, mathematics and statistics, psychology and sociology, economics, and so on.

Applications of the theory range from abstract speculations, relating to ideally rational individuals, to the resolution of specific decision-making problems. Decision theorists investigate the logical consequences of different decision-making rules or explore the logical-mathematical aspects of different descriptions of rational behaviors. On the contrary, applied researchers are interested in analyzing how decisional processes take place in practical situations.

In this perspective, we may distinguish decision theory in *normative decision theory* and *descriptive decision theory*[1]. Normative decision analysts define behavioral rules concerning how decisions should be taken to maximize the well-being of the decision maker. Descriptive decision analysts study how

---

[1]A further specification is that of prescriptive decision theory [13].

decisions are actually taken in practical contexts. This distinction is useful but rather artificial, being the actual way of making decisions certainly relevant for setting theoretical rules, and, in turn, theoretical rules represent an essential element to evaluate the observed behavior of decision makers.

Throughout this book attention will be focused on the essential elements of normative decision theory, whereas descriptive decision theory will be of no direct relevance, as it is the subject of specific disciplines, such as psychology, sociology and, in some respects, economics. Moreover, a series of constraints and conditioning that emerge from the analysis of actual decisional processes will be also taken into account (see Sections 3.5 and 3.6). This specific development of the decision theory that merges the normative and the descriptive approaches is known as *prescriptive decision theory*.

Another relevant distinction in decision theory is that between *individual decisions* and *group decisions*. Individual decisions refer to single individuals as well as to collective bodies (or teams or pools) composed of a multitude of individuals that act as a single one. Indeed, companies, associations, parties, nations, regions, universities, and so on, take individual decisions whenever they aim at achieving a common purpose for the organization. Instead, group decisions are involved when single individuals belonging to the same organisation (e.g., company, association, party, nation, region, university) express different opinions and different purposes with respect to the group's goals or priorities.

The most important part of the research related to group decision theory is traditionally focused on the development of common strategies to manage the components of the group and to distribute resources within the group. In such a context, ethical and moral aspects often assume great relevance. Indeed, in the context of group decisions, it is always possible for an ideally rational member of the group to find it convenient to violate the strategy common to the other members of the group in order to achieve his/her own specific interests. Instead, in the individual decision theory ethical and moral aspects have no relevance and research focuses on the problem of how individuals (defined in the broad sense specified above) can favor their own interests.

A particularly relevant topic related to the ethical aspects of decision theory concerns the notion of rational individual and the relation between moral decisions and rational decisions: in other words, are moral actions also rational actions? Do rational decision makers build ethical societies? Philosophers have failed to provide a satisfactory answer to these questions, until, within the framework of modern decision theory (Section 1.3), specific models of rationality and specific principles of social choice have been developed.

From what illustrated above it is now clear why some philosophers have been fascinated by decision theory: the theory at issue is not limited to applications in traditional philosophical problems, but the theory itself is imbued with philosophical problems (even if philosophers are more interested in the application of decision theory to philosophical problems rather than in the analysis of internal philosophical problems). In this book the philosophical aspects involved in the decision theory are neglected and the main interest focuses on the individual decision theory, leaving the group decision theory with its ethical questions a matter for philosophers rather than statisticians.

1.3   Value theory and utility theory

The foundations of *modern decision theory*, more commonly known as *value theory* or, more generally, *utility theory*, date back to the work of 1947 by Von Neumann and Morgenstern [213]. As will be deeply illustrated in Chapter 3, the two authors introduce a set of axioms concerning the behavior of a rational decision maker and show how, on the basis of these axioms, a real value function, known as *value function* or *utility function* in accordance with the decisional setting, may be defined to represent the decision maker's preferences. The maximum well-being of the decision maker is achieved through the action corresponding to the maximum of the value or utility function.

Several criticisms have been addressed to the utility theory, mainly summarized in the following two points:

1. the rational behavior axioms on which the modern decision theory is built are often empirically violated;

2. the elicitation of a significant value or utility function is anything but simple and many difficulties are often encountered in the practice.

As concerns point 1 above, the problem at issue can be faced through the introduction of alternative utility theories, which relax the original formulation of some problematic axioms in favor of a less constrained axiomatic basis. The topic will be analyzed in depth in Chapter 3.

Also criticisms related to the elicitation of the utility function (point 2 above) have received answers that, in most cases, solve the problem in a satisfactory manner. These criticisms, and related answers, resemble those addressed by the subjective Bayesian approach to statistical inference (Section 2.10), where qualitative knowledge from experts has to be expressed in quantitative terms (i.e., prior probabilities). Similarly, in the setting of utility theory, qualitative preferences for certain actions and their consequences have to be expressed in quantitative terms, as it will be clarified in Chapter 4.

The decision-making process is characterized by three main elements:

- a finite set of alternatives or actions $\mathcal{A}' = (a_1, a_2, \ldots, a_i, \ldots, a_m)$, among which the decision maker has to select the optimal one, according to some rational criterion;

- a discrete or continuous set of states of nature Θ. If the state of nature is discrete we have $\Theta' = (\theta_1, \ldots, \theta_j, \ldots, \theta_k)$, which represents the context in which the decision-making process takes place (operating context);

- a finite set of consequences or results or outcomes (in the presence of a discrete state of nature) $\mathcal{C}' = (c_{11}, c_{12}, \ldots, c_{ij}, \ldots, c_{mk})$ that depend both on the action a_i chosen by the decision maker and on the true state of nature θ_j, that is,

$$c_{ij} = f(a_i, \theta_j) \quad i = 1, \ldots, m; \ j = 1, \ldots, k.$$

Decisions are made up of actions, states and consequences, with the latter ones depending on the action and the state of nature.

The analysis of a decision problem requires that the analyst, which can be the same decision maker or, alternatively, an expert with background in statistics and probability that helps the decision maker, identifies the relevant set of actions, states and consequences to adequately characterise the problem. To make the decision-making process easier a representation through a decision table and a decision tree is recommended. The *decision table* (Table 1.1) is a sort of matrix that displays the (discrete) set $\mathcal{A}$ of actions on the rows and the (discrete) set Θ of states of nature on the columns; consequences c_{ij} are displayed in each cell.

TABLE 1.1
Decision table.

Action	State of nature					
	θ_1	θ_2	...	θ_j	...	θ_k
a_1	c_{11}	c_{12}	...	c_{1j}	...	c_{1k}
a_2	c_{21}	c_{22}	...	c_{2j}	...	c_{2k}
$\vdots$	...	...	$\vdots$	...	$\vdots$	...
a_m	c_{m1}	c_{m2}	...	c_{mj}	...	c_{mk}

The *decision tree* is a diagram composed of nodes that are connected one to the other by ties (Figure 1.1). We may distinguish two types of node. The *decision node* denotes the time point when an action is chosen by the decision maker, whereas the *chance node* is the time point when the result of an uncertain event becomes known (i.e., the state of nature becomes evident). As it will be illustrated in Chapter 5, the decision tree may contain further information, such as the probabilities associated with the states of nature and the utilities or losses associated with the consequences.

After the identification of actions, states, and consequences involved in the decisional process, the decision maker may proceed with the specification and analysis of the decisional problem.

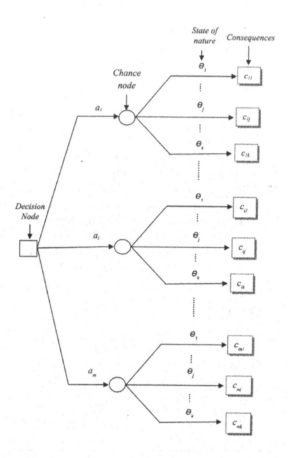

FIGURE 1.1
Decision tree.

1.4 Decisions and informational background

A relevant aspect that the decision maker has to take into account in the decision-making process concerns the level of knowledge about the states of nature. In detail, decisions may be distinguished according to the informational background in which the decision maker operates:

- decisions in situations of *certainty*, when the state of nature is known;

- decisions in situations of *risk*, when the state of nature is unknown, but a probability distribution of the set Θ of states of nature is given or the decision maker is able to assign a subjective level of probability to the elements of Θ;

- decisions in situations of *uncertainty*, when the decision maker cannot or does not want to proceed with the measurement of the plausibility of the states of nature.

It is worth noting that according to some authors (mainly the Bayesian subjectivists, Section 2.10), the distinction between risky and uncertain situations does not make sense. Assessing the plausibility of the states of nature is a necessary condition to solve a decisional problem in an optimal way.

Every decision problem implies a set of consequences c_{ij}, whose values, together with the information on the states of nature, drive the choice of the optimal action. In such a context, the *principle of dominance* suggests excluding all the actions whose consequences are worse than the consequences of some other actions, whatever the state of nature. If an alternative dominates all the others, the principle of dominance leads to choosing such an alternative and the decision problem is solved in an optimal way. Unfortunately, such cases are very uncommon in actual situations. In most cases, there does not exist a dominant alternative and the selection of the optimal action requires a careful analysis of the informational background about the states of nature.

For the sake of clarity, let us assume that consequences c_{ij} are expressed in monetary terms ($c_{ij} = g(a_i, \theta_j)$) and the well-being of the decision maker is exclusively affected by the monetary value, in such a way that a higher monetary value produces a higher well-being.[2] What follows may be immediately generalized to non-monetary consequences, assuming that the decision maker is rationally able to express in quantitative terms his/her preferences about the consequences.

When the state of nature is known, say $\Theta = \theta_j$ (decisions under certainty), the problem of choice reduces to comparing the m consequences $c_{1j}, c_{2j}, \ldots, c_{mj}$ and the decisional problem is easily solved: the decision table (Table 1.1) reduces to a single column (the one corresponding to θ_j) and the best action a^* is the one with the highest monetary value, that is,

$$a^* = \arg\left[\max_i c_{ij}\right] = \arg\left[\max_{a_i} g(a_i, \theta_j)\right].$$

Note that, under certainty, the *rational decision* in favour of a^* is the *right decision*, that is, the decision that maximizes the well-being of the decision maker. Therefore, if the decision maker had a complete knowledge of the future, he/she could simply refer to the principle "take the right decision".

---

[2]In Chapter 3 this assumption will be removed with the introduction of the concept of utility of a consequence.

Unfortunately, most decisions are taken in situations of risk or uncertainty; then they are based on what is believed to happen and not on what actually happens. Therefore, it is usually impossible to take the right decision: at most, the decision maker may act in such a way that a rational decision will be made, evaluating with attention the (partial) set of information available on the possible states of nature. In other words, in contexts of risk or uncertainty the equivalence "rational decision = right decision" is not obvious at all.

In situations of risk, each state of nature θ_j $(j = 1, \ldots, k)$ is associated with a probability $\pi(\theta_j)$ $(\pi(\theta_j) > 0; \sum_j \pi(\theta_j) = 1)$ that refers to its plausibility. In such a context, if the decision-maker fullfills certain rules (axioms) of rational behavior, the optimal action is the one that maximizes the sum of consequences c_{ij} weighted by the related probabilities $\pi(\theta_j)$ (commonly called *expected value*, as it will clarified in Chapter 2, Definition (2.11)), that is,

$$a^* = \arg\left\{\max_i \left[\sum_{j=1}^k c_{ij}\pi(\theta_j)\right]\right\} = \arg\left\{\max_{a_i} \left[\sum_{j=1}^k g(a_i, \theta_j)\pi(\theta_j)\right]\right\}.$$

In practice, the k columns of the decision table (Table 1.1) reduce to a single one, whose m values represent a synthesis of each action. Then, the maximum of these m values is selected and the corresponding action represents the optimal choice.

In Section 3.4 we will see how this criterion, consisting in maximizing the expected monetary value, was criticized since the XVIII century. In 1738, Bernoulli [18] stressed how individuals often do not choose the action that entails the highest monetary advantage but the action with the highest expected "moral value" or, using the modern term due to von Neumann and Morgenstern [213], the highest expected "utility". More in general, in Section 3.4.1 it will be illustrated how, in the case of generic (i.e., monetary or non-monetary) consequences, if the decision maker satisfies some rules of rational behavior, there exists and can be elicited a real-value function $u_{ij} = u(a_i, \theta_j)$, called utility function, associated with the consequences c_{ij} and depending on the actions a_i and the states of nature θ_j, which represents the preferences of the decision maker. In such a context, the optimal action corresponds to the one that maximizes the sum of utilities u_{ij} (instead of consequences) weighted by the related probabilities $\pi(\theta_j)$, that is,

$$a^* = \arg\left\{\max_i \left[\sum_{j=1}^k u_{ij}\pi(\theta_j)\right]\right\} = \arg\left\{\max_{a_i} \left[\sum_{j=1}^k u(a_i, \theta_j)\pi(\theta_j)\right]\right\}.$$

In situations of uncertainty the decisional process gets more problematic. Here we illustrate some criteria[3]:

- *Max-min criterion or Wald's criterion* (extreme pessimistic perspective). The max-min criterion or Wald criterion consists of choosing the action a^*

---

[3]We will come back to the criteria for decisions under uncertainty in Section 5.3, when they will be generalized to account for losses and utilities of consequences.

which corresponds to the maximum of the minimum (monetary) amount

$$a^* = \arg\left[\max_i(\min_j c_{ij})\right].$$

This criterion depicts an attitude of extreme pessimism, because the decision maker operates as if, whatever action he/she chooses, the state of nature will occur (in terms of structural, political, and economic conditions) that will provide him/her with the minimum well-being. Thus, the decision maker protects himself/herself against nature by trying to achieve the maximum among the minimum benefits.

- *Max-max criterion* (extreme optimistic perspective). Differently from the extreme pessimist, the extreme optimist believes that, whatever the chosen action is, nature will be so benign as to grant the maximum well-being. Then, the optimal choice results in the one that maximizes the maximum (monetary) reward:

$$a^* = \arg\left[\max_i(\max_j c_{ij})\right].$$

- *Hurwicz's criterion*. The Hurwicz's criterion represents a compromise between the attitude of the extreme pessimist and the attitude of the extreme optimist. Indeed, the optimal action is chosen in accordance with the following formula:

$$a^* = \arg\left\{\max_i\left[(1-\alpha)\min_j c_{ij} + \alpha\max_j c_{ij}\right]\right\}, \quad 0 \le \alpha \le 1,$$

where α provides a measure of the level of the decision maker's optimism. The max-min criterion is obtained for $\alpha = 0$ and the max-max criterion for $\alpha = 1$.

- *Savage's criterion or min-max regret criterion*. To apply the Savage criterion, the concept of *regret* is introduced as the difference between each element of the decision table (Table 1.1) and the maximum value on the related row, that is,

$$r_{ij} = \max_i c_{ij} - c_{ij}, \quad j = 1, \ldots, k.$$

Then, the action that minimizes the maximum regret is selected:

$$a^* = \arg\left[\min_i(\max_j r_{ij})\right],$$

In other words, through the min-max regret criterion the decision maker limits the damages of a wrong decision.

These four decision criteria may be justified in light of different arguments, which all have a certain character of acceptability. However, the most problematic aspect concerns their application: in fact, if they are adopted in the same decision problem, a choice of four different optimal actions may result. This fact has been used by some authors to state that one or more criteria must necessarily be wrong. However, we think that sufficient reasons do not exist to support this point of view. On the other hand, we reasonably state that all the proposed criteria have a limited applicability, due to the poor information about the states of nature they use, and the most suitable criterion must be chosen from time to time, according to the circumstances.

A further criterion that can be used in situations of uncertainty is the so-called *Laplace criterion*, also known as *insufficient reason criterion*, which identifies as optimal action as the one that corresponds to the maximum sum

$$a^* = \arg \left(\max_i \sum_j c_{ij} \right).$$

The Laplace criterion implicitly attributes the same probability to the states of nature, as there is not sufficient reason to believe that the probability distribution of the states of nature is different from the uniform one. However, the reason why the state of complete ignorance necessarily implies an equal probability of the states of nature is not clear at all. To be honest, even in the other criteria a sort of indirect assessment of the probabilities is carried out. For instance, if we consider the max-max criterion, a degenerate probability distribution is implicitly introduced that assigns value 1 to the probability of the state of nature corresponding to the most favourable consequence, and 0 otherwise.

1.5   Statistical inference and decision theory

As outlined in Section 1.2 above, this book focuses on the normative decision theory, that is, the theory that concerns how decisions should be made by individuals in order to maximize their own well-being. Therefore, in what follows we do not refer to actually observable behaviors but to "ideally rational" behaviors. In particular, a set of rules of "rational behavior" to which a "rational individual" must conform will be deduced in Chapter 3, through axioms and theorems.

A necessary premise to understand the reality, whatever its nature (e.g., social, economic, business, physical, biological) and whatever the purpose (purely cognitive or operational), consists in equipping the phenomenon of interest with a suitable synthetic representation, in order to easily identify relevant characteristics and distinctive elements. To capture the multiform variability or mutability of a phenomenon[4] it may be useful to compute average values to understand what is typical or constant, synthetic indices of variability to measure the tendency to dispersion, and interrelation and dependency indices to highlight the relationships linking two or more phenomena. Moreover, the representation of phenomena and their relationships may be made easier through the specification of analytical models.

As will be illustrated in depth in Section 2.7, the scientific discipline that deals with the issues above is *statistics* in its two main connotations of *descriptive statistics*, when all the manifestations of the phenomenon of interest are available, and *statistical inference* (or inductive statistics), when the set of manifestations is only partially observed (random sample).

In turn, statistical inference distinguishes between the *classical or frequentist approach* (see Section 2.9 for details), if the inductive process is based on sampling data only, and the *Bayesian approach* (see Section 2.10 for details), when the knowledge process from sample to population is based both on sample data (objective knowledge) and prior information (usually, subjective knowledge).

It is worth noting that the distinctive element of the Bayesian approach does not lie so much in the use of prior information *per se*, which is used, even if only informally, also in the classical approach, but in the different use of prior information. From the classical perspective, the states of nature are unknown (or only partially known) and the sample information is the only element that explicitly enters in the inferential process. On the other hand, from the Bayesian perspective a probabilistic knowledge of the states of nature is available (i.e., prior information having an objective or subjective nature) and sample data is used to "update" or "revise" this knowledge in a formal way, according to the Bayes' rule (Section 2.10 for details).

Both the frequentist and the Bayesian approaches deal with the extension of knowledge about a sample of observations to the entire population from which the sample was collected. In both cases, an analytical but synthetic representation of the phenomenon of interest is required. For this aim, a suitable probabilistic model, called *random variable* (see Sections 2.4 and 2.5 for details), is usually introduced that associates a real number (or a sequence of real numbers) with the manifestations of the phenomenon, together with a measure of their plausibility (probability). Random variables distinguish one from the other by the shape of the probability function and by one or more characteristic elements, called *parameters*, whose value is usually unknown. In *parametric* statistical inference the analytic form of the probability function is known, whereas in *non-parametric* statistical inference the analytic

---

[4]A phenomenon is said to be *variable* when its manifestations assume different numerical values that can be enumerated or measurable (quantitative characters), whereas a phenomenon is *mutable* when its manifestations are described by non-numerical attributes (qualitative characters) that have, or not, a natural ordering. The quantitative or qualitative nature of a phenomenon strongly affects the phases of a statistical analysis, from the data collection to the final processing.

form is unknown (Section 2.7). In addition, in the context of classical statistical inference, parameters are unknown but constant values, whereas in the Bayesian setting, they are considered as random variables and, as such, are associated with a probability function, called *prior distribution* (or prior probability distribution). Then, in Bayesian statistical inference prior knowledge is informally used to define the probabilistic model representing the phenomenon of interest, as it happens in classical inference, and is formally used to define the probability distribution of the parameters of such a model.

After having specified a suitable model to describe the phenomenon of interest, the knowledge process from sample to population is achieved through typical inferential instruments, such as estimation and hypothesis testing (see Sections 2.9 and 2.10 for details) and specification of complex models that link different phenomena (Sections 2.11 and 2.12). Results of the inferential process can be used for decision-making purposes, as will be clear in Chapters 5 and 6.

Availability of prior information and/or sample information distinguishes decision theory in (see also Table 1.2):

- *classical decision theory*, when decisions are taken in the absence of any type of information on the probability of the states of nature (Sections 1.4 and 5.3 for details);

- *Bayesian decision theory*, when decisions are taken in the presence of only prior information (prior probability distribution) about the states of nature (Chapter 3);

- *classical statistical decision theory*, when the decision-making process is supported only by sample information; prior information is not formally taken into account (Section 5.4);

- *Bayesian statistical decision theory*, when both sample information and prior information about the states of nature are available and the decision-making process is based on the posterior information obtained updating the prior information on the basis of the sample information (Section 5.5).

TABLE 1.2

Decision theory settings and type of information available on the states of nature for the decision-making process.

Setting	Type of information
classical decision theory	none
Bayesian decision theory	prior probabilities
classical statistical decision theory	sample information
Bayesian statistical decision theory	prior probabilities, sample information

To summarize, knowledge about a phenomenon may rely only on sample information (classical statistical inference) or on both sample and prior information (Bayesian statistical inference). Then, the acquired knowledge may be used to satisfy a mere cognitive need or to solve specific decisional problems.

In the decisional setting, the choice among actions can involve a range of possibilities: from the most naive daily choices (e.g., to reach the work place one can use public transportation, such as bus or train, or the personal car or call a taxi), to more complex choices concerning, for example, the management of a small business (e.g., to proceed or not to restructure a commercial activity; to carry out promotional activities; to extend the range of products offered) or really complex choices, such as those involving the management of medium and large firms as well as the public policies that have impact on a large number of citizens.

To solve in a satisfactory way a decisional problem a clear definition of the problem itself is required, which consists of the identification of all the elements involved in the choice and the possible relationships that connect them. For example, let us assume that a firm want to evaluate the opportunity of producing and selling a new product. Then, an in-depth analysis of the following aspects is required: potential demand for the new product, presence of competitive products on the market, production costs and marketing costs, availability of raw materials, machinery and know-out, availability of monetary resources and the cost of money. However, many other aspects have to be taken into account to solve the decision-making process in an optimal way, where the term "optimal" refers to the achievement of a certain objective through the minimum use of resources or, even, to the achievement of the maximum benefit (e.g., monetary profit) given a certain amount of resources. As already outlined above, the decisional process requires the acquisition of complete or, more often, partial information about the state of the nature in which one operates. For instance, complete information may be gathered with reference to the presence of competitive products on the market and their coverage, while only partial information may be gathered on the buyers and the number of potential new customers.

The logical framework of reference and the available information represent essential *ingredients* of every decision-making process, whereas decision theory and statistical theory, together with methods and models developed in these disciplinary fields, are essential and necessary *tools* for an optimal development of the decision-making process.

1.6 The decision-making approach to statistics

As already outlined, information represents the fundamental element to acquire knowledge in an empirical context. However, it should be outlined that, on one hand, available information is often insufficient to satisfy the cognitive needs and, on the other hand, in some situations, despite the availability of relevant (actual or potential) information, the decision maker does not have clear the aims he/she intends to pursue or he/she lacks adequate methodological background and, consequently, he/she is unable to draw from the available data the specific information of interest and understand its usefulness.

Statistics is the discipline that deals with methods of representing the reality in an essential and simplified way in order to transform raw data (i.e., the manifestations of the phenomenon of interest) into useful information. It fully enters in every decision-making process that requires the use of information, in which three sets of related elements are involved:

- a set of *inputs* (i.e., raw data) consisting of statistical data that represent qualitative or quantitative aspects related to a specific phenomenon;

- a set of *transformations and processing procedures*, which are carried out through the use of suitable statistical approaches and methods;

- a set of *outputs* or *products* (i.e., final information) consisting of numerical indices, graphical representations, tabulations, models, whose meaning and interpretation depend on the interaction between the two previous elements;

- a set of consequences associated with the outputs. For instance, if we deal with a set of observed data and we aim at detecting the main features of a phenomenon, consequences can be described by the loss (of information) determined by the process of synthesis. Further, consequences can be identified in terms of monetary (or other) losses or gains and may be linked to errors occurring during the data transformation process.

In the setting outlined above, we refer to a decisional problem as a schematization that provides the list of all possible consequences and the choice of the optimal one according to the criterion of minimization of losses or maximization of gains (utilities in the following).

In such a setting, it is always possible to formulate every statistical problem in decisional terms. However, some authors do not share this opinion and consider the decision-making logic rather simplistic and reductive. Other authors think the decision-making approach applicable only to problems with operational purposes, and some others consider the decision-making perspective applicable to descriptive and inferential problems, but only according to specific modalities.

In the opinion of authors of this book, elements supporting the decision-making approach to statistics are innumerable. Indeed, the dual purpose - cognitive and operational - traditionally assigned to statistics as a scientific discipline, with the consequent attribution of decisional problems to the second purpose, can be further assigned into the specification of the following dual types of product: (*i*) decisions expressed in terms of actions to be taken and to be realized in practice; (*ii*) decisions expressed in terms of statements that, on the basis of the empirical evidence, define the estimate of an unknown quantity (point and interval estimation; see Sections 2.9.1, 2.9.2, 2.9.3, and 2.10.3) or the compliance with one or more theoretical hypotheses (hypothesis testing; see Sections 2.9.4 and 2.10.4), or the specification of a theoretical model (see Sections 2.11 and 2.12).

The cognitive and operational perspective of statistics always resolves into a decision that can be oriented toward "what to say" or "what to do", according to the specific situation at issue. In other words, the above conceptual distinction between cognitive and operational purposes of statistics is purely terminology and is artificial, mainly if we think that any action can be considered as the result of the statement: "Is the decision the best one among all the possible decisions?"

2

Probability and statistical inference

CONTENTS

2.1 Introduction

In this chapter we shall provide the basic knowledge concerning probability theory and statistical inference, from the frequentist and the Bayesian point of views. We do not claim to be exhaustive about these huge topics; however, as they are essential for the comprehension of the following chapters of the book, we recommend the reader with weak statistical competencies carefully study the present chapter. For insights, we suggest looking up specialized texts, such as [158], [32], [185] for the frequentist approach to statistical inference and [24] and [27] for the Bayesian inference.

We first examine the foundations of the probability theory, focusing on the definitions of some basic concepts, such as random experiment, sample space, event, probability. After an illustration of the most known definitions of probability (classical, frequentist, and subjective definitions), we discuss the Kolmogorov's axioms and some relevant theorems, among which is the Bayes' rule, which provides a useful instrument to implement in practice the calculus of probability.

Next, attention focuses on the concept of random variable. Univariate and multivariate random variables are defined and the related concepts of probability and density functions as well as cumulative probability function are illustrated; the expected value and the moment generating function are illustrated, too. Examples of the most relevant random variables used in the practice are briefly illustrated with special attention to the exponential family.

The remainder of the chapter is devoted to statistical inference. We first deal with the difference between descriptive and inferential statistics and, then, we show the link between inference and probability theory, through the sample distributions. Finally, problems of point estimation, interval estimation, hypothesis testing, and linear regression modeling are illustrated both from the frequentist and the Bayesian points of view.

2.2   Random experiments, events, and probability

Probability theory was born in the XVII century in the context of bets and gambles and nowadays represents the foundation upon which statistics and statistical decision theory are built. Probability theory is an abstract mathematical subject, which is built on some basic concepts, such as random experiment, sample space, and event.

Definition 2.1. *A random experiment is any activity (phenomenon) whose outcome cannot be predicted with certainty.*

Definition 2.2. *The sample space is the set of all possible, exhaustive and mutually exclusive, outcomes of a random experiment. Sample spaces are classified according to the number of elements they contain: a discrete sample space contains a (finite or infinite) countable number of elements, whereas a continuous sample space contains an infinite uncountable number of elements.*

Example 2.1. Finite discrete sample space. *Suppose that a random experiment consists of tossing a coin; then the sample space contains two elements*

$$\Omega = \{H, T\} = \{\omega_1, \omega_2\},$$

with $H = \omega_1$ denoting the outcome "head" and $T = \omega_2$ denoting the outcome "tail". If we assume that the coin could also stay balanced on the edge, the sample space is

$$\Omega = \{H, T, E\} = \{\omega_1, \omega_2, \omega_3\},$$

where $E = \omega_3$ stays for "coin balanced on the edge".

Similarly to the coin toss, several other experiments can be framed in the binary scheme: "success" and "failure". For instance, in the context of bank financing (random experiment), the possible outcomes are: the firm gets the loan (ω_1, "success") or the firm does not get the loan (ω_2, "failure"). In the context of manufacturing quality control (random experiment), a piece from a certain batch can be compliant with the quality specifications (ω_1, "success") or not (ω_2, "failure").

Now, assume that an urn contains n identical balls, up to the number printed on it (from 1 to n); then the sample space is

$$\Omega = \{\omega_1, \omega_2, \ldots, \omega_n\},$$

where ω_i ($i = 1, 2, \ldots, n$) denotes the ball with number i. For instance, in the lottery game ($n = 90$) the sample space is

$$\Omega = \{1, 2, \ldots, 90\} = \{\omega_1, \omega_2, \ldots, \omega_{90}\}.$$

Example 2.2. Infinite discrete sample space. *If the random experiment consists of counting the number of accesses to an internet web site or counting the number of stars in the Universe, the sample space is defined by all non-negative integer numbers, that is,*

$$\Omega = \{0, 1, 2, \ldots, \infty\} = \{\omega_1, \omega_2, \ldots, \infty\}.$$

Example 2.3. Continuous sample space. *If the random experiment consists in testing the duration of a tire or of a light bulb, the sample space is defined by all non-negative real numbers, that is,*

$$\Omega = \{0, \infty\}.$$

In both of the last two examples, the superior extreme of Ω is ∞, as it is not possible to establish in advance the maximum number of internet accesses or the maximum duration of a light bulb, even if, in practice, they cannot be infinite.

Definition 2.3. *In the presence of a discrete sample space, an event E is any collection of elements of Ω. In the presence of a continuous sample space, an event E is any measurable collection of elements of Ω[1].*

---

[1]It is worth outlining that, in a continuous sample space, not any collection of elements of Ω is an event. This is the case of the non-measurable subsets of Ω, that is, all those

In practice, an event is a collection of outcomes (sample points) of random experiments. There exist two special cases: event Ω (*sure event*) that corresponds to the whole sample space and event ϕ (*impossible event*) that denotes an event that cannot be observed. Moreover, single sample points $E_i = (\omega_i)$, $i = 1, 2, \ldots, n$ are named *elementary events* (or atomic events or simple events). Finally, another type of event is represented by the *conditional event* $E_1 | E_2$ (read: E_1 given E_2) that denotes event E_1 after having observed the occurrence of event E_2. In other words, the conditioning of events consists of a new definition of the sample space, from Ω to the conditioning event E_2.

Two other important concepts are those of σ-algebra and probability.

Definition 2.4. *A σ-algebra $\mathcal{B}$ (or σ-field or complete Boolean algebra) is any collection of subsets of Ω that satisfies the following properties:*

- $\phi \in \mathcal{B}$,

- *if $E \in \mathcal{B}$, then $\bar{E} \in \mathcal{B}$ (closing under complementation), where $\bar{E}$ denotes the complementary event of E, that is, the event that is observed when E is not observed,*

- *if $E_1, E_2, \ldots, E_n \in \mathcal{B}$, then $\bigcup_{i=1}^{n} \bar{E}_i \in \mathcal{B}$ (closing under countable unions).*

Obviously, as operations of difference and intersection derive from negation and union, respectively, a σ-algebra is closed also under these two operations.

Example 2.4. *The class of all events associated with an experiment is the event space. An event will always be a subset of the sample space but, as said, the class of all subsets of the sample space will not necessarily correspond to the event space.*

Let us consider the tossing of a die. The die can land with any one of the six faces up, such that there are six possible outcomes of the experiment. Let $\omega_i = i$ spots up ($i = 1, 2, \ldots, 6$). Each ω_i is an elementary event. For this experiment the sample space $\Omega = \{\omega_1, \omega_2, \omega_3, \omega_4, \omega_5, \omega_6\}$ is finite; hence the event space is made of all subsets of Ω; there are $2^6 = 64$ events in $\mathcal{B}$ (including ϕ and Ω) and 6 (out of 64) elementary events.

Concepts of random experiment, sample space, and event are kept together by the concept of probability. In the literature, there exist at least three main definitions of probability: (i) the classical definition, originated in the context of games of chance, (ii) the frequentist definition, to which the frequentist

---

subsets whose structure is so complex that it is not possible to assign it a measure. The formulation of the probability theory due to Kolmogorov [133] is restricted to the so called Borel sets, that is, those sets of real numbers formed from open sets (or closed sets) through the operations of countable union, countable intersection, and relative complement. The first example of non-measurable subset is due to [212]. An example of non-measurable subset is represented by the semi-open sets of type $(a \; ; \; b] = \{x \in \mathcal{R} : a < x \le b\}$ or $[a \; ; \; b) = \{x \in \mathcal{R} : a \le x < b\}$. Throughout the book we will consider only measurable collections of Ω.

approach to statistical inference mainly refers, and (*iii*) the subjective definition, which represents the foundation of the Bayesian approach to statistical inference. All these three definitions comply with the so-called axiomatic definition of probability, from which many useful properties and rules derive to implement in practice the calculus of probabilities.

Definition 2.5. Classical probability. *Probability of an event, $P(E)$, is the ratio between the number n_E of events favorable to it, and the number of all events, given that all events are equally plausible to occur.*

The classical definition of probability is affected by several weaknesses. First, the definition is circular: saying that "all events are equally plausible" is the same as "all events have the same probability". Second, this definition cannot be applied neither when the possible events have not the same probability to occur nor when they are not countable.

Definition 2.6. Frequentist probability. *The probability of an event, $P(E)$, is the limit to which the ratio of the number n_E of occurrences of E to the number n of trials tends, when n goes to $+\infty$:*

$$P(E) = \lim_{n \to +\infty} \frac{n_E}{n}.$$

In other words, according to the frequentist definition the relative frequency[2] of a certain event is a proxy of its probability (approximation for finite n). Even the frequentist definition has been criticized. First, the limit of the number of trials ($+\infty$) is unattainable. Second and more relevant, there exist several contexts where an experiment cannot be repeated under the same conditions or, even, cannot be repeated at all (e.g., economic events).

Definition 2.7. Subjective probability. *The probability of an event, $P(E)$, is the individual's level of belief about its occurrence. This personal judgement relies on opinions and past experience of the individual.*

Detractors of the subjective definition of probability complain against the possibility that the level of probability assigned to a certain event may differ a lot from person to person, according to personal judgement. To this criticism we reply that the subjective probability has to be intended conditionally on the level of knowledge: given the same level of knowledge about a certain event, two individuals will provide the same level of belief about its occurrence. Another criticism is about the difficulty of translating the level of belief in quantitative (numerical) terms. This issue is usually addressed by proposing a sort of gamble to the individual: for instance, asking how much he/she is willing to bet to win a given sum of money if the event occurs.

The "primitive" concepts of random experiment, event, and probability above introduced are linked one to another according to the following statement: *the random experiment produces the event with a given probability*, and

---

[2]The relative frequency is the proportion of statistical units characterized from a certain modality of the character of interest.

represent the foundation upon which the axiomatic definition of probability is built.

Definition 2.8. Axiomatic definition of probability. *Given a sample space Ω and an associated σ-algebra $\mathcal{B}$, a real value function $P(\cdot)$ with domain $\mathcal{B}$ is a probability function if the following axioms are satisfied:*

Axiom 1 *Events build a σ-algebra.*

Axiom 2 *The probability of any event is unique.*

Axiom 3 *The probability of an event is non-negative, that is, $P(E) \geq 0$.*

Axiom 4 *The probability of the sure event equals 1, that is, $P(\Omega) = 1$.*

Axiom 5 *If events $E_1, E_2, \ldots, E_k$ are pairwise disjoint, that is, $E_i \cap E_j = \phi$ $(i, j = 1, \ldots, k)$, then the probability of their union is equal to the sum of probabilities of each event E_i:*

$$P(E_1 \cup E_2 \ldots \cup E_k) = P(E_1) + P(E_2) + \ldots + P(E_k)$$

or

$$P\left(\bigcup_{i=1}^{k} E_i\right) = \sum_{i=1}^{k} P(E_i).$$

Axiom 6 *The probability of the conditional event $E_1|E_2$ (conditional probability) equals the ratio of the probability of intersection of E_1 and E_2 to the probability of the conditioning event E_2:*

$$P(E_1|E_2) = \frac{P(E_1 \cap E_2)}{P(E_2)} \quad \text{if } P(E_2) > 0.$$

The last relation is equivalent to:

$$P(E_1 \cap E_2) = P(E_2)P(E_1|E_2) = P(E_2 \cap E_1) = P(E_1)P(E_2|E_1).$$

More in general, in the presence of any number of events the above relation is extended as follows:

$$P(E_1 \cap E_2 \cap E_3 \cap \ldots) = P(E_1)P(E_2|E_1)P(E_3|E_1 \cap E_2)\ldots$$

The above axioms represent the foundation upon which some relevant theorems are built, which are particularly helpful for the calculus of probabilities.

Theorem 2.1. $P(E) \leq 1$

Theorem 2.2. $P(\phi) = 0$

Theorem 2.3. $E_1 \subset E_2 \Longrightarrow P(E_1) < P(E_2)$

Theorem 2.4. $P(E_1 \cup E_2) = P(E_1) + P(E_2) - P(E_1 \cap E_2)$ *and, in the presence of k events,*

$$P\left(\bigcup_{i=1}^{k} E_i\right) = \sum_i P(E_i) - \sum_i \sum_j P(E_i \cap E_j) + $$
$$+ \sum_i \sum_j \sum_h P(E_i \cap E_j \cap E_h) + \ldots + (-1)^{k+1} I_{i=1}^{k} P(E_i)$$

This theorem reduces to Axiom 5 when the k events are pairwise disjoint.

It is also worth outlining that, as concerns the conditional events, the axioms and the cited relations are always satisfied. Moreover, the occurrence of a given event E_2 does not necessarily modify the probability of any other event, say E_1. In other words, it may happen that

$$P(E_1|E_2) = P(E_1).$$

In such a case E_1 and E_2 are called *statistically independent* (or stochastically independent or independent in probability) and Axiom 6 reduces to

$$P(E_1 \cap E_2) = P(E_1)P(E_2).$$

More in general, given any number of events $E_1, E_2, \ldots, E_k$, they are statistically mutually independent if and only if, *for each subset of events*, the probability of their intersection equals the simple product of probabilities of single events.

2.3 Bayes' rule

As outlined in Chapter 1, the Bayes' rule represents the theoretical fundamental element for the statistical decision theory. This rule allows us to update the prior knowledge of the decision maker through the knowledge coming from sample experiments. The resulting posterior knowledge is more complete and, therefore, more useful than the prior knowledge for the decisional process.

Theorem 2.5. *Let $E_1, E_2, \ldots, E_i, \ldots, E_k$ be a partition of the sample space Ω and let E be a set belonging to Ω. Then, for $i = 1, 2, \ldots, k$,*

$$P(E_i|E) = \frac{P(E_i)P(E|E_i)}{P(E)} = \frac{P(E_i)P(E|E_i)}{\sum_{j=1}^{k} P(E_j)P(E|E_j)}.$$

In practice, the Bayes' rule is a generalization of Axiom 6. As $E_1, E_2, \ldots, E_i, \ldots, E_k$ represent a partition of the sample space Ω, then

$E_i \cap E_j = \phi$ for $i \neq j$; $i, j = 1, 2, \ldots, k$ and $\bigcup_{i=1}^{k} E_i = \Omega$ (i.e., the k events E_i are necessary and disjoint). Therefore, for any E belonging to Ω the following relations hold

$$E = E \cap \Omega = E \cap \bigcup_{i=1}^{k} E_i = \bigcup_{i=1}^{k} (E \cap E_i)$$

and also

$$P(E) = P \left[\bigcup_{i=1}^{k} (E \cap E_i) \right] = \sum_{i=1}^{k} P(E \cap E_i).$$

Finally, from $P(E \cap E_i) = P(E_i)P(E|E_i) = P(E_j \cap E) = P(E)P(E_i|E)$ (Axiom 6), the Bayes' rule is immediately obtained.

To make easier the interpretation of the Bayes' rule we may think of the k events E_i ($i = 1, \ldots, k$) in terms of plausible causes of event E. In such a context, $P(E_i)$ is the *prior probability* of cause E_i, $P(E_i|E)$ denotes the *posterior probability* of cause E_i, and $P(E|E_i)$ is the so called *likelihood* of event E. In practice, the Bayes' rule formalizes the process of learning from experience: observing the occurrence of a certain event E leads to the prior knowledge of a phenomenon E_i (expressed in terms of prior probabilities $P(E_i)$ with $i = 1, \ldots, k$) being updated according to how likely is the occurrence of E given E_i (expressed by the likelihood $P(E|E_i)$). This updated knowledge is expressed in terms of posterior probability $P(E_i|E)$.

Example 2.5. *During a quality control on the productive process, Alpha firm discovers that the 40% of defective pieces is due to mechanical errors, whereas the remaining 60% of defective pieces is due to human errors. Moreover, it is known that defects due to mechanical errors are detected with an accuracy of 90% and defects due to human errors are detected with an accuracy of 50%. Let us now assume that, during quality control, a defective piece is discovered. What is the probability that this defect is due to human error?*
Solution.

- $P(E_1) = 0.40$: *prior probability that a defective piece is due to mechanical error;*

- $P(E_2) = 0.60$: *prior probability that a defective piece is due to human error;*

- $P(D|E_1) = 0.90$: *probability of discovering a defective piece due to mechanical error;*

- $P(D|E_2) = 0.50$: *probability of discovering a defective piece due to human error;*

- $P(E_1|D) =?$: *probability that the discovered defective piece is due to mechanical error.*

On the basis of the Bayes' formula we obtain

$$P(E_1|D) = \frac{P(E_1 \cap D)}{P(D)}$$

$$= \frac{P(E_1)P(D|E_1)}{P(E_1)P(D|E_1) + P(E_2)P(D|E_2)}$$

$$= \frac{0.40 \cdot 0.90}{0.40 \cdot 0.90 + 0.60 \cdot 0.50} = 0.55.$$

Then, the probability that the discovered defective piece is due to mechanical error is equal to 55%, and, consequently, the probability that the discovered defective piece is due to human error equals 45% (= 1- 0.55). Steps to apply the Bayes' formula are shown in Table 2.1.

TABLE 2.1
Application of the Bayes' formula.

Cause E_i	Prior prob. $P(E_i)$	Cond. prob. $P(D|E_i)$	Joint prob. $P(E_i)P(D|E_i)$	Post. prob. $P(E_i|D)$
Mechanical ($i = 1$)	0.40	0.90	0.36	0.55
Human ($i = 2$)	0.60	0.50	0.30	0.45
Total	1.00		0.66	1.00

2.4 Univariate random variables

In many contexts it is useful to deal with probabilistic models that provide a analytic description of the stochastic nature of a certain phenomenon of our interest. Such probabilistic models are commonly named random variables.

Definition 2.9. *A Random Variable (RV) is a function $X(\cdot)$ that maps the outcomes of a random experiment to real numbers, under suitable conditions such that the structure of B (the σ-algebra on Ω) is preserved.*

More formally, the univocal function $X(\cdot)$ defined on the sample space Ω is a RV if

$$A = \{\omega \in \Omega | X(\omega) \leq x\} \in B,$$

that is, if the set A of elementary events ω, such that the value of $X(\cdot)$ is less or equal to any real number x, belongs to the algebra B.

RVs can be characterized as

- discrete RVs, when the function takes a finite number or an infinite countable number of real values;

- continuous RVs, when the function takes a continuous (i.e., uncountable) number of real values.

Definition 2.10. *The cumulative distribution function of RV X, denoted by $F(x)$, is defined by*

$$F(x) = P(X \le x) \quad \forall\, x.$$

More formally, the cumulative distribution function is the probability of A, that is,

$$P(A) = P\left[\omega \in \Omega | X(\omega) \le x\right] = P\left[X(\omega) \le x\right] = P(X \le x).$$

The properties of a cumulative distribution function are:

- $0 \le F(x) \le 1$;

- $\lim_{x \to -\infty} F(x) = 0$;

- $\lim_{x \to +\infty} F(x) = 1$;

- $F(x)$ is a non-decreasing function of x;

- $F(x)$ is right-continuous in the discrete case (the discontinuity points are located in $x_1, x_2, \ldots, x_k$) and it is absolutely continuous (i.e., uniformly continuous and almost anywhere differentiable) in the continuous case.

Let X be a discrete RV and let $x_1 < x_2 < \ldots < x_k$ be modalities of X with related probabilities $p_1, p_2, \ldots, p_k$, then

$$F(x_i) = P(X \le x_i) = \sum_{j=1}^{i} P(X = x_j) = \sum_{j=1}^{i} p_j$$

with $p_j = P(X = x_j)$.

Function $f(x_i)$ defined by $f(x_i) = F(x_i) - F(x_{i-1})$ is known as (mass) probability function and provides the probability p_i that RV X has value x_i:

$$f(x_i) = F(x_i) - F(x_{i-1}) = P(X \le x_i) - P(X \le x_{i-1}) = P(X = x_i) = p_i$$
$$i = 1, 2, \ldots, k.$$

If X is a continuous RV and $F(x)$ is an absolutely continuous function, then it will be always possible to define the derivative $f(x)$ of $F(x)$ with respect to x,

$$f(x) = \frac{dF(x)}{dx} \ge 0,$$

which is known as (probability) density function.

Similarly to the discrete case, the cumulative probability function is attained by

$$\int_{-\infty}^{x} f(y)dy = F(x).$$

Moreover

$$f(x)dx = dF(x) = P(x \le X \le x + dx)$$

denotes the probability that the continuous RV X takes values in an infinitesimal neighborhood of x.

The cumulative distribution functions and the corresponding probability functions (in the discrete case) and density functions (in the continuous case) are characterized by some quantities, known as parameters, that completely identify a specific RV according to the values they assume. As far as this point, the following notation is used

$$F(x; \theta_1, \theta_2, \ldots, \theta_m); \quad f(x; \theta_1, \theta_2, \ldots, \theta_m) = f(x; \boldsymbol{\theta}),$$

with $\boldsymbol{\theta}' = (\theta_1, \theta_2, \ldots, \theta_m)$ denoting the vector of parameters characterizing the probabilistic model (i.e., the RV).

To summarize, to model a phenomenon of interest we may proceed by introducing a suitable RV X with its related probability function $f(x_i)$ (in the discrete case) or density function $f(x)$ (in the continuous case)

$$f(x; \boldsymbol{\theta}) \quad x \in \Omega; \; \boldsymbol{\theta} \in \boldsymbol{\Theta},$$

with: Ω representing the sample space of x, that is, the support of all plausible values of X; $\boldsymbol{\theta}$ vector of parameters characterizing the probabilistic model; $\boldsymbol{\Theta}$ parametric space, that is, the space of all plausible values for (usually unknown) vector $\boldsymbol{\theta}$.

After the specification of the probabilistic model through X and $f(x; \boldsymbol{\theta})$, it is often useful to compute one or more numbers that summarize the typicalities of X. For this purpose we introduce the concept of expected value.

Definition 2.11. *Let X be a RV with cumulative probability function $F(x)$ and let $g(X)$ be a transformation of X. The expected value of $g(X)$, denoted by $E[g(X)]$, is*

$$E[g(X)] = \int_{L-S} g(x)dF(x) = \begin{cases} \sum_{i=1}^{k} g(x_i)f(x_i) & \text{if } X \text{ is discrete} \\ \int_a^b g(x)f(x)dx & \text{if } X \text{ is continuous,} \end{cases}$$

(2.1)

where $\int_{L-S} g(x)dF(x)$ is the Lebesgue-Stieltjes integral [187], $f(x_i)$ is the probability function of X in the discrete case, assuming value x_i with probability $f(x_i)$ ($i = 1, 2, \ldots, k$), whereas $f(x)$ is the density function of X in the continuous case, defined in the interval $[a; b]$.

It is worth noting that the definition of $E(\cdot)$ does not require the probability or density function of $g(X)$; moreover, the linearity property holds. Let X be a RV with probability function $f(x_i)$ (in the discrete case) or $f(x)$ (in the continuous case), let a, b be two constants, and let $g_1(X)$ and $g_2(X)$ be two RVs obtained as transformations of X. Then,

$$E[ag_1(X) + bg_2(X)] = aE[g_1(X)] + bE[g_2(X)].$$

For specific transformations $g(X)$ of X some popular expected values are obtained.

- $g(X) = X^r$, $r = 0, 1, 2, \ldots$

$$\mu_r = E[g(X)] = E(X^r) = \begin{cases} \sum_{i=1}^{k} x_i^r f(x_i) & \text{if } X \text{ is discrete} \\ \int_a^b x^r f(x) dx & \text{if } X \text{ is continuous,} \end{cases}$$

The expected value μ_r is known as r-th moment about the origin or moment of order r about the origin. A special relevance is assumed by the moment of order $r = 1$,

$$\mu = \mu_1 = E[g(X)] = E(X),$$

denoting the arithmetic mean of the RV X.

- $g(X) = (X - \mu)^r$, $r = 0, 1, \ldots$, with $\mu = \mu_1 = E(X)$ and $(X - \mu)$ known as *deviation*

$$\bar{\mu}_r = E[g(X)] = E[(X-\mu)^r] = \begin{cases} \sum_{i=1}^{k} (x_i - \mu)^r f(x_i) & \text{if } X \text{ is discrete} \\ \int_a^b (x - \mu)^r f(x) dx & \text{if } X \text{ is continuous,} \end{cases}$$

denoted as r-th central moment or central moment of order r. Moment of order 1 of the deviation is always null.

A special relevance is assumed by the central moment of order 2,

$$\bar{\mu}_2 = E[g(X)] = E[(X - \mu)^2] = \mu_2 - \mu_1 = \sigma^2,$$

denoting the variance of RV X.

- $g(X) = \left(\frac{X-\mu}{\sigma}\right)^r$, $r = 0, 1, \ldots$, $\mu = \mu_1 = E(X)$ and $\sigma = +\sqrt{\bar{\mu}_2} = +\sqrt{\sigma^2}$

$$\bar{\bar{\mu}}_r = E[g(X)] = E\left[\left(\frac{X - \mu}{\sigma}\right)^r\right] = \begin{cases} \sum_{i=1}^{k} \left(\frac{x_i - \mu}{\sigma}\right)^r f(x_i) & \text{if } X \text{ is discrete} \\ \int_a^b \left(\frac{x-\mu}{\sigma}\right)^r f(x) dx & \text{if } X \text{ is continuous,} \end{cases}$$

denoted as r-th standardized moment or standardized moment of order r. The most relevant standardized moments are obtained for $r = 3$ and $r = 4$:

$$\bar{\bar{\mu}}_3 = E[g(X)] = E\left[\left(\frac{X - \mu}{\sigma}\right)^3\right] = \gamma_3 \quad \text{skewness index,}$$

$$\bar{\bar{\mu}}_4 = E[g(X)] = E\left[\left(\frac{X - \mu}{\sigma}\right)^4\right] = \gamma_4 \quad \text{kurtosis index.}$$

- $g(X) = e^{tX}$, $-h < t < h$, $h > 0$

$$m_x(t) = E[g(X)] = E(e^{tX}) = \begin{cases} \sum_{i=1}^{k} e^{tx_i} f(x_i) & \text{if } X \text{ is discrete} \\ \int_a^b e^{tx} f(x) dx & \text{if } X \text{ is continuous,} \end{cases}$$

is the so called *moment generating function* of X. As suggested by the name, moments of a RV X are defined through function $m_x(t)$, evaluating its derivative of order r in $t = 0$:

$$\mu_r = \frac{d^r}{dt^r} m_x(t)|_{t=0} \quad r = 1, 2, \ldots.$$

The central moment generating function and the standardized moment generating function can be defined in a similar way. It has to be noted that, when the moment generating function exists, there is a biunivocal correspondence between it and the cumulative distribution function.

- $g(X) = e^{itX}$, $-h < t < h$; $h > 0$ and $i = \sqrt{-1}$

$$m_x(it) = E[g(X)] = E(e^{itX}) = \begin{cases} \sum_{j=1}^{k} e^{itx_j} f(x_j) & \text{if } X \text{ is discrete,} \\ \int_a^b e^{itx} f(x)dx & \text{if } X \text{ is continuous,} \end{cases}$$

which is known as the *characteristic function* of X. Similar to the moment generating function, also the characteristic function allows us the computation of the moments of X, however it has the further advantage of being always defined.

In the following we provide a brief summary of some commonly used univariate RVs. For each RV, we describe the main fields of application, the shape of the probability or density function, the mean and variance values, and the moment generating function (when it exists).

Example 2.6. Binomial distribution. *Let us consider n independent trials (sampling with replacement), each of which generates two mutually exclusive and exhaustive outcomes (success and failure). In any trial, probability of success, denoted by p, and probability of failure, dented by $q = 1 - p$, are constant. If $X(\omega) = X = 0, \ldots, n$ is the finite number of successes throughout the n trials of the random experiment, the resulting probability function, known as binomial distribution $X \sim \text{Binom}(n, p)$, is*

$$f(x) = f(x; n, p) = \binom{n}{x} p^x q^{n-x}, \tag{2.2}$$

where n (number of trials) and p (probability of success) are the parameters and x is the number of successes observed in n independent trials.

Arithmetic mean, variance and moment generating function of the binomial RV are:

$$E(X) = \sum_{x=0}^{n} x \binom{n}{x} p^x q^{n-x} = np,$$

$$Var(X) = E[(X - \mu)^2] = \sum_{x=0}^{n} (x - np)^2 \binom{n}{x} p^x q^{n-x} = npq,$$

$$m_x(t) = E(e^{tX}) = \sum_{x=0}^{n} e^{tx} \binom{n}{x} p^x q^{n-x} = (pe^t + q)^n.$$

For $n = 1$ a special case of binomial RV turns out, known as Bernoulli RV.

Example 2.7. Poisson distribution. *Let $X(\omega) = X = 0, 1, 2, \ldots, +\infty$ denote the number of occurrences of a certain event (number of successes) in a given time or spatial interval (e.g., in an hour, in a certain geographical area, $\ldots$,) and let these events be independent in non-overlapping intervals. Then, the probability function of $X(\omega)$, known as Poisson distribution $X \sim P(\lambda)$, is*

$$f(x) = f(x; \lambda) = \frac{\lambda^x e^{-\lambda}}{x!}, \tag{2.3}$$

where $\lambda > 0$ is the distribution parameter, $e \simeq 2.718\ldots$ is the Euler's constant, and x in the number of successes.

Arithmetic mean, variance and moment generating function of the Poisson RV are:

$$E(X) = \sum_{x=0}^{+\infty} x \frac{\lambda^x e^{-\lambda}}{x!} = \lambda,$$

$$Var(X) = \sum_{x=0}^{+\infty} (x - \lambda)^2 \frac{\lambda^x e^{-\lambda}}{x!} = \lambda,$$

$$m_x(t) = e^{tX} = \sum_{x=0}^{+\infty} \frac{e^{tx} e^{-\lambda} \lambda^x}{x!} = e^{\lambda(e^t - 1)}.$$

Thus, parameter λ in the Poisson distribution coincides with its mean and variance $\lambda = \mu = \sigma^2$.

Example 2.8. Negative binomial distribution.

Description: *number of failures required to get a fixed number of successes in independent trials; each trial may only generate a success or a failure.*

Parameters: *probability of success in any trial p; fixed number of successes, k.*

Support: $x = 0, 1, 2, \ldots$.

Probability function:

$$f(x) = f(x; k, p) = \binom{k + x - 1}{x} p^k (1 - p)^x = \binom{-k}{x} p^k (p - 1)^x,$$

with $\binom{-k}{x} = \frac{-k(-k-1)(-k-2)\cdots}{x_i!(n-x)!}$.

Mean and variance:

$$E(X) = \frac{k(1 - p)}{p},$$

$$Var(X) = \frac{k(1 - p)}{p^2}.$$

Moment generating function:

$$m_x(t) = \left(\frac{p}{1 - (1 - p)e^t} \right)^k.$$

Example 2.9. Normal distribution. *The continuous random variable* $X(\omega) = X : -\infty \leq x \leq +\infty$ *with density function*

$$f(x) = f(x; \mu, \sigma^2) = \frac{1}{\sqrt{2\pi\sigma^2}} e^{-\frac{1}{2}\left(\frac{x-\mu}{\sigma}\right)^2} \tag{2.4}$$

is defined normal distribution, or gaussian distribution, or distribution of random errors, $X \sim N(\mu, \sigma^2)$.

The normal RV is the most important continuous distribution, especially for the following reasons:

- *numerous random experiments can be associated with a RV whose distribution is approximately normal;*

- *relatively simple transformations of some random variables, which are not normally distributed, are normally distributed;*

- *some relatively complicated distributions can be approximated sufficiently well from the normal distribution;*

- *some RVs, which are the basis of inferential procedures (e.g., confidence intervals and hypothesis testing), are normally distributed or derive from such a distribution.*

However, it must be outlined that in the past the importance of the normal distribution has been too much emphasized. Indeed, the fundamental role that the distribution played in the "theory of random errors" prompted several scholars to believe that it could cover almost all natural phenomena. In reality, the theoretical justification of the relevant role that normal distribution plays in scientific research lies mostly in the "central limit theorem" (Theorem 2.7, Section 2.8).

Arithmetic mean, variance and moment generating function of the normal distribution are

$$E(X) = \int_{-\infty}^{+\infty} x \frac{1}{\sqrt{2\pi\sigma^2}} e^{-\frac{1}{2}\left(\frac{x-\mu}{\sigma}\right)^2} dx = \mu,$$

$$Var(X) = E[(X-\mu)^2] = \int_{-\infty}^{+\infty} (x-\mu)^2 \frac{1}{\sqrt{2\pi\sigma^2}} e^{-\frac{1}{2}\left(\frac{x-\mu}{\sigma}\right)^2} dx = \sigma^2,$$

$$m_x(t) = E(e^{tX}) = \int_{-\infty}^{+\infty} e^{tx} \frac{1}{\sqrt{2\pi\sigma^2}} e^{-\frac{1}{2}\left(\frac{x-\mu}{\sigma}\right)^2} dx = e^{t\mu+\sigma^2 t^2/2}.$$

Arithmetic mean and variance coincide with the two parameters, μ and σ^2, that characterize the normal distribution.

Example 2.10. Gamma distribution.

Description: it is a very flexible continuous distribution; it is often used to describe waiting-for-occurrence phenomena.

Parameters: number of successes, α, affecting the peakedness of the density function; occurrence time rate, β^{-1}, affecting the spread of the density function.

Support: $x \in (0, +\infty)$.

Density function:

$$f(x; \alpha, \beta) = \frac{1}{\Gamma(\alpha)\beta^\alpha} x^{\alpha-1} e^{-\frac{x}{\beta}}. \tag{2.5}$$

where $\alpha > 0, \beta > 0$, with $\Gamma(\alpha) = (\alpha-1)!$; $f(x)$ monotonically decreases from 0 for $\alpha = 1$ and it is positively skewed for $\alpha > 1$.

Mean and variance:

$$E(X) = \alpha\beta,$$
$$Var(X) = \alpha\beta^2.$$

Moment generating function:

$$m_x(t) = (1 - \beta t)^{-\alpha}.$$

A variant of Gamma distribution is represented by the inverse Gamma distribution, whose density function is

$$f(x; \alpha, \beta) = \frac{1}{\Gamma(\alpha)\beta^\alpha} x^{-\alpha-1} e^{-\frac{1}{x\beta}}. \tag{2.6}$$

Moreover, a special case of the Gamma RV is the exponential RV that is obtained for $\alpha = 1$ and is used to model lifetimes.

Example 2.11. Chi-squared χ^2 distribution.

Description: *it is a special case of Gamma distribution obtained for $\alpha = n/2$ ($n > 0$) and $\beta = 2$; it can be also derived from the sum of n independent squared standard normal distributions.*

Parameters: *the number of degrees of freedom, n.*

Support: $x \in (0, +\infty)$.

Density function:

$$f(x; n) = \frac{1}{2^{n/2}\Gamma(n/2)} x^{n/2-1} e^{-\frac{x}{2}}. \tag{2.7}$$

The density function is positively skewed with kurtosis greater than 3; for $n \to +\infty$, it resembles the normal distribution.

Mean and variance:

$$E(X) = n,$$
$$Var(X) = 2n.$$

Moment generating function:

$$m_x(t) = (1 - 2t)^{-n/2}.$$

Example 2.12. Beta distribution.

Description: *it is a flexible continuous distribution used to model inferior and superior limited RVs; it is also used to describe the distribution of the parameter estimators of some RVs.*

Parameters: α *and* β, *as the Gamma distribution.*

Support: $x \in (a, b)$.

Density function:

$$f(x; a, b, \alpha, \beta) = \frac{1}{B(\alpha, \beta)} \frac{(x-a)^{\alpha-1}(b-x)^{\beta-1}}{(b-a)^{\alpha+\beta-1}}. \qquad (2.8)$$

with $B(\alpha, \beta) = \int_0^1 x^{\alpha-1}(1-x)^{\beta-1}dx$; the shape of $f(x)$ changes with α and β: when $\alpha = \beta$, $f(x)$ is symmetric around $1/2$; when $\alpha > 1, \beta > 1$, $f(x)$ is concave with a single mode; when $\alpha < 1, \beta < 1$, $f(x)$ is U-shaped; when $(\alpha - 1)(\beta - 1) \leq 0$, $f(x)$ is J-shaped.

Mean and variance:

$$E(X) = \frac{\alpha}{\alpha + \beta},$$

$$Var(X) = \frac{\alpha\beta}{(\alpha + \beta)^2(\alpha + \beta + 1)}.$$

Moment generating function: *not useful.*

Example 2.13. Continuous uniform distribution.

Description: *it is obtained as a special case of Beta distribution and it is characterized by a uniformly spread probability in a certain interval.*

Parameters: *a and b that define the interval of variability of X.*

Support: $x \in (a, b)$.

Density function:

$$f(x; a, b) = \frac{1}{b-a}. \qquad (2.9)$$

The density function consists in a segment running parallel to the x-axis.

Mean and variance:

$$E(X) = \frac{a+b}{2},$$

$$Var(X) = \frac{(b-a)^2}{12}.$$

Moment generating function: *not useful.*

Example 2.14. Student's t distribution.

Description: *It is obtained as the ratio between a standard normal distribution and the square root of a χ^2 distribution divided by the number of degrees of freedom; these two components are independent.*

Parameters: *the number of degrees of freedom, n.*

Support: $x \in (-\infty, +\infty)$.

Density function:

$$f(x;n) = \frac{\Gamma(\frac{n+1}{2})}{\Gamma(\frac{n}{2})\sqrt{n\pi}} \frac{1}{(1+x^2/n)^{\frac{n+1}{2}}}.$$

The density function is symmetric around 0 and it resembles the standard normal distribution with $n \to +\infty$.

Mean and variance:

$$E(X) = 0 \quad n \geq 2,$$
$$Var(X) = \frac{n}{n-2} \quad n \geq 3.$$

Moment generating function: *it does not exist.*

Example 2.15. Fisher and Snedecor's F distribution

Description: *It is obtained as the ratio between two independent χ^2 distributions, each of them divided by its degrees of freedom.*

Parameters: *the degrees of freedom of the two χ^2 components, n_1 and n_2.*

Support: $x \in (0, +\infty)$.

Density function:

$$f(x;n_1,n_2) = \frac{\Gamma\left(\frac{n_1+n_2}{2}\right) n_1^{n_1/2} n_2^{n_2/2}}{\Gamma\left(\frac{n_1}{2}\right)\Gamma\left(\frac{n_2}{2}\right)} \frac{x^{n_1/2-1}}{(n_1 x + n_2)^{\frac{n_1+n_2}{2}}}.$$

Mean and variance:

$$E(X) = \frac{n_2}{n_2-2}, \quad n_2 \geq 2,$$
$$Var(X) = \frac{2n_2^2(n_1+n_2-2)}{n_1(n_2-2)^2(n_2-4)}, \quad n_2 > 4.$$

Moment generating function: *it does not exist.*

2.5   Multivariate random variables

Definition 2.12. *A k-dimensional RV is a function $\boldsymbol{X}(\omega) = (x_1, x_2, \ldots, x_k) = \boldsymbol{x}'$ that maps a sample space $\boldsymbol{\Omega}$ into $\mathcal{R}^k$ (euclidean k-dimensional space), such that*

$$A = [\omega \in \boldsymbol{\Omega} | \boldsymbol{X}(\omega) \le \boldsymbol{x}] \in \mathcal{B} \quad \forall \boldsymbol{x} \in \mathcal{R}^k.$$

In other words, a k-dimensional RV is a function with k components that associates k ordered real numbers to each event. Moreover, as $A \in \mathcal{B}$, with $\mathcal{B}$ built on the events $\omega \in \boldsymbol{\Omega}$, the cumulative distribution function of $\boldsymbol{X}' = (X_1, X_2, \ldots, X_k)$ is

$$P(A) = P[\omega \in \boldsymbol{\Omega} | \boldsymbol{X}(\omega) \le \boldsymbol{x}] = P[\boldsymbol{X}(\omega) \le \boldsymbol{x}] = P(\boldsymbol{X} \le \boldsymbol{x})$$
$$= P(X_1 \le x_1, X_2 \le x_2, \ldots, X_k \le x_k) = F(x_1, \ldots, x_k) = F(\boldsymbol{x}).$$

The multivariate RV $\boldsymbol{X}' = (X_1, X_2, \ldots, X_k)$ is discrete if all its components take a finite number or an infinite countable number of real numbers; is continuous if all its components take an uncountable number of real values; is mixed whenever some components are discrete and some others are continuous.

Function $F(\boldsymbol{x})$ satisfies properties that are similar to those that hold in the univariate case:

- the following limits hold:

$$\lim_{x_i \to -\infty} F(x_1, \ldots, x_k) = 0 \quad i = 1, \ldots, k$$

$$\lim_{x_1, \ldots, x_k \to +\infty} F(x_1, \ldots, x_k) = 1$$

$$\lim_{x_i \to +\infty} F(x_1, \ldots, x_i, \ldots, x_k) = F(x_1, \ldots, x_{i-1}, x_{i+1}, \ldots, x_k) \quad i = 1, \ldots, k$$

$$\lim_{x_i \to +\infty, x_j \to +\infty} F(x_1, \ldots, x_i, \ldots, x_k) =$$
$$= F(x_1, \ldots, x_{i-1}, x_{i+1}, \ldots, x_{j-1}, x_{j+1}, \ldots, x_k) \quad i \ne j = 1, \ldots, k,$$

where $F(x_1, \ldots, x_{i-1}, x_{i+1}, \ldots, x_k)$ is the distribution function of the $(k-1)$-dimensional RV;

- $F(\boldsymbol{x})$ is non-decreasing with respect to all the components;

- $F(\boldsymbol{x})$ is right-continuous with respect to all the components in the discrete case and is absolutely continuous (i.e., uniformly continuous and almost anywhere differentiable) in the continuous case.

Similarly to the univariate case, a one-to-one correspondence exists between the cumulative probability function and the probability function,

$$f(x_{1i}, x_{2j}, \ldots, x_{ks}) = P[X_1 = x_{1i}, X_2 = x_{2j}, \ldots, X_k = x_{ks}],$$

in the discrete case, where x_{1i} is the i-th modality of variable X_1, x_{2j} is the j-th modality of variable X_2, and x_{ks} is the s-th modality of variable X_k, and the density function,

$$f(x_1, x_2, \ldots, x_k) = f(\boldsymbol{x}) = \frac{\partial^k}{\partial x_1 \partial x_2 \ldots \partial x_k} F(x_1, x_2, \ldots, x_k),$$

in the continuous case; and vice-versa. Both functions completely identify the multivariate RV $\boldsymbol{X}' = (X_1, X_2, \ldots, X_k)$.

The univariate RVs $X_1, X_2, \ldots, X_i, \ldots, X_k$ that compose the multivariate RV $\boldsymbol{X}' = (X_1, X_2, \ldots, X_k)$ are statistically independent (or independent in probability) if

$$F(x_1, x_2, \ldots, x_k) = F(x_1) F(x_2) \ldots F(x_k)$$

or, alternatively, if

$$f(x_{1i}, x_{2j}, \ldots, x_{ks}) = f(x_{1i}) f(x_{2j}) \ldots f(x_{ks}), \quad \text{in the discrete case,}$$
$$f(x_1, x_2, \ldots, x_k) = f(x_1) f(x_2) \ldots f(x_k), \quad \text{in the continuous case.}$$

Let us now consider the case $k = 2$, that generates the bivariate or 2-dimensional RV $(X_1, X_2) = (X, Y)$, where, to make the notation easier, we put $X_1 = X$ and $X_2 = Y$. The bivariate RV (X, Y) is completely defined by the cumulative distribution function

$$F(x, y) = P(X \leq x, Y \leq y)$$

or, alternatively, by the probability function in the discrete case ($i = 1, 2, \ldots, h$ and $j = 1, 2, \ldots, k$, where h and/or k may tend to $+\infty$)

$$f(x_i, y_j) = F(x_i, y_j) - F(x_i, y_{j-1}) - F(x_{i-1}, y_j) + F(x_{i-1}, y_{j-1})$$
$$= P(X = x_i, Y = y_j)$$

and by the density function in the continuous case ($a \leq x \leq b$ and $c \leq y \leq d$, where a and/or c may tend to $-\infty$ and b and/or d may tend to $+\infty$)

$$f(x, y) = \frac{\partial^2}{\partial x \partial y} F(x, y).$$

The following relations hold:

$$0 \leq f(x_i, y_j) \leq 1, \qquad \sum_{i=1}^{h} \sum_{j=1}^{k} f(x_i, y_j) = 1, \quad \text{in the discrete case,}$$

$$f(x, y) \geq 0, \qquad \int_a^b \int_c^d f(x, y) dx dy = 1, \quad \text{in the continuous case.}$$

Moreover, probability or density functions of the univariate RVs (known as *marginal RVs*) that compose the bivariate RV can be obtained from $f(x_i, y_j)$

in the discrete case or $f(x,y)$ in the continuous case summing up or integrating out the nuisance variable (X or Y) as follows:

$$\sum_{i=1}^{h} f(x_i, y_j) = f(y_j), \quad j = 1, \ldots, k,$$

$$\sum_{j=1}^{k} f(x_i, y_j) = f(x_i), \quad i = 1, \ldots, h,$$

$$\int_{a}^{b} f(x,y)dx = f(y), \quad c \le y \le d,$$

$$\int_{c}^{d} f(x,y)dy = f(x), \quad a \le x \le b.$$

The two RVs X and Y, composing the bivariate RV (X, Y), are statistically independent (or independent in probability) if

$$f(x_i, y_j) = f(x_i)f(y_j), \quad \text{in the discrete case,}$$
$$f(x,y) = f(x)f(y), \quad \text{in the continuous case.}$$

On the other hand, if X and Y are not independent, it may be of interest to analyze the conditional RVs that, in the discrete case, are defined as

- conditional RV $X|\cdot$

$$(X|Y = y_j) = X|y_j, \quad j = 1, \ldots, k,$$

with conditional probability function

$$f(x_i|y_j) = \frac{f(x_i, y_j)}{f(y_j)}, i = 1, 2, \ldots, h; \ j = 1, 2, \ldots, k;$$

- conditional RV $Y|\cdot$

$$(Y|X = x_i) = Y|x_i, \quad i = 1, \ldots, h,$$

with

$$f(y_j|x_i) = \frac{f(y_j, x_i)}{f(x_i)}, i = 1, 2, \ldots, h; \ j = 1, 2, \ldots, k.$$

Overall, there are k conditional RVs $X|y_j$ (as many as are the categories of the conditioning RV Y) and h conditional RVs $Y|x_i$ (as many as are the categories of the conditioning RV X). In the continuous case, the definition of conditional RV and conditional density function is easily extended; in such a case, the conditional RVs $(X|Y = y) = X|y$ and $(Y|X = x) = Y|x$ are infinite.

It is worth outlining that the independence between X and Y does not imply the independence between $X|Z$ and $Y|Z$, where Z is a conditioning RV, that is,

$$f(x,y) = f(x)f(y) \quad \not\Rightarrow \quad f(x,y|z) = f(x|z)f(y|z),$$

and vice-versa. Moreover, if two RVs X and Y are not independent

$$f(x,y) \neq f(x)f(y),$$

the conditional RVs $X|Z$ and $Y|Z$ may be independent, that is,

$$f(x,y|z) = f(x|z)f(y|z).$$

As already outlined for the univariate RVs, the cumulative probability function or the probability or density functions provide a complete definition of the bivariate RV and, also, of the marginal and conditional RVs. However, it is often useful to consider a synthetic (and, therefore, partial) description of the RVs. Similar to the univariate case, the expected value may be defined for suitable transformations $g(\cdot)$ of a bivariate RV (X, Y):

$$E[g(X,Y)] = \begin{cases} \sum_{i=1}^h \sum_{j=1}^k g(x_i, y_j) f(x_i, y_j), & \text{if } (X,Y) \text{ is discrete,} \\ \int_a^b \int_c^d g(x,y) f(x,y) dx dy, & \text{if } (X,Y) \text{ is continuous.} \end{cases}$$

More precisely, some interesting moments are obtained as follows

- $g(X,Y) = X^r Y^s$, $r, s = 0, 1, 2, \ldots$

$$\mu_{rs} = E[g(X,Y)] = E(X^r Y^s) =$$

$$\begin{cases} \sum_{i=1}^h \sum_{j=1}^k x_i^r y_j^s f(x_i, y_j), & \text{if } (X,Y) \text{ is discrete} \\ \int_a^b \int_c^d x^r y^s f(x,y) dx dy, & \text{if } (X,Y) \text{ is continuous.} \end{cases}$$

The expected value μ_{rs} is known as rs-th mixed moment or mixed moment ! of order rs about the origin. Note that for $r = 1, s = 0$ and for $r = 0, s = 1$ the arithmetic means of X and Y are obtained, that is, $\mu_{10} = \mu_x = E(X)$ and $\mu_{01} = \mu_y = E(Y)$.

- $g(X,Y) = (X - \mu_x)^r (Y - \mu_y)^s$, $r, s = 0, 1, 2, \ldots$

$$\bar{\mu}_{rs} = E[g(X,Y)] = E\left[(X - \mu_x)^r (Y - \mu_y)^s\right] =$$

$$\begin{cases} \sum_{i=1}^h \sum_{j=1}^k (x_i - \mu_x)^r (y_j - \mu_y)^s f(x_i, y_j), & \text{if } (X,Y) \text{ is discrete,} \\ \int_a^b \int_c^d (x - \mu_x)^r (y - \mu_y)^s f(x,y) dx dy, & \text{if } (X,Y) \text{ is continuous,} \end{cases}$$

is known as rs-th central mixed moment or mixed moment about the mean of order rs. For $r = 2$ and $s = 0$ the central mixed moment $\bar{\mu}_{20}$ is defined, corresponding to σ_x^2, whereas for $r = 0$ and $s = 2$ $\bar{\mu}_{02}$ is obtained, corresponding to σ_y^2.

Another relevant central mixed moment is $\bar{\mu}_{11}$

$$\bar{\mu}_{11} = E[g(X,Y)] = E\left[(X - \mu_x)(Y - \mu_y)\right] =$$

$$\begin{cases} \sum_{i=1}^{h}\sum_{j=1}^{k}(x_i - \mu_x)(y_j - \mu_y)f(x_i, y_j), & \text{if } (X,Y) \text{ is discrete,} \\ \int_a^b \int_c^d (x - \mu_x)(y - \mu_y)f(x,y)dxdy, & \text{if } (X,Y) \text{ is continuous,} \end{cases}$$

which is known as *covariance* and is usually denoted by σ_{xy} (or σ_{yx}).

The covariance is an absolute measure of association between X and Y and is positive, negative, or null. The covariance is positive, when X and Y tend to vary along the same direction, that is, large values of X tend to be observed with large values of Y and small values of X with small values of Y. In such a case, positive (negative) deviations $(X - \mu_x)$ correspond to positive (negative) deviations $(Y - \mu_y)$ and products $(X - \mu_x)(Y - \mu_y)$ are mainly positive.

The covariance is negative when X and Y tend to vary in opposite directions, that is, when large values of one variable tend to be observed with small values of the other variable (and vice-versa). In such a case, the products $(X - \mu_x)(Y - \mu_y)$ are mainly negative.

- $g(X,Y) = \left(\frac{X-\mu_x}{\sigma_x}\right)^r \left(\frac{Y-\mu_y}{\sigma_y}\right)^s$, $r, s = 0, 1, 2, \ldots$

$$\bar{\bar{\mu}}_{rs} = E[g(X,Y)] = E\left[\left(\frac{X - \mu_x}{\sigma_x}\right)^r \left(\frac{Y - \mu_y}{\sigma_y}\right)^s\right] =$$

$$\begin{cases} \sum_{i=1}^{h}\sum_{j=1}^{k}(\frac{x_i-\mu_x}{\sigma_x})^r(\frac{y_j-\mu_y}{\sigma_y})^s f(x_i, y_j), & \text{if } (X,Y) \text{ is discrete,} \\ \int_a^b \int_c^d (\frac{x-\mu_x}{\sigma_x})^r(\frac{y-\mu_y}{\sigma_y})^s f(x,y)dxdy, & \text{if } (X,Y) \text{ is continuous,} \end{cases}$$

which is known as standardized mixed moment of order rs.

The most relevant standardized mixed moment is the *Bravais-Pearson correlation coefficient*, which is obtained when $r = 1$ and $s = 1$:

$$\bar{\bar{\mu}}_{11} = E\left[\left(\frac{X - \mu_x}{\sigma_x}\right)\left(\frac{Y - \mu_y}{\sigma_y}\right)\right] = \frac{\sigma_{xy}}{\sigma_x \sigma_y} = \rho.$$

Given the following relation[3]:

$$\sigma_{xy}^2 \leq \sigma_x^2 \sigma_y^2 \implies |\rho| \leq 1,$$

---

[3]It is worth noting that what follows derives from the Cauchy-Schwarz inequality:

$$|\langle a', b\rangle|^2 \leq \langle a', a\rangle\langle b', b\rangle,$$

where $\langle a', b\rangle$ is the inner scalar product between vectors a and b. If the dimension of a and b is n, then the Cauchy-Schwarz inequality is expressed as

$$(\sum_{i=1}^{n} a_i b_i)^2 \leq \sum_{i=1}^{n} a_i^2 \sum_{i=1}^{n} b_i^2.$$

the correlation coefficient assumes values in the range $(-1; +1)$ and it achieves the extremes -1 or $+1$ when X and Y are perfectly linearly related (perfect correlation). The sign $-$ or $+$ of ρ is driven by the covariance σ_{xy} and depends on the slope of the straight line that approximates observations on (X, Y): if the line is positively inclined, then $\rho > 0$; on the other hand, if the line is negatively inclined, then $\rho < 0$. When $\rho = 0$ (i.e., $\sigma_{xy} = 0$) the two components X and Y of the bivariate RV (X, Y) are linearly independent (uncorrelated). Note that, linear independency does not exclude a strong non-linear relationship between X and Y (e.g., a parabolic link such as $Y = a + bX^2$): in such a situation ρ is again null. In other words, the linear independence does not imply the statistical independence. In what follows we will see the relevant exception provided by the bivariate normal RV.

Obviously, the opposite is always true, that is, the statistical independence implies the linear independence as well as any other type of independence. Indeed, if X and Y are statistically independent, that is, $f(x, y) = f(x)f(y)$, then (without loss of generality we refer to the continuous case)

$$
\begin{aligned}
\sigma_{xy} &= \int_a^b \int_c^d (x - \mu_x)(y - \mu_y) f(x, y) dx dy \\
&= \int_a^b \int_c^d (x - \mu_x)(y - \mu_y) f(x) f(y) dx dy \\
&= \int_a^b (x - \mu_x) f(x) dx \int_c^d (y - \mu_y) f(y) dy = \bar{\mu}_{10} \bar{\mu}_{01} = 0.
\end{aligned}
$$

Note that what was mentioned above about the moments of a bivariate RV (X, Y) holds also for the conditional RVs $Y|x$ and $X|y$, adjusting the expected value $E(\cdot)$ for the conditional probability or density function. For instance, in the continuous case mean and variance of conditional RVs $Y|x$ and $X|y$ are obtained as:

$$
\mu_{y|x} = E(Y|x) = \int_{-\infty}^{+\infty} y f(y|x) dy,
$$

$$
\mu_{x|y} = E(X|y) = \int_{-\infty}^{+\infty} x f(x|y) dx,
$$

$$
\sigma_{y|x}^2 = E\left[(Y|x - \mu_{y|x})^2 \right] = \int_{-\infty}^{+\infty} (y - \mu_{y|x})^2 f(y|x) dy,
$$

$$
\sigma_{x|y}^2 = E\left[(X|y - \mu_{x|y})^2 \right] = \int_{-\infty}^{+\infty} (x - \mu_{x|y})^2 f(x|y) dx.
$$

To conclude, the transformation $g(X, Y) = e^{t_x X + t_y Y}$ ($-h_x < t < h_x$, $-h_y < t < h_y$, $h_x, h_y > 0$) defines the moment generating function of a bivariate RV

$$
m_{x,y}(t_x, t_y) = E(e^{t_x X + t_y Y}).
$$

Similar to the univariate case, when the moment generating function exists, it fully identifies the RV (X, Y) and allows us to easily derive both the marginal distributions:

$$\lim_{t_y \to 0} m_{x,y}(t_x, t_y) = \lim_{t_y \to 0} E(e^{t_x X + t_y Y}) = E(e^{t_x X}) = m_x(t_x),$$

$$\lim_{t_x \to 0} m_{x,y}(t_x, t_y) = \lim_{t_x \to 0} E(e^{t_x X + t_y Y}) = E(e^{t_y Y}) = m_y(t_y),$$

and the mixed moments about the origin, such as,

$$\frac{d}{dt_x} m_{x,y}(t_x, t_y)|_{t_x=0, t_y=0} = \frac{d}{dt_x} E(e^{t_x X + t_y Y})|_{t_x=0, t_y=0} = \frac{d}{dt_x} E(e^{t_x X})|_{t_x=0}$$

$$= \mu_{10} = \mu_x,$$

$$\frac{d}{dt_y} m_{x,y}(t_x, t_y)|_{t_x=0, t_y=0} = \frac{d}{dt_y} E(e^{t_x X + t_y Y})|_{t_x=0, t_y=0} = \frac{d}{dt_y} E(e^{t_y Y})|_{t_y=0}$$

$$= \mu_{01} = \mu_y,$$

$$\frac{d^2}{dt_x dt_y} m_{x,y}(t_x, t_y)|_{t_x=0, t_y=0} = \frac{d^2}{dt_x dt_y} E(e^{t_x X + t_y Y})|_{t_x=0, t_y=0} = \mu_{11}.$$

What was mentioned in this section as far as the bivariate RVs is easily extended to the general case of a k-dimensional RV, with $k > 2$. More precisely, the moment generating function is

$$m_{x_1, \ldots, x_k}(t_{x_1}, \ldots, t_{x_k}) = E(e^{t_{x_1} X_1 + \ldots + t_{x_k} X_k}),$$

from which the marginal distributions, the conditional distributions, and moments of any order can be derived.

In the following examples we provide a brief summary of some commonly used multivariate RVs. For each RV, we describe the main fields of application, the shape of the probability or density function, the mean and variance values, and the moment generating function (when it exists).

Example 2.16. Multinomial distribution.

Description: *number of $k+1$ different types of success in n independent trials (sampling with replacement); each trial may generate only one of the $k + 1$ possible outcomes; it generalizes the binomial RV.*

Parameters: *$p_1, \ldots, p_k$ and $p_{k+1} = 1 - \sum_{i=1}^{k} p_i$ with $\sum_{i=1}^{k+1} p_i = 1$ and p_i denoting the probability of outcome ω_i in any trial; fixed number of trials, n.*

Support: *$x_i = 0, 1, 2, \ldots, n$ ($i = 1, \ldots, k$) under the constraint $\sum_{i=1}^{k} x_i \leq n$.*

Probability function:

$$f(x_1, \ldots, x_k; n, p_1, \ldots, p_k) = \frac{n!}{x_1! \ldots x_k! \left(n - \sum_{i=1}^{k} x_i\right)!} \cdot$$

$$\cdot p_1^{x_1} \ldots p_k^{x_k} \left(1 - \sum_{i=1}^{k} p_i\right)^{n - \sum_{i=1}^{k} x_i} \tag{2.10}$$

Characteristics indices:

$$E(X_i) = \mu_{x_i} = np_i,$$
$$Var(X_i) = np_i(1 - p_i),$$
$$\sigma_{x_i x_j} = -np_i p_j,$$
$$\rho_{x_i x_j} = -\sqrt{\frac{p_i p_j}{(1 - p_i)(1 - p_j)}}.$$

Moment generating function:

$$m_{x_1, \ldots, x_k}(t_1, \ldots, t_k) = (p_1^{x_1} e^{t_1} + \ldots + p_k^{x_k} e^{t_k} + 1 - p_1 - \ldots - p_k)^n.$$

Example 2.17. Bivariate normal distribution.

Description: *numerous random experiments may be described through a bivariate normal RV; whenever suitably transformed, some RVs have an approximated normal distribution; by virtue of the central limit theorem the normal RV provides a good approximation of several RVs.*

Parameters: μ_x and μ_y, denoting the mean of X and Y, respectively; σ_x^2 and σ_y^2, denoting the variance of X and Y, respectively; ρ, denoting the correlation coefficient of (X, Y).

Support: $-\infty < x < +\infty$; $-\infty < y < +\infty$.

Density function:

$$f_{X,Y}(x, y; \mu_x, \mu_y, \sigma_x, \sigma_y, \rho_{yx}) =$$

$$\frac{1}{2\pi \sigma_x \sigma_y \sqrt{1 - \rho_{yx}^2}} e^{-\frac{1}{2(1 - \rho_{yx}^2)}\left[\left(\frac{x - \mu_x}{\sigma_x}\right)^2 - 2\rho_{yx}\left(\frac{x - \mu_x}{\sigma_x}\right)\left(\frac{y - \mu_y}{\sigma_y}\right) + \left(\frac{y - \mu_y}{\sigma_y}\right)^2\right]}.$$

We observe that, if $\rho = 0$, the two RVs X and Y are statistically independent. Moreover, it turns out that both marginal and conditional distributions are *normal:*

$$f_X(x) = \int_{-\infty}^{+\infty} f(x,y)dy = \frac{1}{\sqrt{2\pi}\sigma_x}e^{-\frac{1}{2\sigma_x^2}(x-\mu_x)^2},$$

$$f_Y(y) = \int_{-\infty}^{+\infty} f(x,y)dx = \frac{1}{\sqrt{2\pi}\sigma_y}e^{-\frac{1}{2\sigma_y^2}(y-\mu_y)^2},$$

$$f_{X|Y}(x|y) = \frac{f_{X,Y}(x,y)}{f_Y(y)} =$$

$$= \frac{1}{\sqrt{2\pi}\sigma_x\sqrt{1-\rho^2}}e^{-\frac{1}{2(1-\rho^2)\sigma_x^2}\left\{(x-\mu_x)-\left[\rho\frac{\sigma_x}{\sigma_y}(y-\mu_y)\right]\right\}^2},$$

$$f_{Y|X}(y|x) = \frac{f_{X,Y}(x,y)}{f_X(x)} =$$

$$= \frac{1}{\sqrt{2\pi}\sigma_y\sqrt{1-\rho^2}}e^{-\frac{1}{2(1-\rho^2)\sigma_y^2}\left\{(y-\mu_y)-\left[\rho\frac{\sigma_x}{\sigma_y}(x-\mu_x)\right]\right\}^2}.$$

Obviously, if $\rho = 0$, the conditional distributions are equal to the marginal ones.

Characteristic indices:

$$\mu_{10} = E(X) = \mu_x,$$
$$\mu_{01} = E(Y) = \mu_y,$$
$$\bar{\mu}_{20} = E\left[(X-\mu_x)^2\right] = \sigma_x^2,$$
$$\bar{\mu}_{02} = E\left[(Y-\mu_y)^2\right] = \sigma_y^2,$$
$$\bar{\mu}_{11} = E\left[\left(\frac{X-\mu_x}{\sigma_x}\right)\left(\frac{Y-\mu_y}{\sigma_y}\right)\right] = \frac{\mu_{11}-\mu_{01}\mu_{10}}{\sigma_x\sigma_y} = \rho.$$

Moment generating function:

$$m_{x,y}(t_x,t_y) = e^{t_xX+t_yY} = e^{t_x\mu_x+t_y\mu_y+\frac{1}{2}(t_x^2\sigma_x^2+2\rho t_x t_y \sigma_x \sigma_y + t_y^2 \sigma_y^2)}.$$

Example 2.18. Multivariate normal distribution

Description: *it is the generalization of bivariate normal RV to the case of $k > 2$ RVs.*

Parameters: *vector of means of $\boldsymbol{X}' = (X_1, \ldots, X_k)$*

$$\boldsymbol{\mu}' = (\mu_1, \mu_2, \ldots, \mu_k)$$

and variances and covariances matrix

$$\boldsymbol{\Sigma} = \begin{pmatrix} \sigma_1^2 & \sigma_{12} & \cdots & \sigma_{1k} \\ \sigma_{21} & \sigma_2^2 & \cdots & \sigma_{2k} \\ \vdots & \vdots & \cdots & \vdots \\ \sigma_{k1} & \sigma_{k2} & \cdots & \sigma_{k^2} \end{pmatrix},$$

with σ_i^2 variance of X_i and σ_{ij} covariance of X_i and X_j $(i, j = 1, \ldots, k)$.

Support: $-\infty < x_i < +\infty$ for $i = 1, \ldots, k$.

Density function:

$$f(x; \mu, \Sigma) = \frac{1}{(2\pi)^{n/2}\sqrt{|\Sigma|}} e^{-\frac{1}{2}(x-\mu)'\Sigma^{-1}(x-\mu)},$$

where $(x - \mu)'\Sigma^{-1}(x - \mu)$ is known as Mahalanobis distance (or generalized distance). Alternatively, the density function may be expressed as

$$f(x; \mu, D, R) = \frac{1}{(2\pi)^{n/2}\sqrt{|DRD|}} e^{-\frac{1}{2}(x-\mu)'(DRD)^{-1}(x-\mu)},$$

where D is the dispersion matrix and R is the correlation matrix:

$$D = \begin{pmatrix} \sigma_1^2 & 0 & \cdots & 0 \\ 0 & \sigma_2^2 & \cdots & 0 \\ \vdots & \vdots & \cdots & \vdots \\ 0 & 0 & \cdots & \sigma_k^2 \end{pmatrix}$$

and

$$R = \begin{pmatrix} 1 & \sigma_{12} & \cdots & \rho_{1k} \\ \rho_{21} & 1 & \cdots & \rho_{2k} \\ \vdots & \vdots & \cdots & \vdots \\ \rho_{k1} & \rho_{k2} & \cdots & 1 \end{pmatrix},$$

with $\rho_{ij} = \frac{\sigma_{x_i x_j}}{\sigma_{x_i}\sigma_{x_j}}$.

Characteristic indices: μ and Σ

Moment generating function:

$$m_x(t) = e^{t'\mu + \frac{1}{2}t'\Sigma t}.$$

Similarly to the bivariate normal RV, each marginal and conditional component of $X' = (X_1, \ldots, X_k)$ is normally distributed. The opposite is not always true: only under the independence of $X_1, \ldots, X_k$, the normality of each component X_i implies the normality of the multivariate RV $X' = (X_1, \ldots, X_k)$.

Example 2.19. Dirichlet distribution.

Description: *the Dirichlet RV is a generalization of the Beta distribution to the multivariate case; it is often used in the Bayesian inferential statistics as a prior distribution of the multinomial RV.*

Parameters: $\boldsymbol{\alpha}' = (\alpha_1, \ldots, \alpha_{k+1})$.

Support: $x_1, \ldots, x_{k+1} > 0$ *with* $x_{k+1} = 1 - \sum_{i=1}^{k} x_i$, $\sum_{i=1}^{k} x_i < 1$.

Density function:

$$f(x_1, \ldots, x_k; \boldsymbol{\alpha}) = \frac{\Gamma\left(\sum_{i=1}^{k+1} \alpha_i\right)}{\prod_{i=1}^{k+1} \Gamma(\alpha_i)} \prod_{i=1}^{k+1} x_i^{\alpha_i - 1}. \tag{2.11}$$

A special case is given by the symmetric Dirichlet distribution, which is obtained when $\alpha_i = \alpha$ *for all* $i = 1, \ldots, k + 1$:

$$f(x_1, \ldots, x_k; \alpha) = \frac{\Gamma\left[(k+1)\alpha\right]}{[\Gamma(\alpha)]^{k+1}} \prod_{i=1}^{k+1} x_i^{\alpha - 1}.$$

Mean and variance:

$$E(X_i) = \frac{\alpha_i}{\sum_{i=1}^{k+1} \alpha_i},$$

$$Var(X_i) = \frac{\alpha_i(\sum_{i=1}^{k+1} \alpha_i - \alpha_i)}{(\sum_{i=1}^{k+1} \alpha_i)^2 (\sum_{i=1}^{k+1} \alpha_i + 1)}.$$

Moment generating function: not useful.

2.6   The exponential family

A multivariate RV $\boldsymbol{X}' = (X_1, \ldots, X_k)$ with parameter vector $\boldsymbol{\theta}' = (\theta_1, \ldots, \theta_r)$ and probability or density function $f(\boldsymbol{x}; \boldsymbol{\theta})$ belongs to the *exponential family* if $f(\boldsymbol{x}; \boldsymbol{\theta})$ may be written as (standard formulation)

$$f(\boldsymbol{x}; \boldsymbol{\theta}) = a(\boldsymbol{\theta})h(\boldsymbol{x})e^{\sum_{j=1}^{r} \varphi_j(\boldsymbol{\theta})t_j(\boldsymbol{x})}, \quad a(\boldsymbol{\theta}) \geq 0, \ h(\boldsymbol{x}) \geq 0, \ \boldsymbol{\theta} \in \Theta,$$

where $\boldsymbol{x}' = (x_1, \ldots, x_k)$ is the vector of observed values of X, $\boldsymbol{\theta}$ is the *standard parameter* and Θ is the *standard parameter space*. Alternatively, the probability or density function of the exponential family may be expressed as

$$f(\boldsymbol{x}; \boldsymbol{\theta}) = h(\boldsymbol{x})e^{\sum_{j=1}^{r} \varphi_j(\boldsymbol{\theta})t_j(\boldsymbol{x}) - d(\boldsymbol{\theta})}, \text{ or}$$

$$f(\boldsymbol{x}; \boldsymbol{\theta}) = e^{\sum_{j=1}^{r} \varphi_j(\boldsymbol{\theta})t_j(\boldsymbol{x}) - d(\boldsymbol{\theta}) + g(\boldsymbol{x})},$$

with $-d(\boldsymbol{\theta}) = \log[a(\boldsymbol{\theta})]$ and $g(\boldsymbol{x}) = \log[h(\boldsymbol{x})]$. The exponential family is regular if the support of $\boldsymbol{X}$ is independent of $\boldsymbol{\theta}$, irregular otherwise. The expression of $f(x; \theta)$ is immediately specified for the univariate case (substitute $\boldsymbol{x}$ with x) and for the case with only one parameter (substitute $\boldsymbol{\theta}$ with θ).

Usually, the exponential family is not represented as above described (standard formulation), but in terms of the so called natural or canonical formulation, which is obtained from the standard one through a suitable reparameterization; the new parameters $\boldsymbol{\varphi}(\boldsymbol{\theta}) \in \boldsymbol{\Phi}$ (where $\boldsymbol{\Phi}$ is the natural parametric space) are the natural or canonical parameters. In the multivariate and multi-parametric case, the natural formulation of the density function of a member of the exponential family is

$$f(\boldsymbol{x}; \boldsymbol{\varphi}) = c(\boldsymbol{\varphi})h(\boldsymbol{x})e^{\sum_{j=1}^{r} \varphi_j t_j(\boldsymbol{x})},$$

or, alternatively,

$$f(\boldsymbol{x}; \boldsymbol{\varphi}) = h(\boldsymbol{x})e^{\sum_{j=1}^{r} \varphi_j t_j(\boldsymbol{x})-d(\boldsymbol{\varphi})}, \text{ or}$$
$$f(\boldsymbol{x}; \boldsymbol{\varphi}) = e^{\sum_{j=1}^{r} \varphi_j t_j(\boldsymbol{x})-d(\boldsymbol{\varphi})+g(\boldsymbol{x})}.$$

Through suitable specifications of $c(\cdot)$, $h(\cdot)$, $\varphi(\cdot)$, and $t(\cdot)$ many common distributions are obtained, among which are binomial (when n is assumed to be known), negative binomial (when k is assumed to be known), multinomial (given n), Poisson, Gamma, Beta, normal, and multivariate normal (for details on these RVs see examples in Sections 2.4 and 2.5). On the other hand, the exponential family does not include all distributions whose support depends on the parameters characterizing the probability or density function (such as, uniform when the extremes are unknown, Student's t, and Fisher-Snedecor F; Section 2.4 for details), unless they are assumed to be known.

In what follows, some examples of exponential families are illustrated, showing the natural formulation of the probability or density function.

Example 2.20. Binomial distribution. *The probability function of the binomial RV is characterized by parameters n and p. Usually, n is known, then the main interest is on $\theta = p$. It can be shown that the binomial RV belongs to the exponential family. Indeed,*

$$f(x_i; n, p) = \binom{n}{x_i}p^{x_i}(1-p)^{n-x_i} = \binom{n}{x_i}(1-p)^n e^{\log(\frac{p}{1-p})x_i}$$
$$= h(x_i)a(p)e^{\varphi(p)t(x_i)}$$

where $a(p) = (1-p)^n \geq 0$, $h(x_i) = \binom{n}{x_i} \geq 0$, $\varphi(p) = \log\left(\frac{p}{1-p}\right)$, and $t(x_i) = x_i$.

Example 2.21. Poisson distribution. *The probability function of the Poisson RV (Eq. (2.3)) is characterized by $\theta = \lambda$. It can be shown that the Poisson RV belongs to the exponential family. Indeed,*

$$f(x_i) = \frac{\lambda^{x_i}e^{-\lambda}}{x_i!} = \frac{1}{x_i!}e^{-\lambda}e^{x_i \log \lambda} = h(x_i)a(\lambda)e^{\varphi(\lambda)t(x_i)},$$

where $h(x_i) = 1/x_i!$, $a(\lambda) = e^{-\lambda}$, $\varphi(\lambda) = \log \lambda$, and $t(x_i) = x_i$.

Example 2.22. Normal distribution. *The probability function of the normal RV is characterized by* $\boldsymbol{\theta} = (\mu, \sigma^2)$. *It can be shown that the normal RV belongs to the exponential family. Indeed,*

$$f(x) = f(x; \mu, \sigma^2) = \frac{1}{\sqrt{2\pi\sigma^2}} e^{-\frac{1}{2}\left(\frac{x-\mu}{\sigma}\right)^2} = \frac{1}{\sqrt{2\pi\sigma^2}} e^{-\frac{\mu^2}{2\sigma^2}} e^{-\frac{1}{2\sigma^2}x^2 + \frac{\mu}{\sigma^2}x},$$

where $a(\mu, \sigma^2) = \frac{1}{\sqrt{2\pi\sigma^2}} e^{-\frac{\mu^2}{2\sigma^2}}$, $h(x) = 1$, $\varphi_1 = -\frac{1}{2\sigma^2}$, $\varphi_2 = \frac{\mu}{\sigma^2}$, $t_1(x) = x^2$, *and* $t_2(x) = x$.

Example 2.23. Multinomial distribution. *The probability function of the k-dimensional multinomial RV (Eq. (2.10)) is characterized by* $k + 1$ *parameters,* n *and* $p_1, p_2, \ldots, p_k$. *Usually,* n *is known and, then, the parameters of interest are* $\boldsymbol{p}' = (p_1, p_2, \ldots, p_k)$. *It can be shown that the multinomial RV belongs to the exponential family. Indeed,*

$$f(\boldsymbol{x}; \boldsymbol{p}) = \frac{n!}{\prod_{i=1}^{k} x_i! \left(n - \sum_{i=1}^{k} x_i\right)!} \prod_{i=1}^{k} p_i^{x_i} \left(1 - \sum_{i=1}^{k} p_i\right)^{n - \sum_{i=1}^{k} x_i}$$

$$= \frac{n!}{\prod_{i=1}^{k} x_i! \left(n - \sum_{i=1}^{k} x_i\right)!} e^{n \log(1 - \sum_{i=1}^{k} p_i)} e^{\sum_{i=1}^{k} x_i \log\left(\frac{p_i}{1 - \sum_{i=1}^{k} p_i}\right)}$$

where $h(\boldsymbol{x}) = \frac{n!}{\prod_{i=1}^{k} x_i! (n - \sum_{i=1}^{k} x_i)!}$, $h(\boldsymbol{p}) = e^{n \log(1 - \sum_{i=1}^{k} p_i)}$, $\varphi_i = \log\left(\frac{p_i}{1 - \sum_{i=1}^{k} p_i}\right)$, *and* $t_i(\boldsymbol{x}) = x_i$.

2.7   Descriptive statistics and statistical inference

Statistical theory distinguishes according to the availability of observations on the whole population of interest or only on a sample (i.e., a subset of the population units). In the first case, we move within *descriptive statistics*, whereas, in the second case, the topics of *statistical inference* are involved. Probability theory, whose fundamentals have been illustrated in the first part of this chapter, represents the link between descriptive statistics and statistical inference, because it provides the essential instruments to study phenomena when only a subset of manifestations is available.

Let F denote the phenomenon of interest and let P be the population, that is, the full set of all the possible manifestations of F (Figure 2.1). Usually, P has a huge dimension and suitable instruments (or functions) are necessary to attain a synthetic representation of P and, then, clear information about F. In such a frame, descriptive statistics is a set of all functions, say

$s_1(\cdot), \ldots, s_i(\cdot), \ldots, s_h(\cdot)$, such as frequencies, averages (arithmetic mean, median, quartiles, mode), deviations (variance, coefficient of variation, range), graphs (histograms, pie plots, box plots), which take elements of P as input and give as output a set R (definitely smaller than P) of synthetic representations of P that helps us to acquire knowledge of F. This process is illustrated in a schematic way in Figure 2.1.

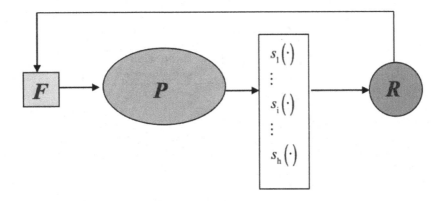

FIGURE 2.1
Graphical representation of the knowledge process in the frame of descriptive statistics.

Usually, the study of phenomena cannot rely on the observations of the whole population, but just a sample of units belonging to the population is available. In such a context, theoretical results and instruments proper to statistical inference are necessary to generalize to the whole population the knowledge acquired from the observed sample. More precisely Figure 2.2, let Ω be a sample of units drawn from population P; let also $t_1(\cdot), \ldots, t_i(\cdot) \ldots, t_k(\cdot)$ a set of functions (not necessarily equal to $s_1(\cdot), \ldots, s_i(\cdot), \ldots, s_h(\cdot)$) that take as input the elements of Ω and give as output a set R_Ω of synthetic representations of Ω. The aim of statistical inference is to detect functions and procedures such that R_Ω is as close as possible to R, the set of synthetic representations of population P.

For instance, let us assume that we are interested in a synthetic index θ that can be obtained through function $s(\cdot)$ applied on P (e.g., θ is the population mean). However, if we cannot observe P but only Ω, we need an alternative function, say $t(\cdot)$, that applied on Ω gives a value $\hat{\theta}$ as close as possible to θ. In other words, we are not interested in the sample characteristics *per sé*, but we are interested in the sample observations to say something about the population from which the sample was drawn. The cognitive process from the sample to the generating population is known as *inductive process* or

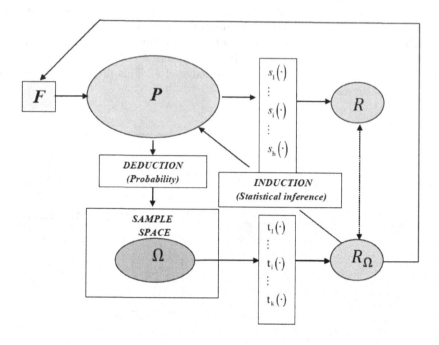

FIGURE 2.2
Graphical representation of the process underlying the classical statistical inference.

inferential process. As opposed to the inductive process, the *deductive process* makes it possible to predict the output of sampling from the population, in virtue of the probability theory (Figure 2.2).

Statistical inference deals with two main issues: estimation and hypothesis testing. Statistical estimation concerns the *measure* of a certain unknown synthetic index θ, whereas hypothesis testing is about the *plausibility of a certain assumption* about θ. In both cases, probability theory provides the theoretical framework to detect instruments (i.e., estimators in the estimation setting and test statistics in the hypothesis testing) that satisfy desirable properties in terms of closeness of the elements of R_Ω to the elements of R.

The complex nature of real phenomena often arises in the formulation of *probabilistic models*, $f(x; \theta)$ (Figure 2.3). In such a context, two main situations of lack of knowledge can be distinguished. The first one is when the analytic form of $f(x; \theta)$ is known (e.g., it is assumed that $f(\cdot)$ is normally distributed), whereas θ is partially or completely unknown. This is the frame of reference of the *parametric statistical inference*. The second situation is encountered when no assumption on the distributive form of $f(x; \theta)$ is introduced: both $f(\cdot)$ and θ are unknown. This latter case defines the frame of reference of the *non-parametric statistical inference* or *distribution-free statistical inference.* Finally, a third and intermediate situation, which is known as *semi-parametric statistical inference*, results when partial model specifications are introduced (e.g., it is assumed that $f(\cdot)$ belongs to the exponential family, but the exact distribution is not specified).

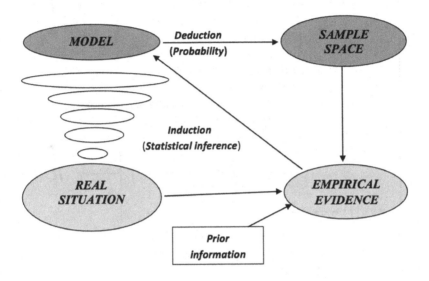

FIGURE 2.3
Graphical representation of the relationship between probability (classical and Bayesian), statistical inference, empirical evidence and model.

In what follows we will deal with the problems of estimation and hypothesis testing in the context of parametric statistical inference, distinguishing the *classical or frequentist approach* and the *Bayesian approach*. In the classical approach to statistical inference the process of knowledge acquisition explicitly makes use only of sample data, whereas in the Bayesian approach to inference the information from the sample data is integrated with the prior knowledge, which is formalized in a suitable way. The knowledge acquisition process illustrated in Figure 2.2 under the classical approach modifies as shown in Figure 2.4 under the Bayesian perspective.

2.8 Sample distributions

Throughout the book we will refer to the simple random sampling with replacement (bernoullian sampling).

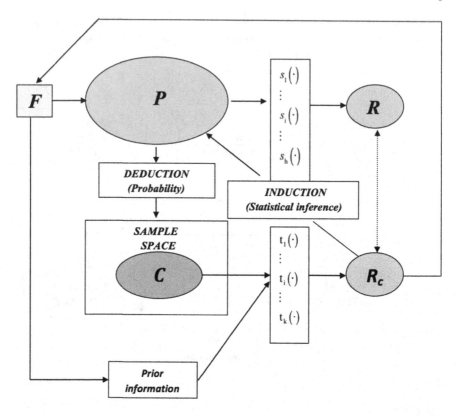

FIGURE 2.4
Graphical representation of the knowledge acquisition process in the frame of
Bayesian statistical inference.

Definition 2.13. *Let X be a RV with probability or density function equal
to $f(x; \boldsymbol{\theta})$, where $\boldsymbol{\theta}$ is the vector of parameters characterizing $f(\cdot)$. A simple
random sample of size n is any sequence $X_1, X_2, \ldots, X_n$ of RVs drawn with
replacement from X. $X_1, X_2, \ldots, X_n$ are i.i.d. and the joint probability or
joint density function is*

$$f(x_1, \ldots, x_n; \theta_1, \ldots, \theta_r) = f(\boldsymbol{x}; \boldsymbol{\theta}) = f(x_1; \boldsymbol{\theta}) \ldots f(x_n; \boldsymbol{\theta}) = \prod_{i=1}^{n} f(x_i; \boldsymbol{\theta}).$$

*Moreover, the sample point $\boldsymbol{X}' = (X_1, X_2, \ldots, X_n)$ is defined in the n-
dimensional sample space (or sample universe) Ω (i.e., $\boldsymbol{X} \in \Omega$).*

In the above formula we denoted by $f(x_i; \boldsymbol{\theta})$ $(i = 1, 2, \ldots, n)$ the probability or density function of the i-th sample unit. Because of the independence
of the sample units, the distribution of each RV X_i equals the distribution of
the generating population X, that is, $f(x_i; \boldsymbol{\theta}) = f(x; \boldsymbol{\theta})$.

Example 2.24. *Let X be a normal RV with mean μ and variance σ^2. Then, the density function of a random sample of size n drawn from X is*

$$f(x_1,\ldots,x_n;\mu,\sigma^2) = \prod_{i=1}^{n} f(x_i;\mu,\sigma^2)$$

$$= \prod_{i=1}^{n} \frac{1}{\sqrt{2\pi\sigma^2}} e^{-\frac{1}{2}\left(\frac{x_i-\mu}{\sigma}\right)^2} = \frac{1}{(2\pi\sigma^2)^{n/2}} e^{-\frac{1}{2\sigma^2}\sum_{i=1}^{n}(x_i-\mu)^2}.$$

Similarly, if X has a Poisson distribution with parameter λ (Eq. (2.3)), the probability function of a random sample of size n is

$$f(x_1,\ldots,x_n;\lambda) = \prod_{i=1}^{n} f(x_i;\lambda) = \prod_{i=1}^{n} \frac{\lambda^{x_i}}{x_i!} e^{-\lambda}.$$

The two functions $f(x_1,\ldots,x_n;\mu,\sigma^2)$ and $f(x_1,\ldots,x_n;\lambda)$ above introduced and, more in general, each sample probability or density function $f(x_1,\ldots,x_n;\boldsymbol{\theta}) = f(\boldsymbol{x};\boldsymbol{\theta}) = \prod_{i=1}^{n} f(x_i;\boldsymbol{\theta})$, may be interpreted in terms of *likelihood function*:

$$\mathcal{L}(\boldsymbol{\theta}) = \mathcal{L}(\boldsymbol{\theta}|\boldsymbol{X} = \boldsymbol{x}) = \prod_{i=1}^{n} f(\boldsymbol{\theta}|\boldsymbol{x}), \tag{2.12}$$

where $\boldsymbol{X}' = (X_1,\ldots,X_n)$ is the n-dimensional RV associated with the n sample units and $\boldsymbol{x}' = (x_1,\ldots,x_n)$ is the sample point, that is, a specific determination of $\boldsymbol{X}$ (i.e., one of the elements of the n-dimensional sample space Ω).

From a formal point of view $f(\boldsymbol{x};\boldsymbol{\theta})$ and $\mathcal{L}(\boldsymbol{\theta})$ are perfectly equivalent; however, from a substantial point of view, the likelihood function has to be interpreted as a function of the (usually unknown) vector $\boldsymbol{\theta}$ for a given observed sample point, whereas the sample probability or density function $f(\boldsymbol{x};\boldsymbol{\theta})$ depends both on $\boldsymbol{\theta}$ and the RVs $X_1,\ldots,X_n$, both of them being unknown. In other words, the likelihood function describes how likely are the possible values of $\boldsymbol{\theta}$.

Example 2.25. *Let X be a Bernoulli RV with parameter $\boldsymbol{\theta} = p$ and let $(X_1,\ldots,X_n)$ be a random sample drawn from X. The sample probability function is*

$$f(X_1,\ldots,X_n,p) = \prod_{i=1}^{n} f(X_i,p) = \prod_{i=1}^{n} p^{X_i}(1-p)^{1-X_i} = p^{\sum_{i=1}^{n} X_i}(1-p)^{n-\sum_{i=1}^{n} X_i},$$

whereas the likelihood function is

$$\mathcal{L}(p|x_1,\ldots,x_n) = \prod_{i=1}^{n} f(x_i,p) = \prod_{i=1}^{n} p^{x_i}(1-p)^{1-x_i} = p^{\sum_{i=1}^{n} x_i}(1-p)^{n-\sum_{i=1}^{n} x_i}.$$

We remind that $x_1,\ldots,x_n$ is a sequence of real numbers, then the specification of $\mathcal{L}(p|x_1,\ldots,x_n)$ depends on the observed sample. For instance:

- *if we observe* $\boldsymbol{x} = (1, 0, 1, 1, 1, 1)$ *(n = 6 and* $\sum_{i=1}^{6} x_i = 5$*), then*

$$\mathcal{L}(p|\boldsymbol{x}) = p^5(1-p);$$

- *if we observe* $\boldsymbol{x} = (1, 1, 0, 1, 1, 1, 1, 1, 1, 1, 0, 1)$ *(n = 12 and* $\sum_{i=1}^{12} x_i = 10$*), then*

$$\mathcal{L}(p|\underline{x}) = p^{10}(1-p)^2;$$

- *if we observe a sample of size n = 36 with* $\sum_{i=1}^{36} x_i = 30$*, then*

$$\mathcal{L}(p|\underline{x}) = p^{30}(1-p)^6.$$

In Figure 2.5 the shape of these three likelihood functions is shown.

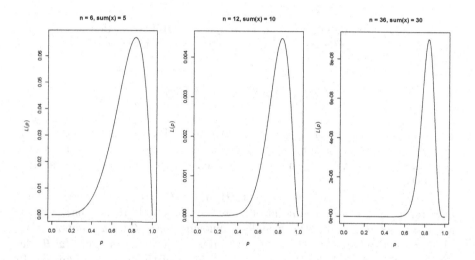

FIGURE 2.5
Likelihood function for three random samples drawn from a Bernoulli RV.

Definition 2.14. *A statistic is a non-constant function* $T(\cdot)$ *of the sample elements* $X_1, \ldots, X_n$, *which does not depend on the unknown parameters* $\boldsymbol{\theta}$. *As* $X_1, \ldots, X_n$ *are RVs, then also any statistic* $T(\boldsymbol{X}) = T(X_1, \ldots, X_n)$ *is a RV and, consequently, has a probability distribution, known as* sample distribution, *whose specification depends on the sample probability or sample density function of* $X_1, \ldots, X_n$.

Let $\boldsymbol{X}' = (X_1, \ldots, X_n)$ be a simple random sample drawn from X and let $E(X^r) = \mu_r$ and $E[(X - \mu_1)^2] = \sigma^2$ be the moment about the origin of order r and the variance of X, respectively. It can be shown that the following relevant relations hold:

- $T(\boldsymbol{X}) = \bar{X}_r = T_r(X_1, X_2, \ldots, X_n) = \frac{1}{n} \sum_{i=1}^{n} X_i^r$ $(r = 1, 2, \ldots,)$ is the sample (or empirical) moment about the origin of order r, known as *sample mean*, and denoted by $\bar{X}$, when $r = 1$:

$$\bar{X} = \frac{1}{n} \sum_{i=1}^{n} X_i.$$

The expected value of $\bar{X}_r$ coincides with the r-th moment about the origin μ_r:

$$E(\bar{X}_r) = E\left(\frac{1}{n} \sum_{i=1}^{n} X_i^r\right) = \frac{1}{n} \sum_{i=1}^{n} E(X_i^r) = \mu_r$$

and, also, $E(\bar{X}) = \mu$.

- $T(\boldsymbol{X}) = S_*^2 = T_r(X_1, X_2, \ldots, X_n) = \frac{1}{n} \sum_{i=1}^{n} (X_i - \bar{X})^2$ is the *sample variance* (or sample moment about the mean of order 2). Its expected value is:

$$E(S_*^2) = E\left(\frac{1}{n} \sum_{i=1}^{n} (X_i - \bar{X})^2\right) = \frac{n-1}{n} \sigma^2. \tag{2.13}$$

- $T(\boldsymbol{X}) = S^2 = T_s(X_1, X_2, \ldots, X_n) = \frac{1}{n-1} \sum_{i=1}^{n} (X_i - \bar{X})^2$ is the *unbiased sample variance*, whose expected value coincides with variance σ^2 of population X:

$$E(S^2) = E\left(\frac{1}{n-1} \sum_{i=1}^{n} (X_i - \bar{X})^2\right) = \sigma^2. \tag{2.14}$$

More in general, the following theorem provides theoretical results that are relevant for the development of the statistical inference.

Theorem 2.6. *If $X_1, \ldots, X_n$ are elements of a random sample drawn from a normal population, with mean μ and variance σ^2, then the following relations hold:*

- *the sample mean $\bar{X}$ is normally distributed with mean μ and variance σ^2/n and, then, the standardized sample mean $Z = \frac{\bar{X}-\mu}{\sigma/\sqrt{n}}$ is distributed as a standard normal, that is,*

$$\bar{X} \sim N(\mu, \frac{\sigma^2}{n}) \quad \text{and} \quad Z \sim N(0,1);$$

- *the squared sum of standardized sample elements $\frac{X_1-\mu}{\sigma}, \frac{X_2-\mu}{\sigma}, \ldots, \frac{X_i-\mu}{\sigma}, \ldots,$ $\frac{X_n-\mu}{\sigma}$ is distributed as a χ^2 RV (Eq. (2.7)), that is,*

$$Y = \frac{1}{\sigma^2} \sum_{i=1}^{n} (X_i - \mu)^2 = \sum_{i=1}^{n} \left(\frac{X_i - \mu}{\sigma}\right)^2 \sim \chi_n^2;$$

and, similarly,

$$V = \frac{(n-1)S^2}{\sigma^2} = \frac{\sum_{i=1}^n (X_i - \bar{X})^2}{\sigma^2} \sim \chi^2_{n-1},$$

with S^2 the unbiased sample variance above defined;

- *$\bar{X}$ and V are independent;*

- *the ratio of the standardized sample mean $Z = \frac{\bar{X} - \mu}{\sigma/\sqrt{n}}$ to the square root of $V = \frac{(n-1)S^2}{\sigma^2}$ divided by its degrees of freedom is distributed as a Student's t, that is,*

$$W = \frac{Z}{\sqrt{V/(n-1)}} = \frac{\frac{\bar{X}-\mu}{\sigma/\sqrt{n}}}{\sqrt{\frac{(n-1)S^2}{\sigma^2}}} = \frac{\bar{X}-\mu}{S/\sqrt{n}} \sim t_{n-1}.$$

Whenever the normality of the generating population X does not hold, the normal distribution is anyway a good proxy for the sample mean (under mild conditions), in virtue of the central limit theorem.

Theorem 2.7. Central limit theorem *(version of Lindberg-Levy). Let $X_1, X_2, \ldots, X_n, \ldots$ be a sequence of independent and identically distributed RVs with mean μ and finite variance $\sigma^2 > 0$. Let also $\bar{X} = \frac{1}{n}\sum_{i=1}^n X_i$ with $E(\bar{X}) = \mu$ and $Var(\bar{X}) = \frac{\sigma^2}{n}$. Then, the standardized RV*

$$Z_n = \frac{\bar{X} - \mu}{\sigma/\sqrt{n}}$$

tends to the standard normal distribution when $n \to +\infty$.

In other words, according to the central limit theorem, if $X' = (X_1, X_2, \ldots, X_n)$ is a simple random sample of size n drawn from a population with mean μ and finite variance σ^2, the probability or density function of the sample mean converges to a normal distribution with mean μ and variance σ^2/n, as n approaches to infinity. In practice, the approximation is often good for finite values of n that are only moderately high. The goodness of the approximation depends on the generating population. For instance, symmetric distributions are better approximated than skewed distributions.

2.9   Classical statistical inference

In Section 2.8 we described the sample mean and the sample variance, under the assumption of normal populations. Similar considerations may be extended to other statistics $T(\cdot)$ and to non-normal populations. The general idea consists of collapsing sample information through suitable functions of sample elements $T(X_1, X_2, \ldots, X_n)$ so that the knowledge acquired from the empirical evidence may be extended to the whole population.

The hard core of statistical inference lies in the definition of *optimality criteria* and in detecting *rules* that make it possible to satisfy these criteria so that the inductive process leads to valid conclusions about the generating population. Obviously, the optimality criteria depend on the specific inductive problem (e.g., point estimation, interval estimation, hypothesis testing). To make the comprehension of these issues easier, the concept of sufficient statistic and the Neyman-Fisher factorization criterion are useful.

Definition 2.15. Sufficiency. *Let $\boldsymbol{X}' = (X_1, \ldots, X_n)$ be a simple random sample drawn from a discrete or continuous RV X, having probability or density function $f(x; \theta)$. A statistic $T(\boldsymbol{X}) = T(X_1, \ldots, X_n)$ is sufficient for θ if and only if the sample distribution conditional on any value of $T(\cdot)$, that is, $f(x_1, x_2, \ldots, x_n | T(\boldsymbol{X}) = t(\boldsymbol{x}))$, is constant for any value of θ. In other words, $T(\boldsymbol{X}) = T(X_1, \ldots, X_n)$ is a sufficient statistic if and only if the conditional distribution of $\boldsymbol{X}' = (X_1, \ldots, X_n)$ given $T(\boldsymbol{X}) = t(\boldsymbol{x})$ is independent of θ.*

From the definition of sufficient statistic it follows that:

Theorem 2.8. *If $p(\boldsymbol{x}|\theta)$ is the joint probability or joint density function of $\boldsymbol{X}$ and $q(t|\theta)$ is the probability of $T(\boldsymbol{X})$, then necessary and sufficient condition in order to that $T(\boldsymbol{X})$ is a sufficient statistic for θ is that*

$$\frac{p(\boldsymbol{x}|\theta)}{q(T(\boldsymbol{x})|\theta)}$$

is independent from θ for all $\boldsymbol{x} \in \boldsymbol{\Omega}$.

Moreover, to easily find a sufficient statistic for parameter θ, one can rely on the Neyman-Fisher factorization criterion [95].

Theorem 2.9. Neyman-Fisher factorization criterion. *Let $\boldsymbol{X}' = (X_1, \ldots, X_n)$ be a simple random sample drawn from X with joint probability or joint density function $f(\boldsymbol{x}; \theta)$. A statistic $T(X_1, X_2, \ldots, X_n)$ is sufficient for θ if and only if the following relation holds:*

$$f(x_1, \ldots, x_n; \theta) = \prod_{i=1}^{n} f(x_i; \theta) = g[T(x_1, \ldots, x_n; \theta)]h(x_1, \ldots, x_n), \quad (2.15)$$

where $h(x_1, \ldots, x_n)$ is a non-negative function of the sample elements and $g[T(x_1, \ldots, x_n; \theta)]$ is a non-negative function depending on θ and on the sample elements through $T(\cdot)$.

The concept of sufficiency and the factorization criterion may be extended to a vector $\boldsymbol{\theta}$ of unknown parameters and to a vector $\boldsymbol{T}(\boldsymbol{X})$ of statistics. In such a case, statistics of vector $\boldsymbol{T}(\boldsymbol{X})$ are named jointly sufficient for $\boldsymbol{\theta}$.

2.9.1   Optimal point estimators

Without loss of generality, let us assume that $\boldsymbol{\theta} = \theta$. The point estimation of an unknown parameter θ consists in identifying a function of sample elements, $\hat{\Theta} = T(X_1, \ldots, X_n)$, such that the real value $\hat{\theta}$ obtained for an observed sample $(x_1, \ldots, x_n)$ is as close as possible to the unknown entity $\boldsymbol{\theta}$. Such a function is named (point) estimator:

Definition 2.16. Point estimator. *Let X be a RV with probability or density function $f(x; \theta)$, with $\boldsymbol{\theta}$ denoting the unknown parameter, and let $\boldsymbol{X'} = (X_1, \ldots, X_n)$ be a simple random sample from X. A (point) estimator $\hat{\Theta}$ of $\boldsymbol{\theta}$ is any statistic $T(X_1, \ldots, X_n)$ whose determinations are used to estimate the unknown parameter θ.*

It is worth outlining that an estimator is a function of the sample elements; then it is a RV and its determinations change with the sample according to the probability distribution of the population which the random sample refers to. On the other hand, the value of $\hat{\Theta}$ obtained for a certain observed sample $(x_1, \ldots, x_n)$ is a scalar and represents the *estimate* $\hat{\theta}$ of θ.

Point estimation implies the reduction of data dimensionality from n (the sample size) to 1 (the real number corresponding to the estimate of the unknown parameter). To avoid losing relevant information through the process of dimensionality reduction, statistic $T(X_1, \ldots, X_n)$ has to be chosen in a suitable way, such that some optimal statistical properties hold. Note that it is improper to evaluate *how much good* is an estimate. Indeed, as $\hat{\theta}$ is a real number its goodness depends on its closeness to θ. However, the goodness of $\hat{\theta}$ cannot be verified, as θ is unknown. Rather, it makes sense to evaluate how much good is an estimation method and, then, an estimator, whose behavior in terms of statistical properties can be always evaluated with reference to the sample space.

In what follows, we briefly illustrate the most relevant optimal properties.

Definition 2.17. Sufficiency. *Estimator $\hat{\Theta} = T(X_1, \ldots, X_n)$ is sufficient if it is based on a sufficient statistic.*

In other words, as function $T(X_1, \ldots, X_n)$ summarizes the information from the sample, a reasonable strategy consists of searching the optimal estimator in the class of sufficient statistics. Out of this class, any other statistic would result in a loss of information.

In the class of sufficient statistics, it is desirable that values of an estimator are as close as possible to θ. In this regard, the concentration and the closeness (according to Pitman) of an estimator can be evaluated.

Definition 2.18. Concentration. *Let $\hat{\Theta}^* = T^*(X_1, \ldots, X_n)$ and $\hat{\Theta} = T(X_1, \ldots, X_n)$ be two estimators of θ. If the following relation holds*

$$p(\theta - \delta \leq \hat{\Theta}^* \leq \theta + \delta) \geq p(\theta - \delta \leq \hat{\Theta} \leq \theta + \delta)$$

for any $\delta > 0$, then $\hat{\Theta}^*$ is more concentrated than $\hat{\Theta}$.

Definition 2.19. Closeness. *Let* $\hat{\Theta}^* = T^*(X_1, \dots, X_n)$ *and* $\hat{\Theta} = T(X_1, \dots, X_n)$ *be two estimators of* θ. *If the following relation holds*

$$p(|\hat{\Theta}^* - \theta| < |\hat{\Theta} - \theta|) \geq 0.5$$

for every θ, *then* $\hat{\Theta}^*$ *is closer (according to Pitman) to* θ *than* $\hat{\Theta}$.

The two above relations represent relative criteria, as the comparison between two particular estimators, $\hat{\Theta}^*$ and $\hat{\Theta}$, is involved. If the above inequalities hold for every $\hat{\Theta}$ alternative to $\hat{\Theta}^*$, then the two properties are interpreted in an absolute way.

Concentration and closeness represent reasonable and desirable properties; however the existence of estimators satisfying these properties is very uncommon; in addition, their evaluation is often analytically complex. In practice, synthetic indices are usually computed, instead of referring to the entire probability distribution, based on the differences between the estimator $\hat{\Theta}$ and the unknown parameter θ, such as $|\hat{\Theta} - \theta|$ and $(\hat{\Theta} - \theta)^2$. In such situations, the concept of efficiency becomes relevant.

Definition 2.20. Efficiency in the mean simple error. *Let* $\hat{\Theta}^* = T^*(X_1, \dots, X_n)$ *and* $\hat{\Theta} = T(X_1, \dots, X_n)$ *be two estimators of* θ *and let*

$$mse(\hat{\Theta}^*) = E(|\hat{\Theta}^* - \theta|); \quad mse(\hat{\Theta}) = E(|\hat{\Theta} - \theta|)$$

be the mean simple errors of $\hat{\Theta}^*$ *and* $\hat{\Theta}$, *respectively. If*

$$mse(\hat{\Theta}^*) \leq mse(\hat{\Theta})$$

for every $\theta \in \Theta$ *and for every estimator* $\hat{\Theta}$ *different from* $\hat{\Theta}^*$, *then* $\hat{\Theta}^*$ *is the most efficient estimator in the mean simple error.*

Definition 2.21. Efficiency in the mean square error. *Let* $\hat{\Theta}^* = T^*(X_1, \dots, X_n)$ *and* $\hat{\Theta} = T(X_1, \dots, X_n)$ *be two estimators of* θ *and let*

$$MSE(\hat{\Theta}^*) = E[(\hat{\Theta}^* - \theta|)^2]; \quad MSE(\hat{\Theta}) = E[(\hat{\Theta} - \theta)^2]$$

be the mean square errors of $\hat{\Theta}^*$ *and* $\hat{\Theta}$, *respectively. If*

$$MSE(\hat{\Theta}^*) \leq MSE(\hat{\Theta})$$

for every $\theta \in \Theta$ *and for every estimator* $\hat{\Theta}$ *different from* $\hat{\Theta}^*$, *then* $\hat{\Theta}^*$ *is the most efficient estimator in the mean square error.*

Usually, the most efficient estimator (in the mean simple error or in the mean square error) does not exist or, if it exists, its detection may be cumbersome. In such situations, it is easier to compare two estimators, say $\hat{\Theta}_1$ and $\hat{\Theta}_2$, of the same parameter θ: estimator $\hat{\Theta}_1$ is more efficient in the mean simple error than $\hat{\Theta}_2$ if $mse(\hat{\Theta}_1) < mse(\hat{\Theta}_2)$; similarly, estimator $\hat{\Theta}_1$ is more efficient in the mean square error than $\hat{\Theta}_2$ if $MSE(\hat{\Theta}_1) < MSE(\hat{\Theta}_2)$.

In all those situations that do not admit an estimator that minimizes the mean square error or the mean simple error for every value of θ, the search for a sub-optimal solution can be limited to constrained classes of estimators, being usually more probable to find an efficient estimator in such a restricted class. The most common constraint is that of unbiasedness; another common constraint, which usually is used in addition to the unbiasedness, is the linearity constraint.

Definition 2.22. Unbiasedness. *An estimator* $\hat{\Theta} = T(X_1, \ldots, X_n)$ *of* θ *is unbiased if* $E(\hat{\Theta}) = \theta$ *for every* $\theta \in \Theta$. *Similarly,* $\hat{\Theta}$ *is unbiased if the quantity* $b = E(\hat{\Theta}) - \theta$, *known as bias, is equal to 0.*

It is worth noting that the mean square error of an estimator can be decomposed in the sum of two elements, the variance of the estimator and its squared bias, as follows

$$
\begin{aligned}
MSE(\hat{\Theta}) &= E[(\hat{\Theta} - \theta)^2] \\
&= E[(\hat{\Theta} - E(\hat{\Theta}) + E(\hat{\Theta}) - \theta)^2] \\
&= Var(\hat{\Theta}) + [E(\hat{\Theta}) - \theta]^2 \\
&= Var(\hat{\Theta}) + b^2.
\end{aligned}
$$

If $\hat{\Theta}$ is an unbiased estimator (i.e., $b = 0$), then the mean square error and the variance coincide, that is,

$$
MSE(\hat{\Theta}) = Var(\hat{\Theta}).
$$

In other words, in the constrained class of unbiased estimators the most efficient estimator in the mean square error is the estimator with the minimum variance.

The minimum value that the variance of an unbiased estimator *can* assume is established by the Cramér-Rao inequality.

Theorem 2.10. Cramér-Rao inequality. *Let* X *be a RV with probability or density function* $f(x; \theta)$, *with* $\theta \in \Theta$ *unknown. Under some mild regularity conditions, the following relation holds:*

$$
Var(\hat{\Theta}) = \frac{1}{E\left\{\left[\frac{d}{d\theta} \log \prod_{i=1}^{n} f(X_i; \theta)\right]^2\right\}} = \frac{1}{nE\left\{\left[\frac{d}{d\theta} \log f(X; \theta)\right]^2\right\}} = \frac{1}{\mathcal{I}(\theta)},
$$

where the denominator $\mathcal{I}(\theta)$ *is known as Fisher's information . The Fisher's information provides a measure of the sample information. Moreover, let*

$$
\ell(\theta) = \log \mathcal{L}(\theta) = \log f(x_i; \theta),
$$

be the natural logarithm of the likelihood function, known as log-likelihood function, *and*

$$
S(\theta) = \frac{d}{d\theta} log\mathcal{L}(\theta) \tag{2.16}
$$

be its first derivative, known as score function. *Then, it can be shown that the Fisher's information is equal to the variance of the score function,*

$$Var[S(\theta)] = E\{[S(\theta)]^2\} = \mathcal{I}(\theta).$$

The Cramér-Rao inequality implies that, if $\hat{\Theta}$ is an unbiased estimator and $Var(\hat{\Theta}) = 1/\mathcal{I}(\theta)$, then $\hat{\Theta}$ is the most efficient estimator in the constrained class of unbiased estimators. On the other hand, if $Var(\hat{\Theta}) > 1/\mathcal{I}(\theta)$, no conclusion about the efficiency of $\hat{\Theta}$ can be drawn: in such a situation another more efficient estimator could be exist. It can be shown that the Cramér-Rao limit is reached, under regularity conditions, only in the RVs belonging to the exponential family, which are characterized by one parameter.

Moreover, as stated by the Rao-Blackwell theorem, the variance of an unbiased estimator depending on a sufficient statistic is smaller than the variance of any other unbiased estimator that does not depend on a sufficient statistic:

Theorem 2.11. Rao-Blackwell theorem. *Let $\hat{\Theta}$ be an unbiased estimator of θ with $E(\hat{\Theta}^2) < +\infty$ for all possible values of θ. Let $T(\mathbf{X}) = t(X_1, \ldots, X_n)$ be sufficient for θ and $\hat{\Theta}^* = f(T)$ and $E(\hat{\Theta}^*) = \theta$. It turns out that*

$$E(\hat{\Theta}^* - \theta)^2 \le E(\hat{\Theta} - \theta)^2 \quad \forall \theta,$$

where the inequality is strict whenever $\hat{\Theta}$ does not depend on the sufficient statistic $T(\mathbf{X})$.

Properties of unbiasedness and efficiency are evaluated with respect to a finite sample size. In addition, analyzing the behavior of an estimator for an increasing sample size (asymptotic behavior) is also of interest to evaluate its goodness. As far as this point, a desirable asymptotic property is represented by the consistency.

Definition 2.23. Weak consistency. *Let $\hat{\Theta}_n = T_n(X_1, \ldots, X_n)$ be an estimator of θ. $\hat{\Theta}_n$ is weakly consistent if*

$$\lim_{n \to +\infty} (p|\hat{\Theta}_n - \theta| < \epsilon) = 1$$

for every $\theta \in \Theta$ and for any $\epsilon > 0$.

Definition 2.24. Strong consistency. *Let $\hat{\Theta}_n = T_n(X_1, \ldots, X_n)$ be an estimator of θ. $\hat{\Theta}_n$ is strongly consistent if*

$$\lim_{n \to +\infty} MSE(\hat{\Theta}_n) = 0$$

or also

$$\lim_{n \to +\infty} mse(\hat{\Theta}_n) = 0$$

for every $\theta \in \Theta$.

Strong consistency implies weak consistency.

2.9.2   Point estimation methods

Several point estimation methods have been proposed in the statistical literature, among which are the method based on the minimization of the MSE
and the maximum likelihood method.

The method based on the minimization of the MSE is directly related to
the concept of efficiency. The estimator that minimizes the MSE is the most
efficient among all the possible estimators and its choice represents the dominant strategy to estimate θ. Such an estimator, if it exists, is usually known
as *the best estimator*. However, an estimator that minimizes the MSE with respect to all the possible values of $\theta \in \Theta$ does not often exist: usually, the MSE
is minimized by different estimators in different subsets of Θ (e.g., the MSE is
minimized by $\hat{\Theta} = \hat{\Theta}_1$ for $\theta < 0$ and by $\hat{\Theta} = \hat{\Theta}_2$ for $\theta \geq 0$). In such a situation
the estimation problem does not admit a dominant strategy. A solution con-
sists in narrowing the class of estimators according to one or more constraints
and, then, verify if it is possible to minimize the MSE in this constrained class.
Obviously, such an estimator, if it exists, is not the most efficient in the absolute sense, but only relative to the restricted class of constrained estimators.
For instance, in the class of unbiased estimators the most efficient estimator
is that one with the minimum variance. Such an estimator is usually known
as *the best unbiased estimator* and it represents the dominant strategy under
the unbiasedness constraint. A further constraint consists in searching for the
most efficient estimator among the statistics that are linear combinations of
the sample elements, such as $T(X_1, \ldots, X_n) = \alpha_0 + \sum_{i=1}^{n} \alpha_i X_i$. In the class of
linear and unbiased estimators, the statistic that minimizes the MSE is known
as *Best Linear Unbiased Estimator* (BLUE). A popular example of the BLUE
estimator is provided by the one that results by applying the least squares
estimation method to the linear regression model (Section 2.11).

The estimation method based on the minimization of the MSE offers a nice
interpretation from a decisional perspective. Indeed, if the MSE is interpreted
in terms of loss (or utility) function and the possible values of the unknown
parameter θ are interpreted in terms of states of nature, the existence of the
most efficient estimator is equivalent to the existence of a dominant action,
that is, an action that minimizes the loss (maximizes the utility) for any state
of nature.

Another relevant and commonly adopted approach to point estimation is
the Maximum Likelihood (ML) method. As outlined in Section 2.8, the likelihood function $\mathcal{L}(\theta)$, defined as in Eq. (2.12), depends only on the unknown
vector of parameters θ, given an observed sample $x' = (x_1, \ldots, x_n)$. In other
words, $\mathcal{L}(\theta)$ provides an assessment of the plausibility of the possible values of
θ defined by Θ. Then, it is reasonable to choose as the point estimation of θ
the most likely value, that is, the value $\tilde{\theta}$ that maximizes the likelihood function. Usually, the log-likelihood function $\ell(\theta)$ is maximized instead of $\mathcal{L}(\theta)$.
As $\ell(\cdot)$ is a monotonic transformation of $\mathcal{L}(\cdot)$, value $\tilde{\theta}$ that maximizes the former function is the same as the value maximizing the latter one. Besides, the

maximization of $\ell(\cdot)$ is easier than the maximization of $\mathcal{L}(\cdot)$, as it does not involve the computation of the product $\prod_{i=1}^{n} f(x_i; \theta)$.

The success of the ML approach is due to the statistical properties satisfied by the ML estimators:

- *Invariance.* If $\tilde{\theta}$ is the ML estimator of θ, then $g(\tilde{\theta})$ is the ML estimator of $g(\theta)$. For instance, in the normal model the ML estimation of σ is provided by the squared root of $\tilde{\sigma}^2$; in the Poisson model (Eq. (2.3)) the ML estimate of $1/\lambda$ is $1/\tilde{\lambda}$.

- *Sufficiency.* If there exist sufficient statistics for θ, then the ML estimator depends on them and is a sufficient estimator itself. This property follows from the Neyman-Fisher factorization criterion: if sufficient estimators exist, then the log-likelihood function is decomposed in the sum of two components, according to Eq. (2.15), one depending only on the unknown parameter and the other one depending on the sample.

- *Efficiency (for finite samples).* If there exists an unbiased estimator whose variance reaches the Cramér-Rao limit, then this estimator is an ML estimator.

- *Asymptotic efficiency.* Under mild regularity conditions, the ML estimator is asymptotically efficient. More in detail, the ML estimator is (*i*) asymptotically unbiased, as $\lim_{n \to +\infty} E(\tilde{\boldsymbol{\theta}}_n) = \boldsymbol{\theta}$; (*ii*) its variance tends to the Cramér-Rao limit that, in turn, tends to $\lim_{n \to +\infty} Var(\tilde{\boldsymbol{\theta}}_n) = 1/\mathcal{I}(\theta)$; (*iii*) as usually $1/\mathcal{I}(\theta)$ goes to 0 for $n \to +\infty$, then strong and weak consistency of the ML estimator follows.

- *Asymptotic normality.* For $n \to +\infty$, $\tilde{\theta}_n$ is normally distributed with mean θ and variance equal to the Cramér-Rao limit, that is $1/\mathcal{I}(\theta)$.

To better outline the last two properties, the acronymous BANE and CANE are used: BANE stays for Best Asymptotically Normal Estimator, and CANE stays for Consistent Asymptotically Normal Estimator.

Example 2.26. Estimation of Bernoulli parameter - *The log-likelihood of a Bernoulli RV is given by*

$$\ell(p) = \sum_{i=1}^{n} \log f(x_i; p) = \sum_{i=1}^{n} \log[p^{x_i}(1-p)^{1-x_i}] = \log p \sum_{i=1}^{n} x_i + \log(1-p)(n - \sum_{i=1}^{n} x_i)$$

and the related score function in Eq. (2.16) is

$$S(p) = \frac{1}{p} \sum_{i=1}^{n} x_i - \frac{1}{1-p}(n - \sum_{i=1}^{n} x_i).$$

Solving $S(p) = 0$ with respect to p, the maximum likelihood estimation of p is obtained as

$$\tilde{p} = \frac{1}{n} \sum_{i=1}^{n} x_i = \bar{x}.$$

Then, the maximum likelihood estimator[4] of p is the sample mean $\tilde{p} = \frac{1}{n}\sum_{i=1}^{n} X_i = \bar{X}$.

Note that each X_i is a Bernoulli RV; then $\sum_{i=1}^{n} X_i$ (sum of successes in n independent trials) has a binomial distribution with parameters n and p and the sample mean $\bar{X} = \frac{1}{n}\sum_{i=1}^{n} X_i$ (proportion of successes) is well approximated by a normal distribution with mean p and variance $p(1-p)/n$, when n goes to infinity (central limit theorem).

Moreover, $\sum_{i=1}^{n} X_i$ (as well as any one-to-one transformation of it) is a sufficient statistic for p; then $\tilde{p} = \bar{X}$ is a sufficient and unbiased estimator of p. In addition, its MSE equals the variance and reaches the Cramér-Rao limit. Indeed,

$$\mathcal{I}(p) = -E\left[\frac{d}{dp}S(p)\right] = -E\left\{\frac{d}{dp}\left[\frac{1}{p}\sum_{i=1}^{n} X_i - \frac{1}{1-p}\left(n - \sum_{i=1}^{n} X_i\right)\right]\right\}$$

$$= E\left[\frac{1}{p^2}\sum_{i=1}^{n} X_i + \frac{1}{(1-p)^2}\left(n - \sum_{i=1}^{n} X_i\right)\right]$$

$$= \frac{np}{p^2} + \frac{n(1-p)}{(1-p)^2} = \frac{n(1-p)+np}{p(1-p)} = \frac{n}{p(1-p)}.$$

Then,

$$[\mathcal{I}(p)]^{-1} = \frac{p(1-p)}{n},$$

which is the variance of $\bar{X}$. Therefore, the maximum likelihood estimator $\tilde{p} = \bar{X}$ is the best estimator of p in the class of unbiased estimators.

Finally, $\tilde{p} = \bar{X}$ is strongly (and, then, also weakly) consistent, as

$$\lim_{n\to+\infty} MSE(\bar{X}_n) = \lim_{n\to+\infty} Var(\bar{X}_n) = \lim_{n\to+\infty} \frac{p(1-p)}{n} = 0.$$

Example 2.27. Estimation of Poisson parameter - *The log-likelihood of the Poisson RV (Eq. (2.3)) is given by*

$$\ell(\lambda) = \sum_{i=1}^{n} \log f(x_i;\lambda) = \sum_{i=1}^{n} \log \frac{\lambda^{x_i}e^{-\lambda}}{x_i!} = \log\lambda\sum_{i=1}^{n} x_i - n\lambda - \sum_{i=1}^{n}\log x_i$$

and the related score function is

$$S(\lambda) = \frac{1}{\lambda}\sum_{i=1}^{n} x_i - n.$$

---

[4]Note that the use of capital letters to denote random variables (e.g., estimators) and of small letters to denote the observed values of a RV (e.g., estimates) is not fulfilled by specific cases: for instance, the same symbol $\tilde{p}$ is used to denote both estimator and estimate of p, $\tilde{\lambda}$ to denote both estimator and estimate of λ, $\tilde{\mu}$ for μ, $\tilde{\sigma}^2$ for σ^2.

Solving $S(\lambda) = 0$ with respect to λ, the maximum likelihood estimation of λ is obtained as

$$\tilde{\lambda} = \frac{1}{n} \sum_{i=1}^{n} x_i = \bar{x}.$$

Then, similarly to the case of the parameter of the Bernoulli RV, the maximum likelihood estimator of λ is the sample mean $\tilde{\lambda} = \bar{X}$.

In virtue of the additive property of the Poisson RV, $\sum_{i=1}^{n} X_i$ has a Poisson distribution with parameter $n\lambda$, which coincides with its mean and variance, and $\tilde{\lambda} = \bar{X}$ is well approximated by a normal distribution with mean λ and variance λ/n, when n goes to infinity (central limit theorem).

Moreover, as $\sum_{i=1}^{n} X_i$ (as well as any one-to-one transformation of it) is a sufficient statistic for λ, then $\tilde{\lambda} = \bar{X}$ is a sufficient and unbiased estimator of λ. In addition, its MSE equals the variance and reaches the Cramér-Rao limit. Indeed,

$$\mathcal{I}(\lambda) = -E\left[\frac{d}{d\lambda}S(\lambda)\right] = -E\left[\frac{d}{d\lambda}\left(\frac{1}{\lambda}\sum_{i=1}^{n} X_i - n\right)\right]$$

$$= -E\left(-\frac{\sum_{i=1}^{n} X_i}{\lambda^2}\right) = \frac{n}{\lambda}.$$

Then,

$$[\mathcal{I}(\lambda)]^{-1} = \frac{\lambda}{n},$$

which is the variance of $\bar{X}$. Therefore, the maximum likelihood estimator $\tilde{\lambda} = \bar{X}$ is the best estimator of λ in the class of unbiased estimators.

Finally, $\tilde{\lambda} = \bar{X}$ is strongly (and, also, weakly) consistent, as

$$\lim_{n \to +\infty} MSE(\bar{X}_n) = \lim_{n \to +\infty} Var(\bar{X}_n) = \lim_{n \to +\infty} \frac{\lambda}{n} = 0.$$

Example 2.28. Estimation of normal parameters - *The log-likelihood of the normal RV is given by*

$$\ell(\mu, \sigma^2) = \sum_{i=1}^{n} \log f(x_i; \mu, \sigma^2) = \sum_{i=1}^{n} \log \frac{1}{\sqrt{2\pi\sigma^2}} e^{-\frac{1}{2\sigma^2}(x_i - \mu)^2}$$

$$= -\frac{n}{2}\log(2\pi) - \frac{n}{2}\log\sigma^2 - \frac{1}{2\sigma^2}\sum_{i=1}^{n}(x_i - \mu)^2.$$

As the log-likelihood function depends on two parameters, four different situations are envisaged: (i) estimation of μ, (ii) estimation of σ^2 with known μ, (iii) estimation of σ^2 with unknown μ, (iv) estimation of both μ and σ^2.

Estimation of μ - *As far as the estimation of μ, the knowledge of σ^2 is useless, as the equation $S(\mu) = 0$ is independent of σ^2, being $S(\mu)$ the score function computed with respect to μ, that is,*

$$S(\mu) = -\frac{1}{2\sigma^2}\sum_{i=1}^{n} 2(x_i - \mu)(-1) = \frac{1}{\sigma^2}\left(\sum_{i=1}^{n} x_i - n\mu\right). \qquad (2.17)$$

Solving $S(\mu) = 0$ with respect to μ, the maximum likelihood estimation of μ is given by

$$\tilde{\mu} = \frac{1}{n}\sum_{i=1}^{n} x_i = \bar{x}.$$

Then, the maximum likelihood estimator of μ is the sample mean $\tilde{\mu} = \bar{X}$.

Variable $\tilde{\mu} = \bar{X}$ is normally distributed with mean μ and variance σ^2/n. Also, $\bar{X}$ is an unbiased and sufficient (as it is based on the sufficient statistic $\sum_{i=1}^{n} X_i$) estimator of μ. Moreover, its MSE equals the variance and reaches the Cramér-Rao limit. Indeed,

$$\mathcal{I}(\mu) = -E\left(\frac{d}{d\mu}S(\mu)\right) = -E(\frac{-n}{\sigma^2}) = n/\sigma^2.$$

Then,

$$[\mathcal{I}(\mu)]^{-1} = \sigma^2/n,$$

which is the variance of $\bar{X}$. Therefore, the maximum likelihood estimator $\tilde{\mu} = \bar{X}$ is the best estimator of μ in the class of unbiased estimators. It is also strongly (and weakly) consistent for μ.

Estimation of σ^2 with known μ - *The score function computed with respect to σ^2 is given by*

$$S(\sigma^2) = -\frac{n}{2\sigma^2} + \frac{1}{2\sigma^4}\sum_{i=1}^{n}(x_i - \mu)^2; \qquad (2.18)$$

then the equation $S(\sigma^2) = 0$ depends both on σ^2 (parameter of interest) and μ (nuisance parameter). If the nuisance parameter is known, then the maximum likelihood estimation of σ^2 is given by

$$\tilde{\sigma}^2 = \frac{1}{n}\sum_{i=1}^{n}(x_i - \mu)^2 = s_{**}^2$$

and the maximum likelihood estimator of σ^2 is $\tilde{\sigma}^2 = S_{**}^2 = \frac{1}{n}\sum_{i=1}^{n}(X_i - \mu)^2$, known as sample variance with known μ.

As shown in the Theorem 2.6, the RV

$$\frac{n\tilde{\sigma}^2}{\sigma^2} = \frac{nS_{**}^2}{\sigma^2} = \sum_{i=1}^{n}\left(\frac{X_i - \mu}{\sigma}\right)^2$$

is distributed as a χ^2 with n degrees of freedom, with mean and variance equal to

$$E\left(\frac{nS_{**}^2}{\sigma^2}\right) = n; \qquad Var\left(\frac{nS_{**}^2}{\sigma^2}\right) = 2n,$$

respectively. Then,

$$E(\tilde{\sigma}^2) = E(S_{**}^2) = \sigma^2; \qquad Var(\tilde{\sigma}^2) = Var(S_{**}^2) = 2\sigma^4/n.$$

Therefore, the sample variance $\tilde{\sigma}^2 = S_{**}^2$ is an unbiased and sufficient (as it is based on the sufficient statistic $\sum_{i=1}^{n}(X_i - \mu)^2$) estimator of σ^2. It is also the most efficient estimator in the class of unbiased estimators, as its variance reaches the Cramér-Rao limit.

$$\mathcal{I}(\sigma^2) = -E\left(\frac{d}{d\sigma^2}S(\sigma^2)\right) = -E\left[\frac{n}{2\sigma^4} - \frac{1}{2\sigma^6}\sum_{i=1}^{n}(x_i - \mu)^2\right] = -\frac{n}{2\sigma^4} + \frac{n}{\sigma^4} = \frac{n}{2\sigma^4}$$

that implies that

$$[\mathcal{I}(\sigma^2)]^{-1} = \frac{2\sigma^4}{n}, \tag{2.19}$$

which corresponds to the variance of $\tilde{\sigma}^2$.

Finally, $\tilde{\sigma}^2 = S_{**}^2$ is strongly (and weakly) consistent for σ^2, as

$$\lim_{n\to+\infty} MSE(\tilde{\sigma}_n^2) = \lim_{n\to+\infty} Var(\tilde{\sigma}_n^2) = \lim_{n\to+\infty} \frac{2\sigma^4}{n} = 0$$

Joint estimation of μ and σ^2 - *If the value of μ is unknown, S_{**}^2 cannot be used as point estimator of σ^2, as it depends on the nuisance parameter μ. In such a case, it is useful to substitute μ in the formula of S_{**}^2 with its estimate. More in general, when both μ and σ^2 are unknown the joint estimation of both of them is obtained by solving the following system of equations*

$$\begin{cases} S(\mu) = 0 \\ S(\sigma^2) = 0, \end{cases}$$

where $S(\mu)$ and $S(\sigma^2)$ are defined in Eq. (2.17) and (2.18), respectively. The resulting maximum likelihood estimates follow as

$$\tilde{\mu} = \bar{x}; \qquad \tilde{\sigma}^2 = \frac{1}{n}\sum_{i=1}^{n}(x_i - \bar{x})^2 = s_*^2.$$

Then, the maximum likelihood estimators of μ and σ^2 are the sample mean and the sample variance, respectively,

$$\tilde{\mu} = \bar{X}; \qquad \tilde{\sigma}^2 = \frac{1}{n}\sum_{i=1}^{n}(X_i - \bar{X})^2 = S_*^2.$$

As concerns the properties of $\tilde{\mu} = \bar{X}$ and $\tilde{\sigma}^2 = S_*^2$, we observe that $\sum_{i=1}^{n} X_i$ and $\sum_{i=1}^{n} X_i^2$ are jointly sufficient for μ and σ^2. Then, also estimators $\tilde{\mu} = \bar{X}$ and $\tilde{\sigma}^2 = S_*^2$ are jointly sufficient, as they are based on jointly sufficient statistics.

As shown in the Theorem 2.6, the RV

$$\frac{nS_*^2}{\sigma^2} = \sum_{i=1}^{n} \left(\frac{X_i - \bar{X}}{\sigma} \right)^2$$

is distributed as a χ^2 with $n - 1$ degrees of freedom, with mean and variance equal to

$$E\left(\frac{nS_*^2}{\sigma^2}\right) = n - 1 \qquad Var\left(\frac{nS_*^2}{\sigma^2}\right) = 2(n-1).$$

Then (see also Eq. (2.13)),

$$E(S_*^2) = \frac{n-1}{n}\sigma^2; \qquad Var(S_*^2) = 2\sigma^4 \frac{n-1}{n^2},$$

respectively. Therefore, S_*^2 is a biased estimator of σ^2 and, consequently, the Cramér-Rao theorem cannot be applied and nothing can be said about the efficiency of S_*^2. However, as $MSE(S_*^2) = \sigma^4(2n-1)/n^2$ and $MSE(S_*^2)$ tends to 0 when n goes to infinity, S_*^2 is a strongly (and weakly) consistent estimator of σ^2.

It is worth outlining that, in virtue of its unbiasedness, statistic $S^2 = \frac{1}{n-1}\sum_{i=1}^{n}(X_i - \bar{X})^2$ is usually preferred to S_*^2 to estimate σ^2, although S^2 is not a maximum likelihood estimator. Indeed, as

$$\frac{(n-1)S^2}{\sigma^2} = \frac{nS_*^2}{\sigma^2},$$

then (see also Eq. (2.14))

$$E\left(\frac{(n-1)S^2}{\sigma^2}\right) = E\left(\frac{nS_*^2}{\sigma^2}\right) = n - 1 \implies E(S^2) = \sigma^2.$$

Moreover, as $Var\left(\frac{(n-1)S^2}{\sigma^2}\right) = Var\left(\frac{nS_*^2}{\sigma^2}\right) = 2(n-1)$, then $Var(S^2) = 2\sigma^4/(n-1)$.

In conclusion, statistic S^2 is a sufficient and unbiased estimator of σ^2, but it is not the most efficient estimator, as it does not reach the Cramér-Rao limit provided in Eq. (2.19):

$$Var(S^2) = 2\sigma^4/(n-1) > 2\sigma^4/n = [\mathcal{I}(\sigma^2)]^{-1}.$$

On the other hand, the difference between $Var(S^2)$ and the Cramér-Rao limit is small and it reduces when n increases; moreover, S^2 is strongly (and weakly) consistent for σ^2.

2.9.3 Confidence intervals

The point estimation provides a single value that represents a measurement of the unknown parameter, but no indication is given about the probability that this value is close to θ. The estimation method based on the confidence intervals overcomes this drawback by providing a range of plausible values for θ together with a measurement, in probabilistic terms, of the reliability of this range of values.

More in detail, the estimation approach based on confidence intervals consists in detecting, on the basis of the sample evidence, two values L_1 (inferior limit) and L_2 (superior limit) such that

$$p(L_1 \leq \theta \leq L_2) = 1 - \alpha \quad 0 < \alpha < 1,$$

where $L_1 = T_1(X_1, \ldots, X_n)$ and $L_2 = T_2(X_1, \ldots, X_n)$ ($L_1 < L_2$) are random variables depending on the n sample elements and $1 - \alpha$ (usually equal to 0.95, 0.99, 0.999) is the *confidence level* that denotes the proportion of random samples containing the true value of θ. In practice, we usually observe a single random sample and, then, obtain a single pair of determinations of L_1 and L_2 denoting the interval $[l_1, l_2]$. The observed interval $[l_1, l_2]$ can be one of $(1 - \alpha)\%$ intervals including θ or one of $\alpha\%$ intervals not including θ. Note that the probability that $[l_1, l_2]$ contains θ is 1 in the former case and 0 in the latter case.

A confidence interval is characterized by two elements:

- the reliability, measured by the confidence level $1 - \alpha$,

- the informational capacity, measured by the interval size $(l_2 - l_1)$.

Theoretically, it would be desirable to identify intervals with a high confidence level and a small size. Unfortunately, this dual aim cannot be pursued, as when $1 - \alpha$ increases, the interval size increases too, and vice-versa. Only when the sample size is not yet specified is it possible to reduce the interval size for a certain confidence level, by increasing the sample size. Given the sample size, the usual approach consists of setting the confidence level and looking for the minimum size interval, in the class of all intervals of level $1 - \alpha$. This approach is based on the concept of *pivotal element* or *pivot*.

Definition 2.25. Pivotal element or pivot. *Let X be a RV with probability or density function $f(x; \theta)$, $\theta \in \Theta$, and let $\boldsymbol{X}' = (X_1, \ldots, X_n)$ be a simple random sample from X. A pivot is a quantity $Q(\boldsymbol{X}, \theta)$ having the following characteristics:*

- $Q(\boldsymbol{X}, \theta)$ *is a function of the sample elements* $\boldsymbol{X}' = (X_1, \ldots, X_n)$,

- *it depends on* θ,

- *it is independent of other unknown parameters (nuisance parameters),*

- *its distribution is known,*

- *it can be inverted with respect to θ.*

In practice, the procedure to determine a confidence interval follows the following steps:

1. we identify a pivot $Q(\boldsymbol{X}, \theta)$, usually starting from the maximum likelihood estimator of θ,

2. we set the confidence level $1 - \alpha$,

3. we find the minimum size interval $[q_1; q_2]$ that satisfies the following relationship:

$$p(q_1 \leq Q(\boldsymbol{X}, \theta) \leq q_2) = 1 - \alpha \qquad (2.20)$$

4. we invert the relation $q_1 \leq Q(\boldsymbol{X}, \theta) \leq q_2$ with respect to θ so that the confidence interval for θ turns out, that is,

$$p[L_1(\boldsymbol{X}, q_1) \leq \theta \leq L_1(\boldsymbol{X}, q_2)] = 1 - \alpha.$$

Example 2.29. *Let $\boldsymbol{X}' = (X_1, \ldots, X_n)$ be a random sample drawn from a normal RV with unknown mean μ and known variance σ^2. The pivot for the confidence interval for μ is*

$$Z = \frac{\bar{X} - \mu}{\sigma/\sqrt{n}} \sim N(0, 1).$$

Indeed, Z depends on the sample elements through the point estimator $\bar{X}$ and on the unknown parameter μ, has a completely known distribution, and is invertible with respect to μ. The minimum size interval is obtained by choosing symmetric values for q_1 and q_2 in Eq. (2.20), that is, $q_2 = -q_1 = z_{\alpha/2}$, with $z_{\alpha/2}$ superior quantile of level $\alpha/2$ of a standard normal distribution. The confidence interval for μ follows:

$$p(\bar{X} - z_{\alpha/2}\sigma/\sqrt{n} \leq \mu \leq \bar{X} + z_{\alpha/2}\sigma/\sqrt{n}) = 1 - \alpha. \qquad (2.21)$$

For instance, the confidence interval at 95% level is obtained by substituting in the above expression $\alpha = 0.05$ and, consequently $z_{\alpha/2} = 1.96$. Then, the inferior and superior limits of the interval for μ are $L_1 = \bar{X} - 1.96\sigma/\sqrt{n}$ and $L_2 = \bar{X} + 1.96\sigma/\sqrt{n}$.

Example 2.30. *If variance σ^2 of the normal RV is unknown too, then the confidence interval in Eq. (2.21) cannot be used, as L_1 and L_2 depend on σ (nuisance parameter). In such a case the unknown σ is substituted with its estimate and an alternative pivot has to be chosen. If σ is estimated through S, the square root of the unbiased sample variance $S^2 = \frac{1}{n-1}\sum_{i=1}^{n}(X_i - \bar{X})^2$, then the RV (Theorem 2.6)*

$$W = \frac{\bar{X} - \mu}{S/\sqrt{n}} \sim t_{n-1}$$

is a pivot as it depends on the sample elements and on the unknown parameter of interest μ, it is independent of any nuisance parameter (i.e., σ) and its distribution is completely known. The minimum size interval follows:

$$p(\bar{X} - t_{\alpha/2}S/\sqrt{n} \leq \mu \leq \bar{X} + t_{\alpha/2}S/\sqrt{n}) = 1 - \alpha,$$

where $t_{\alpha/2}$ is the superior quantile of a Student's t at level $\alpha/2$.

Example 2.31. *The RV*

$$V = \frac{(n-1)S^2}{\sigma^2} = \sum_{i=1}^{n} \left(\frac{X_i - \bar{X}}{\sigma}\right)^2 \sim \chi_{n-1}^2$$

(Theorem 2.6) can be used as a pivot to define the confidence interval for the unknown parameter σ^2. Choosing q_1 and q_2 in Eq. (2.20) in a symmetric way, the following confidence interval results

$$p\left(\frac{(n-1)S^2}{\chi_{\alpha/2}^2} \leq \sigma^2 \leq \frac{(n-1)S^2}{\chi_{1-\alpha/2}^2}\right) = 1 - \alpha,$$

where $\chi_{1-\alpha/2}^2$ and $\chi_{\alpha/2}^2$ are the inferior and superior quantiles, respectively, of the RV at level $\alpha/2$. It is worth observing that this interval is not of minimum size, as the χ^2 distribution is skewed.

2.9.4 Hypothesis testing

Hypothesis testing is an inferential procedure that is strongly related with the decisional process. This is particularly evident whenever rejecting an hypothesis about a certain unknown entity (e.g., a parameter) in favor of another hypothesis related to a certain behavior. In what follows the main concepts connected with hypothesis testing are introduced.

Definition 2.26. Statistical hypothesis. *An hypothesis (in the statistical sense) is a statement that completely (simple hypothesis) or partially (composite hypothesis) specifies the probability distribution of a RV. This statement usually refers to the parameters characterizing the probability distribution of a RV, when the functional form of this distribution is assumed to be known (parametric hypothesis) or when no assumption about the functional form is introduced (distribution-free hypothesis or non-parametric hypothesis); obviously, in this latter case, the hypothesis may involve the parameters as well as the distribution form. The hypothesis is usually denoted by H_0 and it is named null hypothesis.*

In addition to the null hypothesis H_0, an alternative hypothesis H_1 is usually formulated, which is complementary to H_0, in the sense that one between H_0 and H_1 is surely true.

In what follows we refer only to parametric hypotheses. In such a context, a simple hypothesis consists of a single real value that completely identifies the unknown parameter θ, such as $\theta = \theta_0$, whereas a composite hypothesis includes a range of possible values for θ, such as $\theta > \theta_0$, $\theta \neq \theta_0$, or $\theta \in [a; b]$.

Definition 2.27. Hypothesis testing. *A (statistical) hypothesis testing is a rule to decide whether to reject or not reject the null hypothesis, given the sample evidence.*

In other words, hypothesis testing relies on sampling observations from the population of interest. If these observed values are coherent with the null hypothesis one concludes in favor of it (acceptance of H_0 or failing to reject H_0); otherwise one rejects it.

In practice, an hypothesis testing splits the sample space Ω into two disjoint subsets, the *reject region RR* and the *non-reject region* $\overline{RR} = \Omega - RR$ (Figure 2.6). The decisional process underlying the testing procedure consists of two actions: rejecting H_0 if the sample point belongs to RR and failing to reject H_0 if the sample point belongs to $\overline{RR}$.

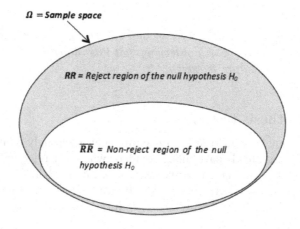

FIGURE 2.6
Sample space partition.

This decisional process implies two types of error, related to the true but unknown state of nature (i.e., H_0 true and H_0 false), as summarized in Table 2.2. A *I type error* is made when H_0 is true and it is rejected, whereas a *II type error* is made when H_0 is false and it is not rejected. Moreover, when H_0 is a simple hypothesis, the probability of a I type error, that is, the probability of rejecting H_0 when H_0 is true, is denoted by

$$\alpha = p(\boldsymbol{X} \in RR|H_0),$$

TABLE 2.2

Decision process underlying hypothesis testing.

	State of world	
	H_0 is true	H_0 is false
Action		
Not reject H_0	right decision	II type error
Reject H_0	I type error	right decision

where α is the *significance level* of the test. In addition, the probability of a II type error, that is, the probability of failing to reject H_0 when H_0 is false, is denoted by

$$\beta(H_1) = p(\boldsymbol{X} \in \overline{RR}|H_1),$$

where H_1 is the alternative hypothesis. Note that $\beta(H_1)$ depends on the specification of H_1. The complement to 1 of $\beta(H_1)$ is the probability $\gamma(H_1) = 1 - \beta(H_1)$, which denotes the *power of the test*, that is, the probability of rejecting H_0 when H_0 is false (i.e., probability of correctly rejecting the null hypothesis). Function $\gamma(H_1)$ depends on (*i*) the specification of the alternative hypothesis, (*ii*) the significance level of the test, and (*iii*) the sample size.

In the case of a composite null hypothesis, the significance level is defined as

$$\alpha = \sup_{H \subset H_0} p(\boldsymbol{X} \in RR|H_0).$$

Theoretically, the optimal test would be a test that minimizes α and $\beta(H_1)$ at the same time, given the sample size. However, as such a test does not exists, a different strategy is usually adopted: we set the significance level and, among all tests of level α, we select the one that minimizes $\beta(H_1)$. As minimizing $\beta(H_1)$ is the same as maximizing the power $\gamma(H_1)$, this test, if it exists, is the Most Powerful (MP) test. Note that a test is the more powerful the higher are the significance level, the distance between θ_0 and θ_1, and the sample size.

In more detail, the MP test is a rule that identifies the *best* reject region RR, that is, the region that satisfies the following two conditions, for $H_0 : \theta = \theta_0$ against $H_1 : \theta = \theta_1$:

$$p(X \in RR|H_0) = \alpha$$
$$p(X \in RR|H_1) \geq p(X \in RR_i|H_1)$$

where RR_i ($i = 2, 3, \ldots$) represents a reject region alternative to RR such that $p(\boldsymbol{X} \in RR_i|H_0) = \alpha$.

In the case of a simple null hypothesis and a simple alternative hypothesis, the Neyman-Pearson theorem states that the best reject region always exists.

Theorem 2.12. Neyman-Pearson theorem. *Let X be a random variable with probability or density function $f(x; \theta)$ and let $\boldsymbol{x}' = (x_1, \ldots, x_n)$ be a random sample from X. Let θ_0 and θ_1 be two possible specifications for θ, that is, $H_0 : \theta = \theta_0$ and $H_1 : \theta = \theta_1$, and let K be any positive real number. Let us assume that RR is a sub-space of Ω such that*

$$\frac{\mathcal{L}(\theta_1)}{\mathcal{L}(\theta_0)} \geq K \implies \boldsymbol{x} \in RR,$$

with $\mathcal{L}(\cdot)$ denoting the likelihood function and K satisfying $p(\boldsymbol{X} \in RR | H_0) = \alpha$. Then, among all the sub-spaces of Ω at level α, sub-space RR is the one with the smallest probability of II type error. Note that the above relation also implies that

$$\frac{\mathcal{L}(\theta_1)}{\mathcal{L}(\theta_0)} < K \implies \boldsymbol{x} \in \overline{RR},$$

where $\overline{RR} = \Omega - RR$.

In practice, the sample space to which one refers is not the space of variability of the random sample $\boldsymbol{X}' = (X_1, \ldots, X_n)$, but the space of variability of a suitable statistic $T(X_1, \ldots, X_n)$, which, in the context of the hypothesis testing, is known as the *test statistic*. For instance, if $\theta = \mu$, the test statistic is the sample mean, that is, $T(X_1, \ldots, X_n) = \bar{X}$, and the sample space of reference is the entire real axis. Figure 2.7 displays a partition of the real axis to test $H_0 : \mu = \mu_0$ against $H_1 : \mu = \mu_1$ with $\mu_1 > \mu_0$. The curve on the left is the probability density function of the sample mean $\bar{X}$ under H_0, whereas the curve on the right is the probability density function of $\bar{X}$ under H_1. Both curves are defined all along the real axis (from $-\infty$ to $+\infty$).

From Figure 2.7 is evident that a decision maker can never be sure if his/her decision is right: he/she can only adopt a rational behavior and decide in favor of the most likely hypothesis, according to the Neyman-Pearson theorem.

The test displayed in Figure 2.7 is an example of a right tail test (or right side test), as the reject region is located on the right extreme tail of the density distribution of the test statistic $T(X_1, \ldots, X_n)$. If $\mu_1 < \mu_0$ then a left tail test results, with the reject region located on the left extreme tail of the distribution of $T(X_1, \ldots, X_n)$.

It is worth outlining that the Neyman-Pearson theorem is valid for any finite number of parameters characterizing the distribution of X and it does not explicitly require the independence of the n sample units. Moreover, it provides the conditions to have an MP test as well as the process to detect the best reject region. Unfortunately, the Neyman-Pearson theorem operates only when H_0 and H_1 are both simple. In all those situations when one or both the hypotheses are composite a more general procedure is provided by the likelihood ratio test. This procedure has the advantage of detecting the best reject region, whenever it exists.

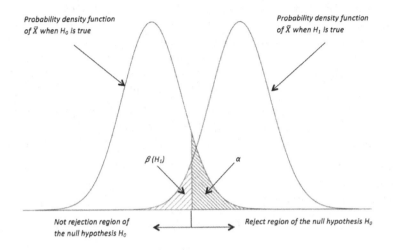

FIGURE 2.7
Hypothesis testing on μ: probability density distribution of the sample statistic $\bar{X}$ under H_0 and under H_1, partition of the sample space in the reject region and acceptance region, probability of the I type error (α) and II type error ($\beta(H_1)$).

Definition 2.28. Likelihood Ratio (LR) test. *Let X be a random variable with probability or density function $f(x;\theta)$ and let $x' = (x_1,\ldots,x_n)$ be a random sample from X. Let $H_0 : \theta = \Theta_0$ and $H_1 : \theta = \Theta_1$ be simple or composite hypotheses, where $\Theta_0 \in \Theta$, $\Theta_1 \in \Theta$, and $\Theta_0 \cup \Theta_1 = \Theta$. Besides, let us denote by $\mathcal{L}(\hat{\Theta}_0)$ the maximum value of likelihood with respect to θ under H_0 and by $\mathcal{L}(\hat{\Theta})$ the maximum value of likelihood with respect to θ in $\Theta_0 \cup \Theta_1$. Then, the reject region is composed of all the sample points such that*

$$LR = \frac{\mathcal{L}(\hat{\Theta}_0)}{\mathcal{L}(\hat{\Theta})} = \frac{\max_\theta \mathcal{L}(\theta|\theta \in \Theta_0)}{\max_\theta \mathcal{L}(\theta|\theta \in \Theta)} < K, \quad 0 \leq K \leq 1,$$

where K is chosen in such a way that the probability of a I type error is α.

It is worth noting that the ratio $\frac{\mathcal{L}(\hat{\Theta}_0)}{\mathcal{L}(\hat{\Theta})}$ is always less than or equal to 1 and its probabilistic distribution provides the value of K at the α significance level.

The probability distribution of LR cannot always be derived in a simple way. However, for a sufficiently high sample size and under some general regularity conditions, the sample variable $W = -2\log LR$ is approximately distributed as a χ^2_{df}, with df denoting the number of equality constraints specified on the parameters under H_0.

Given a simple null hypothesis H_0 and a significance level α, let us assume that a test exists that minimizes $\beta(H_1)$ for all the possible specifications of a composite alternative hypothesis H_1. This test is an LR test. Such a type of test is named a Uniformly Most Powerful (UMP) test. Note that the UMP test may exist only for unidirectional alternative hypotheses (e.g., $H_1 : \theta > \theta_0$ or $H_1 : \theta < \theta_1$). In the case of a bidirectional alternative hypotheses the LR test may be the Uniformly Most Powerful test only in the restricted class of Unbiased tests (Uniformly Most Powerful and Unbiased - UMPU test).

Definition 2.29. Unbiased test. *A test is unbiased if $\gamma(H_1) \geq \alpha$, that is, if the probability of correctly rejecting H_0 is always greater than or equal to the probability of wrongly rejecting H_0 (i.e., rejecting H_0 when H_0 is true).*

Example 2.32. *Let $x' = (x_1, \ldots, x_n)$ be a sample drawn from a normal distribution with unknown mean μ and variance $\sigma^2 = 1$ and let also the two following simple hypotheses*

$$H_0 : \mu = \mu_0$$
$$H_1 : \mu = \mu_1 > \mu_0$$

be given. The best reject region, that is, the region that minimizes $\beta(H_1)$ given α, can be defined through the Neyman-Pearson theorem. The likelihood functions under H_0 and H_1 are

$$\mathcal{L}(\mu_1) = \prod_{i=1}^{n} f(x_i, \mu_1) = (2\pi)^{-n/2} e^{-\frac{1}{2}\sum_i (x_i - \mu_1)^2} \quad \text{and}$$

$$\mathcal{L}(\mu_0) = \prod_{i=1}^{n} f(x_i, \mu_0) = (2\pi)^{-n/2} e^{-\frac{1}{2}\sum_i (x_i - \mu_0)^2},$$

respectively. Then, the best reject region is detected through

$$\frac{\mathcal{L}(\mu_1)}{\mathcal{L}(\mu_0)} = e^{\frac{1}{2}\left[\sum_i (x_i - \mu_0)^2 - \sum_i (x_i - \mu_1)^2\right]} \leq K$$

where K is a constant term depending on α. After some algebra, this inequality becomes

$$\bar{x} \geq \frac{2\log(K) - n(\mu_0^2 - \mu_1^2)}{2n(\mu_1 - \mu_0)} = K^*.$$

As under $H_0 : \mu = \mu_0$ $\bar{X}$ is normally distributed with mean μ_0 and variance $\sigma^2 = 1/n$, it is possible to find that value of K^* such that

$$p\left(\bar{x} \geq \frac{2\log(K) - n(\mu_0^2 - \mu_1^2)}{2n(\mu_1 - \mu_0)} \bigg| H_0\right) = \alpha.$$

In practice, as the right side of the above inequality is a constant function of K, it is sufficient to find that value K^* that satisfies

$$p(\bar{X} \geq K^* | H_0) = \alpha.$$

or

$$p(\frac{\bar{X} - \mu_0}{1/\sqrt{n}} \geq \frac{K^* - \mu_0}{1/\sqrt{n}} | H_0) = \alpha.$$

This last equation is equivalent to solving $p(Z \geq z_{alpha}) = \alpha$, *where* $Z = \frac{\bar{X} - \mu_0}{1/\sqrt{n}}$ *is a standard normal RV and* z_α *is the corresponding quantile whose right side area (i.e., the reject region) is equal to* α.

If the variance is unknown, the test statistic Z *is substituted with* $T = \frac{\bar{X} - \mu_0}{S/\sqrt{n}}$, *which is distributed as a Student's t with* $n - 1$ *degrees of freedom. The corresponding reject region is defined by the interval* $(t_{alpha}, +\infty)$, *where* t_{alpha} *is the superior quantile of a Student's t whose right side area is* α.

In the presence of a composite unidirectional alternative hypothesis, such as

Case a) $H_0 : \mu = \mu_0$

$H_1 : \mu > \mu_0,$

the reject region results from the iterated application of the Neyman-Pearson theorem for each specification of the alternative hypothesis. The reject region is defined by the same interval $(t_\alpha, +\infty)$, with $p(T > t_\alpha) \leq \alpha$, as when H_1 is simple, that is, $H_1 : \mu = \mu_1 > \mu_0$. This test is a UMP test.

Similarly, in the case of a left tail test with a simple null hypothesis and a composite alternative hypothesis with opposite direction, that is,

Case b) $H_0 : \mu = \mu_0$

$H_1 : \mu < \mu_0$

the reject region is defined by the interval $(-\infty, -t_\alpha)$, with $-t_\alpha$ satisfying $p(T < -t_\alpha) \leq \alpha$. Also in this case the test is UMP.

The two one tail tests above mentioned (Case a and Case b) are both UMP. On the other hand, in the case of a two tails test, that is, when the alternative hypothesis is bidirectional, such as

Case c) $H_0 : \mu = \mu_0$

$H_1 : \mu \neq \mu_0,$

then it is not possible to detect the UMP test. Indeed, if the true value of μ were greater than μ_0, the UMP test would be that detected in Case *a*; if the true value of μ were less than μ_0, the UMP test would be that detected in Case *b*. As any additive information is available with respect to hypothesis $H_1 : \mu \neq \mu_0$, it is usually attributed the same weight to the two alternatives

$\mu > \mu_0$ and $\mu < \mu_0$. The resulting reject region is then defined by the two intervals $(-\infty, -t_{\alpha/2})$ and $(t_{\alpha/2}, +\infty)$, where the critical values $-t_{\alpha/2}$ and $t_{\alpha/2}$ $t > 0$ satisfy $p(T < -t_{\alpha/2}) = p(T > t_{\alpha/2}) = \alpha/2$. This test is Uniformly Most Powerful in the constrained class of Unbiased tests (UMPU test).

Alternatively to the approach based on the reject region, the hypothesis testing may equivalently be performed using the p-value.

Definition 2.30. *p-value. Let $T(\cdot)$ be a test statistic and let t_0 be the actual value of $T(\cdot)$ for a given sample. The p-value is the probability that the test statistic would be more extreme than the actual observed value (i.e., greater than t_0 if t_0 is positive and smaller than t_0 if t_0 is negative).*

This definition is commonly accepted when the alternative hypothesis is unidirectional ($H_1 : \theta > \theta_0$ or $H_1 : \theta < \theta_0$). A more questionable situation is when the alternative hypothesis is bidirectional, that is, when $H_1 : \theta \neq \theta_0$. In such a case, as the extreme values include both the left and the right tails of the distribution, the p-value should be doubled according to several authors.

The p-value is useful because it provides a measurement of empirical evidence against the null hypothesis. In practice, the p-value approach consists in rejecting the null hypothesis when the p-value is smaller than the significance level α of the test, as in such a case if H_0 would be true, we should admit that we drew a definitely uncommon sample. Then, we prefer to conclude that the null hypothesis is false and reject it.

A problem related to the definition of p-value concerns the inclusion of the actual value t_0 of $T(\cdot)$ in the computation of p-value, when the test statistic has a discrete distribution. Usually, this value is included so that a *conservative test* is obtained, as the significance level is greater than the nominal value α.

Another questionable aspect is related to the ambiguous interpretation of the p-value. For instance, let us assume to test $H_0 : \mu = 65$ against $H_0 : \mu > 65$ through a sample drawn from a normal population with known variance $\sigma^2 = 1$. Now, let us consider two cases: (*i*) $n = 4$ and $\bar{x} = 66$ and (*ii*) $n = 400$ and $\bar{x} = 65.1$. In both cases the p-value is equal to 0.0228 and, then, we should reject the null hypothesis $\mu = 65$, even if the two situations are completely different. This problem is not only related to the p-value approach: it involves the entire classical hypothesis testing theory. A satisfactory solution is provided in the context of the Bayesian statistical inference, as will be clarified in the next section.

2.10   Bayesian statistical inference

The classical approach to statistical inference recognizes sample data as the unique eligible source to attain an *objective knowledge* of real phenomena. In addition to sample data, in the Bayesian approach a relevant role is covered by the prior information, which is formally included in the process of knowledge through the Bayes' rule[5].

As previously outlined, probabilistic models are usually introduced to make easier the comprehension of real phenomena whose manifestations are dominated by variability. In the Bayesian context, probabilistic models are also used to represent the prior information: in such a case, however, the models do not represent the objective variability in the data, but a virtual variability due to the total or partial lack of knowledge. In practice, different from the frequentist approach, where parameters characterizing the population distribution are unknown but constant values, in the Bayesian approach the population parameters are considered as RVs and, as such, a probabilistic distribution, known as *prior distribution*, is introduced to formalize the prior information about them.

Formulating the prior distribution represents a crucial issue in the Bayesian approach and main criticisms to this approach point against the use of the subjective definition of probability that would undermine the objectivity of the scientific process. For this reason, several authors incline towards *uninformative prior distributions* (also named default distributions, conventional distributions, reference distributions, objective distributions) that do not require a subjective probability elicitation process (see Section 2.10.2 for more details).

In more detail, two main schools of thought exist in the Bayesian field: the *subjectivist Bayesian school* and the *objectivist Bayesian school*. The subjective Bayesians support the subjective choice of prior distribution. The problem of

---

[5]Pompilj [173] writes (translation from the original Italian text): "I will try to illustrate the meaning and the extent of the Bayes formulas by quoting some passages from one of my articles in the magazine Archimede [172].

Daily experience always puts us in front of apparently paradoxical contrasts because in them the parts invoke, as support of the opposing theses, the same facts, on which they perfectly agree.

Why, [...], do the parties agree on the facts and then come to conflicting conclusions? Through which mechanism is each of us persuaded by certain interpretations? What is the subjective and objective component of this persuasion? These are very old problems; and it certainly cannot satisfy the dogmatic explanation of the ancient sophists: man is the measure of all things [...].

In *Six characters in search of an author* [by Pirandello] when the Manager interrupts his stepdaughter's, exclaiming: - Let's come to the point. This is only discussion -, the father character intervenes by clarifying: - Very good, sir! But a fact is like a sack which won't stand up when it is empty. In order that it may stand up, one has to put into it the reason and the sentiment which have caused it to exist -.

This joke of the father contains the true essence of the problem just outlined; because once recognized, according to the Pirandellian image, that a fact is like a sack, we can easily understand how depending on what you put in it can take on one aspect rather than another. [...]"

A similar conclusion is reached by Gini. This author, in addition to being a precursor [81] of what is usually defined as the empirical Bayesian approach to statistical inference [37], in two contributions [79, 80] anticipates much of the criticism addressed, in the following years, to the theory of significance tests by the supporters of the Bayesian approach to statistical inference.

eliciting the prior distribution is similar to the problem of eliciting the utility function, which is deeply treated in Chapter 4; for further details, see [123] and [75]. On the other hand, the objective Bayesians sustain the necessity of an objective assessment of the prior distribution. In turn, objective Bayesians distinguish between empirical Bayesians[6], which justify the Bayesian approach only when the prior empirical evidence is adequate, and those that prefer using uninformative prior distributions (Section 2.10.2). For further considerations on these topics we suggest the following references: [15, 16, 83, 125, 130, 181].

Above considerations and, more in general, critics of the Bayesian approach are usually justified by a supposed objectivity of the frequentist inferential approach. However, nothing could be more wrong than this idea. Indeed, although according to the frequentist approach the knowledge process formally relies only on the sample data, the prior information plays informally a very relevant role, as happens, for instance, when the analytic form of the probabilistic model generating the sample data is assumed to be known, as is usual in the parametric inference.

Prior knowledge about the data generating process drives the choice of the probabilistic model representing the real phenomenon and, as a consequence, it affects the inferential conclusions, even if the observed sample data is the same. To better understand this point, let us assume that event "head" is observed k times in n independent tosses of a coin. This empirical evidence may be represented by two different probabilistic models, which reflect the data generating process: if the number n of tosses is fixed, a suitable probabilistic model is the binomial one, whereas if k is fixed and, then, n depends on the number of tosses to obtain k times the event "head", the optimal probabilistic model is the negative binomial one.

Example 2.33. *Let p denote the probability of head in a coin toss and let k denote the number of successes (heads) observed in n independent trials. If the number n of trials is fixed, then the data generating process is described through a binomial RV (Eq. (2.2)):*

$$f(x) = f(x; n, p) = \binom{n}{x} p^x (1-p)^{n-x}$$

with $E(X) = np$ and $Var(X) = np(1-p)$. On the other hand, if the number k of successes is fixed, the data generating process is described by a negative binomial RV (Example 2.8):

$$f(x) = f(x; k, p) = \binom{k+x-1}{x} p^k (1-p)^x$$

with $n = k+x$ and $E(X) = k(1-p)/p$ and $Var(X) = k(1-p)/p^2$.

---

[6]Note that the term "empirical Bayesian" is here used differently from the current literature, which adopts the expression empirical Bayesian approach with reference to the use of the empirical evidence to make inference both on the likelihood and on the prior distribution.

Let us assume that we are interested in the event "10 heads" in 15 trials. In the binomial case we have $n = 15$ (number of trials) and $x = 10$ (number of successes); then the likelihood function is

$$\mathcal{L}(p|X = 10) = \binom{15}{10}p^{10}(1-p)^{15-10}.$$

In the negative binomial case, $k = 10$ (number of successes) and $x = 5$ (number of failures such that the number of trials is again $n = k + x = 10 + 5 = 15$), then

$$\mathcal{L}(p|X = 10) = \binom{10+5-1}{5}p^{15}(1-p)^5.$$

It can be observed that the two above likelihoods are similar with the exception of the constant term (number of permutations with resampling)

$$\binom{15}{10} \neq \binom{10+5-1}{5} = \binom{14}{5} = \binom{14}{9}.$$

However, the maximum likelihood estimates of parameter p are really different: $\hat{p} = \frac{k}{n} = \frac{10}{15} = 0.67$ in the former case and $\hat{p} = \frac{k}{k+n} = \frac{10}{25} = 0.40$ in the latter case.

This example outlines how the prior information affects statistical data analysis procedures as well as inferential conclusions derived from these analyses. In the opinion of the authors of this book, the restrictive interpretation of objectivity of science based on keeping out any subjective element from the scientific knowledge process is not sufficient to reject the Bayesian approach and the use of subjective probabilities. The true question is not as concerned with using prior information as the nature and the correct use of itself: the nature of prior knowledge depends on the theoretical background of the scientist (or decision maker) and its correct use is provided by the correct use of the Bayes' formula.

In Section 2.9, we illustrated inferential approaches to estimation and hypothesis testing, under the assumption that the probability or density function $f(x; \theta) = f(x|\theta)$ of RV X is known, up to parameter θ that is an unknown constant term. We also assumed to observe a simple random sample $\mathbf{X}' = (X_1, \dots, X_n)$ from X with probability or density function $f(x_1, \dots, x_n; \theta) = f(\mathbf{x}; \theta) = \prod_{i=1}^{n} f(x_i; \theta)$, where $f(x_i; \theta) = f(x; \theta)$.

Under the Bayesian approach the probabilistic model $f(x; \theta) = f(x|\theta)$ is substituted with the joint distribution $f(x, \theta)$. Both elements x and θ are RVs: x is a RV because of the variability characterizing the real phenomenon (random variability) and θ is a RV because of the lack of knowledge about its numerical value (virtual variability).

The joint distribution $f(x, \theta)$ may be expressed as follows:

$$f(x, \theta) = f(x|\theta)f(\theta) = f(x|\theta)\pi(\theta),$$
$$f(x, \theta) = f(\theta|x)f(x) = \pi(\theta|x)f(x),$$

where $f(\theta) = \pi(\theta)$ is the (mass or density) *prior probability function* of the random vector θ and $f(\theta|x) = \pi(\theta|x)$ is the (mass or density) *posterior probability function* of random vector θ. From the two equalities above the analytic expression of the Bayes' formula is derived, that is (under the assumption of a continuous parametric space):

$$\pi(\theta|x) = \frac{f(x|\theta)\pi(\theta)}{f(x)} = \frac{f(x|\theta)\pi(\theta)}{\int_\theta f(x|\theta)\pi(\theta)d\theta} \propto f(x|\theta)\pi(\theta). \qquad (2.22)$$

Moreover, if $f(x|\theta)$ is interpreted in terms of likelihood function (i.e., $f(x|\theta) = \mathcal{L}(\theta)$), the Bayes' formula may be written as

$$\pi(\theta|x) = \frac{\mathcal{L}(\theta)\pi(\theta)}{\int_\theta f(x|\theta)\pi(\theta)d\theta} = \frac{\mathcal{L}(\theta)\pi(\theta)}{f(x)} \propto \mathcal{L}(\theta)\pi(\theta), \qquad (2.23)$$

where $f(x) = \int_\theta f(x|\theta)\pi(\theta)d\theta$ is a normalization constant term and $\mathcal{L}(\theta)\pi(\theta)$ is the *kernel* of the posterior distribution. More in detail, $f(x)$ denotes the marginal prior sample distribution of $X' = (X_1, \ldots, X_n)$ and is known also as *predictive prior distribution* of sample X. Instead, the kernel $\mathcal{L}(\theta)\pi(\theta)$ is that part of probability or density function that depends on RV X. For instance, the kernel associated with the normal RV

$$f(x, \mu|\sigma^2) = \left(\frac{1}{2\pi\sigma^2}\right)^{n/2} e^{-\frac{1}{2\sigma^2}(x-\mu)^2} \propto e^{-\frac{1}{2\sigma^2}(x-\mu)^2}$$

is $e^{-\frac{1}{2\sigma^2}(x-\mu)^2}$.

In addition to the predictive prior distribution, the predictive posterior distribution is defined as

$$f(x^*|x) = \int_\theta f(x^*|\theta, x)\pi(\theta|x)d\theta,$$

and denotes the distribution of a new sample, $X = x^*$, given a previously observed sample $X = x$ drawn from the same RV.

To summarize, functions introduced above have the following probabilistic interpretation (where the term probability has to be intended as density function in the continuous case and probability function in the discrete case):

- $\pi(\theta)$: parameter prior distribution (prior probability);

- $\pi(\theta|x)$: parameter posterior distribution (posterior probability);

- $f(x|\theta)$: conditional sample distribution;

- $\mathcal{L}(\theta)$: likelihood function (is not a probability, that is, $\int \mathcal{L}(\theta)d\theta$ is not necessarily equal to 1 and $\mathcal{L}(\theta)$ may not be finite);

- $f(x)$: marginal prior sample distribution or *predictive prior distribution*;

- $f(\boldsymbol{x}^*|\boldsymbol{x})$: marginal posterior sample distribution or *predictive posterior distribution*.

Usually, it may happen that only a subset of elements of $\boldsymbol{\theta}' = (\theta_1, \ldots, \theta_h, \theta_{h+1}, \ldots, \theta_r)$ is of interest, say, for instance, $\boldsymbol{\theta}'_1 = (\theta_1, \ldots, \theta_h)$ ($h < r$), whereas the remaining elements, $\boldsymbol{\theta}'_2 = (\theta_{h+1}, \ldots, \theta_r)$ are nuisance parameters. In such a situation, the nuisance parameters are integrated out and the posterior distribution of $\boldsymbol{\theta}_1$ turns out to be

$$\pi(\boldsymbol{\theta}_1|\boldsymbol{x}) = \int_{\boldsymbol{\theta}_2} \pi(\boldsymbol{\theta}|\boldsymbol{x})d\boldsymbol{\theta}_2 = \int_{\boldsymbol{\theta}_2} \pi(\boldsymbol{\theta}_1, \boldsymbol{\theta}_2|\boldsymbol{x})d\boldsymbol{\theta}_2 = \int_{\boldsymbol{\theta}_2} \pi(\boldsymbol{\theta}_1|\boldsymbol{\theta}_2, \boldsymbol{x})\pi(\boldsymbol{\theta}_2|\boldsymbol{x})d\boldsymbol{\theta}_2.$$

In the past, the main reasons that interfered with the development of Bayesian inference rely on two main aspects of the Bayes' rule. The first one refers, as is outlined above, to the criticisms against the subjective assessment of prior knowledge and its formalization in terms of probability distribution ($\pi(\boldsymbol{\theta})$). The second aspect refers to the computational difficulties in solving the integral

$$f(\boldsymbol{x}) = \int_{\boldsymbol{\theta}} f(\boldsymbol{x}|\boldsymbol{\theta})\pi(\boldsymbol{\theta})d\boldsymbol{\theta}.$$

as well as in computing the characteristic indices (e.g., moments of RV $\boldsymbol{\theta} = \theta$). Indeed, the computation of the expected value of any function $g(\boldsymbol{\theta})$ requires the solution of two integrals, as is clear from the following expression:

$$E[g(\boldsymbol{\theta})] = \int_{\boldsymbol{\theta}} g(\boldsymbol{\theta})\pi(\boldsymbol{\theta}|\boldsymbol{x})d\boldsymbol{\theta} = \frac{\int_{\boldsymbol{\theta}} g(\boldsymbol{\theta})f(\boldsymbol{x}|\boldsymbol{\theta})\pi(\boldsymbol{\theta})d\boldsymbol{\theta}}{\int_{\boldsymbol{\theta}} f(\boldsymbol{x}|\boldsymbol{\theta})\pi(\boldsymbol{\theta})d\boldsymbol{\theta}}.$$

The introduction of Markov Chain Monte Carlo (MCMC) methods in the statistical field strongly simplified the computational aspects and favored the theoretical development of the Bayesian approach as well as its dissemination among practitioners. MCMC methods include a wide class of algorithms, which aim at approximating a complex distribution through repeated random sampling based on a Markov chain. Among the most commonly used algorithms, we mention the Metropolis-Hastings, Gibbs sampler, slice sampling and perfect sampling, which are implemented in the free software WinBUGS[7].

In addition, a special relevance is covered by the conjugate prior distributions (Section 2.10.1), which are especially useful from a computational point of view, because their use in the Bayes' formula does not involve the computation of the normalization term $f(\boldsymbol{x})$.

Before proceeding, it is worth noting that the prior probability or density distribution $\pi(\boldsymbol{\theta})$ is characterized by one or more parameters $\boldsymbol{\delta}' = (\delta_1, \ldots, \delta_s)$,

---

[7]WinBUGS is a really flexible software, developed in the context of the Bayesian inference Using Gibbs Sampling (BUGS) project. It allows us the Bayesian analysis of complex statistical models through MCMC methods. The project started in 1989 at the MRC Biostatistic Unit of Cambridge University and developed in collaboration with the Imperial College School of Medicine of London.

which are named *hyperparameters*. To make explicit the dependence of the prior distribution on the hyperparameters we adopt the expression $\pi(\boldsymbol{\theta}; \boldsymbol{\delta}) = \pi(\boldsymbol{\theta} \mid \boldsymbol{\delta})$, whereas the posterior distribution becomes

$$\pi(\boldsymbol{\theta} \mid \boldsymbol{x}, \boldsymbol{\delta}) = \frac{f(x \mid \boldsymbol{\theta})\pi(\boldsymbol{\theta}; \boldsymbol{\delta})}{\int_{\boldsymbol{\theta}} f(x \mid \boldsymbol{\theta})\pi(\boldsymbol{\theta}; \boldsymbol{\delta})d\boldsymbol{\theta}} = \frac{f(x \mid \boldsymbol{\theta})\pi(\boldsymbol{\theta}; \boldsymbol{\delta})}{f(x)}$$

$$= \frac{\mathcal{L}(\boldsymbol{\theta})\pi(\boldsymbol{\theta}; \boldsymbol{\delta})}{f(\boldsymbol{x})} \propto \mathcal{L}(\boldsymbol{\theta})\pi(\boldsymbol{\theta}; \boldsymbol{\delta}).$$

Therefore, the predictive distributions follow

$$f(\boldsymbol{x} \mid \boldsymbol{\delta}) = \int_{\boldsymbol{\theta}} f(x \mid \boldsymbol{\theta})\pi(\boldsymbol{\theta}; \boldsymbol{\delta})d\boldsymbol{\theta}$$

and

$$f(\boldsymbol{x}^*, \boldsymbol{\delta} \mid \boldsymbol{x}) = \int_{\boldsymbol{\theta}} f(\boldsymbol{x}^* \mid \boldsymbol{\theta}, \boldsymbol{x})\pi(\boldsymbol{\theta} \mid \boldsymbol{x}, \boldsymbol{\delta})d\boldsymbol{\theta}.$$

2.10.1   Conjugate prior distributions

Definition 2.31. *Given Bayes' formula in Eq. (2.23), if the posterior distribution $\pi(\boldsymbol{\theta} \mid \boldsymbol{x})$ and the prior distribution $\pi(\boldsymbol{\theta})$ belong to the same family of RVs, then $\pi(\boldsymbol{\theta} \mid \boldsymbol{x})$ and $\pi(\boldsymbol{\theta})$ are conjugate distributions and $\pi(\boldsymbol{\theta})$ is the conjugate prior distribution for the likelihood function $\mathcal{L}(\boldsymbol{\theta})$ [178]. Briefly, one can say that a conjugate prior distribution is closed under sampling.*

Conjugate prior distributions are usually really flexible and, consequently, are suitable in several practical situations. However, their use should always be supported by adequate prior knowledge. In what follows, the process of transforming a conjugate prior distribution in a posterior distribution is illustrated with reference to some RVs (i.e., binomial, Poisson, normal): the starting point is usually the kernel of the prior distribution, which is transformed through the Bayes' formula in the kernel of the posterior distribution; finally, the normalization constant term $f(\boldsymbol{x}) = \int_{\boldsymbol{\theta}} f(\boldsymbol{x} \mid \boldsymbol{\theta})\pi(\boldsymbol{\theta})d\boldsymbol{\theta}$ is derived.

Example 2.34. *Let X be a binomial distribution (Eq. (2.2)); the likelihood function is*

$$\mathcal{L}(p; \boldsymbol{x}) \propto p^{\sum_{i=1}^{n} x_i}(1 - p)^{n - \sum_{i=1}^{n} x_i}.$$

Let us also assume that the prior distribution of p ($p \in [0, 1]$) is a Beta RV (Eq. (2.8)) with $a = 0$ and $b = 1$,

$$\pi(p) = \pi(p; 0, 1, \alpha, \beta) = \frac{1}{B(\alpha, \beta)}p^{\alpha-1}(1 - p)^{\beta-1} \propto p^{\alpha-1}(1 - p)^{\beta-1}.$$

The likelihood function is

$$\mathcal{L}(p; \boldsymbol{x}) \propto p^{\sum_{i=1}^{n} x_i}(1 - p)^{n - \sum_{i=1}^{n} x_i}.$$

According to the Bayes' formula, the posterior distribution of p is proportional to the product of the likelihood and the prior distribution

$$\pi(p|x) = \frac{\pi(p)\mathcal{L}(p;\boldsymbol{x})}{f(x)} \propto p^{\alpha-1}(1-p)^{\beta-1}p^{\sum_{i=1}^{n} x_i}(1-p)^{n-\sum_{i=1}^{n} x_i}$$

$$\propto p^{x+\alpha-1}(1-p)^{n+\beta-x-1},$$

which, considering the constant term

$$f(x) = \frac{1}{B(\sum_{i=1}^{n} x_i + \alpha, n+\beta)} = \frac{\Gamma(\sum_{i=1}^{n} x_i + \alpha)\Gamma(n+\beta)}{\Gamma(\sum_{i=1}^{n} x_i + \alpha + n + \beta)} =$$

$$= \frac{(\sum_{i=1}^{n} x_i + \alpha - 1)!(n+\beta-1)!}{(\sum_{i=1}^{n} x_i + \alpha + n + \beta - 1)!},$$

represents a Beta distribution with parameters $\sum_{i=1}^{n} x_i + \alpha$ and $n+\beta$, that is,

$$\pi(p|\boldsymbol{x}) = \frac{1}{B(\sum_{i=1}^{n} x_i + \alpha, n+\beta)} p^{\alpha-1}(1-p)^{\beta-1}p^{\sum_{i=1}^{n} x_i}(1-p)^{n-\sum_{i=1}^{n} x_i}.$$

Similar considerations hold if X is distributed as a multinomial RV with k dimensions and n known (Eq. (2.10)): in this case the conjugate prior distribution is the Dirichlet RV (Eq. (2.11)).

Example 2.35. Let X be a Poisson distribution (Eq. (2.3)); the likelihood function is

$$\mathcal{L}(\lambda;\boldsymbol{x}) = \prod_{i=1}^{n} \frac{\lambda^{x_i} e^{-n\lambda}}{x_i!} \propto e^{-n\lambda} x_i! \, \lambda^{\sum_{i=1}^{n} x_i}.$$

Let us also assume that the prior distribution of λ is a Gamma RV (Eq. (2.5)). Then, the posterior distribution of λ is proportional to the product of the likelihood and the prior distribution

$$\pi(\lambda|\boldsymbol{x}) = \frac{\pi(\lambda)\mathcal{L}(\lambda;\boldsymbol{x})}{f(x)} \propto \lambda^{\sum_{i=1}^{n} x_i + \alpha - 1} e^{-\left(n\lambda + \frac{\lambda}{\beta}\right)},$$

$$\propto \lambda^{\alpha^*-1} e^{-\frac{\lambda}{\beta^*}}$$

which, considering the constant term

$$f(\boldsymbol{x}) = \frac{1}{\Gamma(\alpha)\beta^\alpha},$$

represents a Gamma distribution with parameters $\alpha^* = \sum_{i=1}^{n} x_i + \alpha$ and $\beta^* = \frac{\beta}{\beta n + \lambda}$, that is,

$$\pi(\lambda|\boldsymbol{x}) = \frac{1}{\Gamma(\alpha)\beta^\alpha} \lambda^{\alpha^*-1} e^{-\frac{\lambda}{\beta^*}}.$$

Example 2.36. *In the case of a normally distributed RV (Eq. (2.4)), the likelihood function*

$$\mathcal{L}(\mu, \sigma^2; \boldsymbol{x}) = \frac{1}{(2\pi\sigma^2)^{n/2}} e^{-\frac{1}{2\sigma^2}\sum_{i=1}^n (x_i-\mu)^2}$$

$$\propto e^{-\frac{1}{2\sigma^2}\sum_{i=1}^n (x_i-\mu)^2}$$

$$\propto e^{-\frac{n}{2\sigma^2}(\bar{x}-\mu)^2}$$

has to be factorized with the prior probability of σ^2 when μ is known, with the prior probability of μ when σ^2 is unknown, and with the prior joint probability of (μ, σ^2) when both parameters are unknown.

When μ is known, the inverse Gamma distribution (Eq. (2.6))

$$\pi(\sigma^2|\mu; \alpha, \beta) = \frac{1}{\Gamma(\alpha)\beta^\alpha}(\sigma^2)^{-\alpha-1}e^{-\frac{1}{\sigma^2\beta}}$$

is a conjugate prior distribution for σ^2. Indeed, the resulting posterior distribution is an inverse Gamma with parameters $\alpha^* = \alpha + \frac{n}{2}$ and $\beta^* = \beta + \frac{n}{2}(\bar{x}-\mu)^2$:

$$\pi(\sigma^2|\mu, \boldsymbol{x}; \alpha, \beta) \propto e^{-\frac{n}{2\sigma^2}(\bar{x}-\mu)^2}(\sigma^2)^{-(\alpha+\frac{n}{2})-1}e^{-\frac{\beta}{\sigma^2}}$$

$$\propto (\sigma^2)^{-(\alpha+\frac{n}{2})-1}e^{-\frac{\beta+\frac{n}{2}(\bar{x}-\mu)^2}{\sigma^2}}$$

$$\propto (\sigma^2)^{-\alpha^*-1}e^{-\frac{\beta^*}{\sigma^2}}.$$

When σ^2 is known, the normal distribution

$$\pi(\mu|\sigma^2 = \sigma_0^2; \mu_0) = \frac{1}{(2\pi\sigma_0^2)^{n/2}} e^{-\frac{1}{2}\left(\frac{\mu-\mu_0}{\sigma_0}\right)^2}$$

is a conjugate prior distribution for μ. Indeed, considering a normally distributed predictive prior function $f(\boldsymbol{x})$ with hyperparameters μ_0 and $\sigma^2 + \sigma_0^2$,

$$\boldsymbol{X} \sim N(\mu_0, \sigma^2 + \sigma_0^2), \tag{2.24}$$

the resulting posterior distribution is normal

$$\pi(\mu|\sigma, \boldsymbol{x}; \mu_0, \sigma_0^2) \propto e^{-\frac{1}{2}\left[\frac{n}{\sigma^2}(\bar{x}-\mu)^2 + \frac{1}{\sigma_0^2}(\mu-\mu_0)^2\right]}$$

$$\propto e^{-\frac{1}{2}\frac{\sigma^2+n\sigma_0^2}{\sigma^2\sigma_0^2}\left[\mu-\frac{n\bar{x}\sigma_0^2+\mu_0\sigma^2}{\sigma^2+n\sigma_0^2}\right]^2}$$

$$\propto e^{-\frac{1}{2\sigma_1^2}(\mu-\mu_1)^2},$$

with

$$\mu_1 = \frac{n\bar{x}\sigma_0^2 + \mu_0\sigma^2}{\sigma^2 + n\sigma_0^2}, \tag{2.25}$$

$$\sigma_1^2 = \frac{\sigma^2\sigma_0^2}{\sigma^2 + n\sigma_0^2}. \tag{2.26}$$

The related posterior predictive distribution, $f(\boldsymbol{x}^*|\boldsymbol{x})$, is normal too with mean equal to μ_1 and variance equal to $\sigma^2 + \sigma_1^2$.

When both μ and σ^2 are unknown, the derivation of a conjugate prior distribution is possible under the assumption of dependence between μ and σ^2:

$$\pi(\mu, \sigma^2) = \pi(\mu|\sigma^2)\,\pi(\sigma^2).$$

If one assumes that $\pi(\mu|\sigma^2)$ is a normal distribution

$$\pi(\mu|\sigma^2) = \frac{1}{(2\pi\sigma_1^2)^{n/2}} e^{-\frac{1}{2\sigma_1^2}(\mu-\mu_1)^2},$$

with $\sigma_1^2 = \sigma^2/n_0$ and n_0 to be fixed, and $\pi(\sigma^2)$ is an inverse Gamma distribution

$$\pi(\sigma^2) = \frac{(\sigma^2)^{-\alpha-1}\beta^\alpha e^{-\frac{\beta}{\sigma^2}}}{\Gamma(\alpha)},$$

then a prior normal-Gamma distribution results

$$\pi(\mu, \sigma^2) = \pi(\mu|\sigma^2)\,\pi(\sigma^2) = \frac{e^{-\frac{1}{2\sigma_1^2}(\mu-\mu_1)^2}}{(2\pi\sigma_1^2)^{n/2}}\frac{(\sigma^2)^{-\alpha-1}\beta^\alpha e^{-\frac{\beta}{\sigma^2}}}{\Gamma(\alpha)}.$$

The joint posterior distribution deriving from the Bayes' formula is a normal-Gamma, too. Indeed,

$$\pi(\mu, \sigma^2|\boldsymbol{x}) = \frac{\pi(\mu|\sigma^2)\pi(\sigma^2)\mathcal{L}(\mu,\sigma^2;\boldsymbol{x})}{f(\boldsymbol{x})}$$

$$= \frac{e^{-\frac{1}{2\sigma_1^2}(\mu-\mu_1)^2}}{(2\pi\sigma_1^2)^{n/2}}\frac{(\sigma^2)^{-\alpha-1}\beta^\alpha e^{-\frac{\beta}{\sigma^2}}}{\Gamma(\alpha)}\frac{1}{(2\pi\sigma_1^2)^{n/2}}$$

$$\times\, e^{-\frac{1}{2\sigma^2}[(n-1)s^2+n(\bar{x}-\mu)^2]}/f(\boldsymbol{x})$$

$$\propto (\sigma^2)^{-\frac{1}{2}}e^{-\frac{n+n_0}{2\sigma^2}\left(\mu-\frac{n_0\mu_1}{n+n_0}-\frac{n\bar{x}}{n+n_0}\right)^2}(\sigma^2)^{-\left(\frac{n}{2}+\alpha\right)-1}$$

$$\times\, e^{-\frac{1}{\sigma^2}\left[\beta+\frac{n-1}{2}s^2+\frac{n\,n_0}{2(n+n_0)}(\bar{x}-\mu_1)^2\right]}$$

$$\propto (\sigma^2)^{-\frac{1}{2}}e^{-\frac{1}{2\sigma_*^2}(\mu-\mu_*)^2}(\sigma^2)^{-\alpha_*-1}e^{-\frac{\beta_*}{\sigma^2}},$$

with $\sigma_*^2 = \frac{\sigma^2}{n+n_0}$, $\mu_* = \frac{n_0\mu_1}{n+n_0} - \frac{n\bar{x}}{n+n_0}$, $\alpha_* = \frac{n}{2}+\alpha$, and $\beta_* = \beta + \frac{n-1}{2}s^2 + \frac{n\,n_0}{2(n+n_0)}(\bar{x}-\mu_1)^2$.

Examples above illustrated may be derived through suitable specifications of the regular exponential family (Section 2.6). Let X be a RV belonging to the exponential family with $\boldsymbol{\theta}' = (\theta_1,\ldots,\theta_k)$ and let $\boldsymbol{x}' = (x_1,\ldots,x_n)$ be a random sample from X. The likelihood function for $\boldsymbol{\theta}$ is

$$\mathcal{L}(\boldsymbol{\theta}|\boldsymbol{x}) = \prod_{i=1}^{n} f(x_i, \boldsymbol{\theta}) = [a(\boldsymbol{\theta})]^n \prod_{i=1}^{n} h(x_i)e^{\sum_{j=1}^{r}\psi_j(\boldsymbol{\theta})\sum_{i=1}^{n}t_j(x_i)},$$

whereas a prior distribution for $\boldsymbol{\theta}$ is introduced that belongs to the exponential family

$$\pi(\boldsymbol{\theta}|\alpha, \boldsymbol{\beta}) \propto [a(\boldsymbol{\theta})]^{\alpha} e^{\sum_{j=1}^{r} \psi_j(\boldsymbol{\theta})\beta_j},$$

with $\alpha > 0$ and $\boldsymbol{\beta}' = (\beta_1, \ldots, \beta_r)$. It can be shown that the resulting posterior distribution resembles $\pi(\boldsymbol{\theta}|\alpha, \boldsymbol{\beta})$, namely,

$$\pi(\boldsymbol{\theta}|\boldsymbol{x}, \alpha, \boldsymbol{\beta}) \propto [a(\boldsymbol{\theta})]^{\alpha_*} e^{\sum_{j=1}^{r} \psi_j(\boldsymbol{\theta})\beta_j*},$$

with $\alpha_* = \alpha + n$ and $\beta_j* = \beta_j t_j(\boldsymbol{x})$.

To conclude, the choice of the conjugate prior distribution is usually subjective, whereas its specification involving unknown hyperparameters can be driven both by subjective and objective evaluations. The numerical value assigned to the hyperparameters can derive from evaluations of experts or, alternatively, from results of previous empirical analyses. In both cases the resulting prior distribution is *informative*, in the sense that it contributes in a relevant way to the synthesis of data. In this regard, O'Hagan [160] states that

---

"The most important consideration in the use of prior information is to ensure that the prior distribution honestly reflects genuine information, not personal bias, prejudice, superstition or other factors that are justly condemned in science as *subjectivity*."

O'Hagan (2004)

2.10.2   Uninformative prior distributions

Differently from informative prior distributions, the uninformative prior distributions are dominated by the likelihood function and, therefore, they only marginally affect the posterior distribution. A possible common problem with uninformative prior distributions is when a prior distribution does not integrate to 1, that is, $\int_{\boldsymbol{\theta}} f(\boldsymbol{\theta}) d\boldsymbol{\theta} = +\infty$. In such a case, the prior distribution is called *improper*: as a main consequence, it may turn out that also the posterior distribution is improper and, then, no inference is possible.

A first rule to define an uninformative prior distribution is based on the principle of insufficient reason, according to which if k alternatives are indistinguishable, then each alternative should be assigned a probability equal to $1/k$. With reference to the binomial distribution, this principle translates in assigning a uniform distribution in $[0, 1]$ to parameter p (probability of success). However, using the uniform prior distribution may generate two main problems. First, the uniform distribution is not invariant under re-parameterization. Second, if the parametric space is infinite, then the uniform prior distribution is improper.

Jeffreys [122] proposes a general rule to define an uninformative prior distribution. The Jeffreys prior accounts for the information on the unknown parameters that is provided by the sample units and, consequently, it is defined proportionally to the positive square root of the determinant of the Fisher information matrix. The Jeffreys prior is nowadays commonly used for models with a single parameter $\boldsymbol{\theta} = \theta$, due to its invariance under reparameterization of θ:

$$\pi_{jef}(\theta) \propto \mathcal{I}(\theta)^{1/2} = \left\{ -E \left[\frac{d^2 \log f(\boldsymbol{x}; \theta)}{d\theta^2} \right] \right\}^{1/2}.$$

However, in the presence of multiple parameters, the Jeffreys prior is not invariant under reparameterization and, for several distributions, is improper and violates the principle of likelihood.

Example 2.37. *Jeffreys prior distribution for a binomial RV $X \sim$ Binomial(n, p) is a Beta RV (Eq. (2.8)) with parameters $\alpha = 1/2$ and $\beta = 1/2$:*

$$\pi_{jef}(p) \propto \mathcal{I}(p)^{1/2} = \left\{ -E \left[\frac{d^2 \log f(\boldsymbol{x}; p)}{dp^2} \right] \right\}^{1/2} = \left(\frac{np}{p^2} + \frac{n - np}{(1 - p)^2} \right)^{1/2} =$$

$$= \left(\frac{n}{p(1 - p)} \right)^{1/2} \propto p^{-1/2}(1 - p)^{-1/2}$$

and, as outlined in the Example 2.34, it represents a prior conjugate distribution for the binomial one.

Despite this result, the Jeffreys prior distributions are not generally conjugate priors, as can be verified with the Poisson RV.

Example 2.38. *Jeffreys prior distribution for a Poisson RV $X \sim$ Poisson(λ) is a Gamma RV (Eq. (2.5)) with parameters $\alpha = 0.5$ and $\beta = 0$:*

$$\pi_{jef}(\lambda) \propto \mathcal{I}(\lambda)^{1/2} = \left\{ -E \left[\frac{d^2 \log f(\boldsymbol{x}; \lambda)}{d\lambda^2} \right] \right\}^{1/2} = \left(\frac{1}{\lambda} \right)^{1/2} = \lambda^{-1/2}.$$

Another example of uninformative prior distribution is the *reference prior*, due to Bernardo [17] and afterwards developed with Berger (see [16] and references therein). The reference prior is based on the maximization of a suitable measure of divergence or distance between prior and posterior distributions. More in detail, let $f(x, \theta)$ be the density function of X with a single parameter θ and let $T(X)$ be a sufficient statistic for θ; also assume that there is a one-to-one correspondence between $f(x, \theta)$ and $f(T(x), \theta)$. The reference prior

$\pi_{ref}(\theta)$ is obtained by maximizing the expected value of the Kulback-Leibler distance, which is defined as

$$K_n[\pi(\theta, \boldsymbol{x}_n^*), \pi(\theta)] = \int \pi(\theta, \boldsymbol{x}_n^*) \log[\pi(\theta, \boldsymbol{x}_n^*)/\pi(\theta)]d\theta,$$

where $\boldsymbol{x}_n^*$ is an observed random sample of length n. Denoting with K_n^π the expected value of $K_n[\pi(\theta, \boldsymbol{x}_n^*), \pi(\theta)]$ with respect to $\boldsymbol{X}$

$$K_n^\pi = E_{\boldsymbol{x}_n^*}\{K_n[\pi(\theta, \boldsymbol{x}_n^*), \pi(\theta)]\}$$

$$= \int \int \cdots \int \left\{ \int \pi(\theta, \boldsymbol{x}_n^*) \log[\pi(\theta, \boldsymbol{x}_n^*)/\pi(\theta)]d\theta \right\} dx_1 dx_2 \dots dx_n,$$

the resulting reference prior maximizes

$$K_\infty^\pi = \lim_{n \to \infty} K_n^\pi.$$

In practice, as the above limit is infinite, one computes K_n^π for a sufficiently high number of times: the reference prior is the one that corresponds to the limit of this sequence.

Note that for one dimensional parameters, the reference priors and the Jeffreys priors are equivalent, whereas in the multidimensional case they differ.

Another type of uninformative prior distribution was proposed by Jaynes [119, 120] and it is based on the maximization of *entropy*. Entropy is defined as

$$H(X) = -\sum_{i=1}^{k} f(x_i, \theta) \log f(x_i, \theta), \quad i = 1, \dots, k,$$

in the case of discrete RVs with a single parameter θ and probability function $f(x_i, \theta)$, and

$$H(X) = -\int_{-\infty}^{+\infty} f(x, \theta) \log f(x, \theta)dx,$$

for continuous RVs with density function $f(x)$.

The prior distribution $\pi_{ja}(\theta)$ of parameter θ is obtained by maximizing the entropy under some constraints. In the case of discrete RVs and no constraint (with the exception of the normalization one), the distribution that maximizes the entropy is the uniform one, that is, $\pi_{ja}(\theta) = \frac{1}{k}$. A similar result, that is, a continuous uniform distribution $\pi_{ja}(\theta) = \frac{1}{b-a}$ (Eq. (2.9)), is reached with continuous RVs defined in the finite range $[a, b]$. In addition to the normalization constraint, if the first moment about the origin is known, say $E(\theta) = \beta$, and values of X are non-negative, the prior distribution $\pi_{ja}(\theta)$ is a negative exponential (for other examples see [149]):

$$\pi_{ja}(\theta) = \frac{1}{\beta} e^{-\frac{\theta}{\beta}}.$$

The extensive use of uninformative prior distributions is usually connoted in terms of the *objective Bayesian approach*, since the uninformative priors only marginally affect the posterior distribution. However, the term "objective" is quite improper, as the choice of the uninformative prior is based on subjective considerations and, moreover, results differ according to the specified prior distribution. As far as this point, Bernardo states that "uninformative priors do not exist" [116]. A further interesting point of view is due to Seidenfeld [193], which states

---

"I claim that the twin inductive principles which form the core of objective Bayesianism are unacceptable. Invariance (due to H. Jeffreys) and the rule of maximum entropy (due to E. Jaynes) are each incompatible with conditionalization (Bayes' theorem). I argue that the former principle leads to inconsistent representations of *ignorance*, i.e., so called informationless priors generated by the invariance principle are at odds with Bayes' theorem. I claim that Jaynes' rule of maximizing the entropy of a distribution to represent *partial information* is likewise unacceptable. It leads to precise probability distributions that are excessively aprioristic, containing more information than the evidence generating them allows. Again, the conflict is with Bayes' theorem."

Seidenfeld (1979)

2.10.3 Bayesian point and interval estimation

In the Bayesian context, using just sample data to derive a point estimate of the unknown parameter θ is improper, as it contradicts the Bayesian philosophy. Sample data is used to update the prior distribution of parameter θ and the resulting posterior distribution provides all the information about the true value of θ. Only when the posterior distribution degenerates to 1, then we obtain a single point for θ. More in general, the posterior distribution of θ may be synthesized in a single point through its mode

$$\tilde{\theta} = \text{argmax}_\theta f(\theta|\boldsymbol{x})$$

or, alternatively, its arithmetic mean $\hat{\theta}$

$$\hat{\theta} = E[g(\theta)] = \int_\theta \theta d[f(\theta|\boldsymbol{x})]$$

or median $\overline{\theta}$

$$\int_{\overline{\theta}}^{+\infty} d[f(\theta|x)] = \int_{-\infty}^{\overline{\theta}} d[f(\theta|x)] = \frac{1}{2}.$$

Along the same lines as above, any other moment of interest may be computed through a suitable specification of function $g(\cdot)$ in $E[g(\theta)] = \int_\theta g(\theta)d[f(\theta|x)]$ (Eq. (2.1) for the definition of expected value).

As concerns the confidence intervals, in the Bayesian approach we speak about *credible intervals* in the univariate case ($\boldsymbol{\theta} = \theta$) or, more in general, *credible regions* in the multivariate case. A credible region is a range of values within which parameter $\boldsymbol{\theta}$ falls with a certain probability. In particular, we distinguish between *posterior credibility regions* $C_\alpha(x)$

$$\int_{C_\alpha(\boldsymbol{x})} \pi(\boldsymbol{\theta}|\boldsymbol{x}) = 1 - \alpha$$

and *prior credibility regions* C_α

$$\int_{C_\alpha} \pi(\boldsymbol{\theta}) = 1 - \alpha.$$

Confidence intervals and credible intervals differ in some interpretative points. First, credible intervals treat their limits as fixed and the unknown parameter as a RV, whereas confidence intervals treat their limits as random and the unknown parameter as a constant. Before the random sampling, we say that the confidence interval $[L_1, L_2]$ includes the unknown parameter with probability $1 - \alpha\%$, whereas in the Bayesian context we say that the unknown parameter belongs to the credible interval C_α with probability $1 - \alpha\%$. In addition, as usual in the Bayesian approach, credible intervals rely on the prior distribution of the unknown parameter, other than on the sample evidence, whereas in the frequentist approach confidence intervals rely only on the sample evidence. Finally, the presence of unknown nuisance parameters does not represent a problem in the definition of credible regions, whereas it needs a specific treatment in the frequentist approach. Usually, in the definition of confidence intervals the nuisance parameters are substituted with suitable point estimates; however this approach implies a change in the probability distribution of the limits L_1 and L_2.

2.10.4 Bayesian hypothesis testing

The classical hypothesis testing (Section 2.9.4) is based on two hypotheses about the population parameter vector: the null hypothesis $H_0 : \boldsymbol{\theta} \in \boldsymbol{\Theta}_0$ and the alternative hypothesis $H_1 : \boldsymbol{\theta} \in \boldsymbol{\Theta}_1$, with $\boldsymbol{\Theta}_0 \cup \boldsymbol{\Theta}_1 = \boldsymbol{\Theta}$ and $\boldsymbol{\Theta}_0 \cap \boldsymbol{\Theta}_1 = \phi$. Given a significance level α, the null hypothesis is rejected whenever the sample point lies in the reject region; otherwise the null hypothesis is not rejected; an alternative approach is based on the p-value.

In the Bayesian context, the procedure to test hypotheses is simplified. The general approach consists of computing the posterior probabilities of the unknown parameters under H_0 and H_1, that is,

$$P_{0|\boldsymbol{x}} = P(\boldsymbol{\theta} \in \boldsymbol{\Theta}_0 | \boldsymbol{x})$$
$$P_{1|\boldsymbol{x}} = P(\boldsymbol{\theta} \in \boldsymbol{\Theta}_1 | \boldsymbol{x}),$$

and, then, the most probable hypothesis is chosen. In other words, H_0 is chosen if the ratio

$$\frac{P_{0|\boldsymbol{x}}}{P_{1|\boldsymbol{x}}} = \frac{P(\boldsymbol{\theta} \in \boldsymbol{\Theta}_0 | \boldsymbol{x})}{P(\boldsymbol{\theta} \in \boldsymbol{\Theta}_1 | \boldsymbol{x})},$$

known as *posterior odds*, is greater than 1, whereas H_1 is preferred if $\frac{P_{0|\boldsymbol{x}}}{P_{1|\boldsymbol{x}}}$ is less than 1. Similarly, the *prior odds*

$$\frac{P_0}{P_1} = \frac{P(\boldsymbol{\theta} \in \boldsymbol{\Theta}_0)}{P(\boldsymbol{\theta} \in \boldsymbol{\Theta}_1)}$$

may be calculated.

The above procedure based on the comparison between posterior probabilities of $\boldsymbol{\theta}$ under H_0 and H_1 may have some drawbacks. For instance, in the case of a simple null hypothesis $H_0 : \boldsymbol{\theta} = \boldsymbol{\theta}_0$ versus a bidirectional hypothesis $H_1 : \boldsymbol{\theta} \neq \boldsymbol{\theta}_0$, the above procedure cannot be applied, as the posterior probability of H_0 is zero, that is, $P_{0|\boldsymbol{x}} = P(\boldsymbol{\theta} = \boldsymbol{\theta}_0 | \boldsymbol{x}) = \int_{\boldsymbol{\theta}_0}^{\boldsymbol{\theta}_0} f(x|\boldsymbol{\theta})P(\boldsymbol{\theta})d\boldsymbol{\theta} = 0.$

A possible solution consists of accepting H_0 if $\boldsymbol{\theta}_0$ belongs to the credibility interval for $\boldsymbol{\theta}$ or assigning a prior probability greater than 0 to H_0, that is, $P(\boldsymbol{\theta}_0) = P(\boldsymbol{\theta} = \boldsymbol{\theta}_0) > 0$, by introducing a mixture prior distribution between a discrete and a continuous RV. Obviously, there is no problem when both hypotheses are simple: in this case both prior probabilities are greater than 0 and $P_1 = P(\boldsymbol{\theta} = \boldsymbol{\theta}_1) = 1 - P_0 = 1 - P(\boldsymbol{\theta} = \boldsymbol{\theta}_0)$.

Alternatively, the hypothesis testing may be driven by the value assumed by the *Bayes factor* in favor of H_0, which is defined as the ratio between posterior odds and prior odds

$$B_0 = \frac{P_{0|\boldsymbol{x}}}{P_{1|\boldsymbol{x}}} \bigg/ \frac{P_0}{P_1} = \frac{P_{0|\boldsymbol{x}}P_1}{P_{1|\boldsymbol{x}}P_0} = \frac{P(\boldsymbol{\theta} \in \boldsymbol{\Theta}_0 | \boldsymbol{x})P(\boldsymbol{\theta} \in \boldsymbol{\Theta}_1)}{P(\boldsymbol{\theta} \in \boldsymbol{\Theta}_1 | \boldsymbol{x})P(\boldsymbol{\theta} \in \boldsymbol{\Theta}_0)},$$

whereas the Bayes factor in favor of H_1 is

$$B_1 = 1/B_0 = \frac{P_{1|\boldsymbol{x}}P_0}{P_{0|\boldsymbol{x}}P_1} = \frac{P(\boldsymbol{\theta} \in \boldsymbol{\Theta}_1 | \boldsymbol{x})P(\boldsymbol{\theta} \in \boldsymbol{\Theta}_0)}{P(\boldsymbol{\theta} \in \boldsymbol{\Theta}_0 | \boldsymbol{x})P(\boldsymbol{\theta} \in \boldsymbol{\Theta}_1)}.$$

If both H_0 and H_1 are simple, then $P_{0|x} \propto P_0 f(x|\boldsymbol{\theta}_0)$ and $P_{1|x} \propto P_1 f(x|\boldsymbol{\theta}_1)$ and, then, the Bayes factor B_0 reduces to the ratio between the two likelihood functions:

$$B_0 = \frac{P_{0|\boldsymbol{x}}}{P_{1|\boldsymbol{x}}} \bigg/ \frac{P_0}{P_1} = \frac{P_{0|\boldsymbol{x}}P_1}{P_{1|\boldsymbol{x}}P_0} = \frac{f(x|\boldsymbol{\theta}_0)P_0 P_1}{f(x|\boldsymbol{\theta}_1)P_1 P_0} = \frac{f(x|\boldsymbol{\theta}_0)}{f(x|\boldsymbol{\theta}_1)}.$$

This result is completely unsatisfactory for supporters of the subjective Bayesian approach, as it implies an automatic elimination of prior knowledge. Somewhat different is the situation when prior knowledge loses relevance because more and more empirical evidence is collected. As far as this point is concerned, we note that Bayesian and frequentist results tend to converge when the sample size increases.

The approach based on the Bayes factor is also suitable when the decision maker has to choose between alternative statistical models representing real phenomena. Let M denote any statistical model and Ξ the space of all statistical models representing a certain phenomenon; then hypothesis $H_0 : \theta \in \Theta_0$ can be more generally interpreted as $H_0 : M = M_0$, whereas the alternative hypothesis $H_1 : \theta \in \Theta_1$ is specified by $H_1 : M = M_1$, with $M_0 \in \Xi$ and $M_1 \in \Xi$. In such a situation, the Bayes factor is defined as the weighted ratio between the likelihoods of the two models

$$B_0 = \frac{P(M = M_0|\boldsymbol{x})P(M = M_1)}{P(M = M_1|\boldsymbol{x})P(M = M_0)} = \frac{\int_{\boldsymbol{\theta} \in \boldsymbol{\theta}_0} f(\boldsymbol{x}|\boldsymbol{\theta})P_0(\boldsymbol{\theta})d\boldsymbol{\theta}}{\int_{\boldsymbol{\theta} \in \boldsymbol{\theta}_1} f(\boldsymbol{x}|\boldsymbol{\theta})P_1(\boldsymbol{\theta})d\boldsymbol{\theta}}$$

and provides a measure of how much better model M_0 represents the reality with respect to model M_1. In practice, the Bayes factor is usually interpreted according to the following rule of thumb:

- if $B_0 \geq 1$, prior and sample evidence definitely support model M_0;

- if $10^{-1/2} \leq B_0 < 1$, evidence against model M_0 is small;

- if $10^{-1} \leq B_0 < 10^{-1/2}$ evidence against model M_0 is medium;

- if $10^{-2} \leq B_0 < 10^{-1}$ evidence against model M_0 is high;

- if $B_0 < 10^{-2}$ prior and sample evidence definitely support model M_1.

In the presence of $s > 2$ competing models M_i $(i = 1, \ldots, s)$, the selection of the optimal model requires the computation of $s(s-1)/2$ Bayes factors, that is, one for each pair of models. A faster strategy starts with the comparison between any pair of models, say M_1 and M_2; then the best of these two models is compared with M_3, and so on until model M_s is compared with the model selected at the previous step.

2.11   Multiple linear regression model

In this paragraph we will sketch the basic elements concerning the linear regression model. For a more detailed discussion the reader is referred, among others, to textbooks by Zellner [223], Greene [87], and Gujarati and Porter [89].

2.11.1 The statistical model

Let us introduce a sample of k variables, $Y, X_2, \ldots, X_k$, observed on n independent individuals. We assume that variable Y (dependent variable or response variable) is explained by a linear combination of $X_2, \ldots, X_k$ (independent variables or explanatory variables or covariates) plus an error term:

$$y_i = \beta_1 + \beta_2 x_{i2} + \ldots + \beta_k x_{ik} + e_i, \quad i = 1, \ldots, n,$$

where β_j $(j = 1, \ldots, k)$ are unknown parameters (regression coefficients) that explain the effect of X_j on the expected value of Y, and e_i is the unknown stochastic disturbance or stochastic error term. Using the matrix notation, the model is equivalently written as

$$\boldsymbol{y} = \boldsymbol{X}\boldsymbol{\beta} + \boldsymbol{e}, \tag{2.27}$$

where: $\boldsymbol{y}' = (y_1, \ldots, y_n)$ is the n-dimensional vector of the responses, $\boldsymbol{X}$ is the data matrix of dimension $n \times k$

$$\boldsymbol{X} = \begin{bmatrix} 1 & x_{12} & \cdots & x_{1j} & \cdots & x_{1k} \\ 1 & x_{22} & \cdots & x_{2j} & \cdots & x_{2k} \\ \vdots & \vdots & \cdots & \vdots & \cdots & \vdots \\ 1 & x_{i1} & \cdots & x_{ij} & \cdots & x_{ik} \\ \vdots & \vdots & \cdots & \vdots & \cdots & \vdots \\ 1 & x_{n1} & \cdots & x_{nj} & \cdots & x_{nk} \end{bmatrix},$$

$\boldsymbol{\beta}' = (\beta_1, \beta_2, \ldots, \beta_k)$ is the k-dimensional vector of regression coefficients, $\boldsymbol{e}' = (e_1, e_2, \ldots, e_n)$ is the n-dimensional vector of errors.

The model in Eq. (2.27) is known as the (classical) linear regression model and it relies on the following assumptions:

- matrix $\boldsymbol{X}$ is nonstochastic and of full rank, that is, $rank(\boldsymbol{X}) = k \leq n$;

- $E(\boldsymbol{e}) = \boldsymbol{0}$, where $\boldsymbol{0}$ is the n-dimensional null vector;

- $Var(\boldsymbol{e}) = \boldsymbol{\Sigma}_e = E(\boldsymbol{ee}') = \sigma^2 \boldsymbol{I}_n$, where σ^2 is a usually unknown parameter, $\boldsymbol{I}_n$ is an $n \times n$ identity matrix, and random variables e_i are uncorrelated and homoscedastic; this matrix is called the *variance-covariance matrix*.

From the above assumptions, it turns out that

$$E(\boldsymbol{y}) = E(\boldsymbol{y}|\boldsymbol{X}) = \boldsymbol{X}\boldsymbol{\beta} = \boldsymbol{y}^*,$$
$$Var(\boldsymbol{y}) = Var(\boldsymbol{y}|\boldsymbol{X}) = \boldsymbol{\Sigma}_y = \sigma^2 \boldsymbol{I}_n.$$

2.11.2   Least squares estimator and maximum likelihood estimator

The estimation of unknown parameters $\boldsymbol{\beta}$ and σ^2 is usually performed through two main approaches: the least squares method and the maximum likelihood method.

According to the least squares method, we aim at finding those values of $\boldsymbol{\beta}$ that minimize the (squared) distance between the actual values $\boldsymbol{y}$ and the expected values $\boldsymbol{y}^*$. Formally, let $Q(\boldsymbol{\beta})$ be the residual sum of squares, that is,

$$Q(\boldsymbol{\beta}) = \sum_{i=1}^{n}(y_i - y_i^*)^2 = (\boldsymbol{y} - \boldsymbol{X}\boldsymbol{\beta})'(\boldsymbol{y} - \boldsymbol{X}\boldsymbol{\beta}).$$

Minimizing $Q(\boldsymbol{\beta})$ with respect to the unknown vector $\boldsymbol{\beta}$, we obtain

$$\hat{\boldsymbol{\beta}} = (\boldsymbol{X}'\boldsymbol{X})^{-1}\boldsymbol{X}'\boldsymbol{y},$$

which is the *ordinary least squares (OLS) estimator* for $\boldsymbol{\beta}$.

Given $\hat{\boldsymbol{\beta}}$, the following relations turn out

$$\hat{\boldsymbol{y}}^* = \hat{\boldsymbol{y}} = \boldsymbol{X}\hat{\boldsymbol{\beta}},$$
$$\hat{\boldsymbol{e}} = \boldsymbol{y} - \hat{\boldsymbol{y}} = \boldsymbol{y} - \boldsymbol{X}\hat{\boldsymbol{\beta}},$$
$$\hat{\boldsymbol{e}}'\hat{\boldsymbol{e}} = \boldsymbol{y}'\boldsymbol{y} - \hat{\boldsymbol{\beta}}'\boldsymbol{X}'\boldsymbol{y},$$
$$\hat{\sigma}^2 = \frac{\hat{\boldsymbol{e}}'\hat{\boldsymbol{e}}}{n-k},$$
$$E(\hat{\sigma}^2) = E(\hat{\boldsymbol{e}}'\hat{\boldsymbol{e}})/(n-k) = \sigma^2,$$
$$E(\hat{\boldsymbol{e}}'\hat{\boldsymbol{e}}) = \sigma^2\boldsymbol{I}_n.$$

The OLS estimator $\hat{\boldsymbol{\beta}}$ is linear and unbiased and minimizes the variance in the class of linear unbiased estimators (Gauss-Markov theorem): by virtue of these properties the OLS estimator is *the best linear unbiased estimator* (BLUE). Indeed,

$$E(\hat{\boldsymbol{\beta}}) = \boldsymbol{\beta},$$
$$E(\hat{\boldsymbol{y}}^*) = E(\hat{\boldsymbol{y}}) = E(\boldsymbol{X}\hat{\boldsymbol{\beta}}) = \boldsymbol{X}\boldsymbol{\beta} = \boldsymbol{y}^*,$$
$$Var(\hat{\boldsymbol{\beta}}) = Var(\hat{\boldsymbol{\beta}}|\boldsymbol{X}) = \boldsymbol{\Sigma}_{\hat{\boldsymbol{\beta}}} = \sigma^2(\boldsymbol{X}'\boldsymbol{X})^{-1},$$
$$Var(\hat{\boldsymbol{y}}^*) = \boldsymbol{\Sigma}_{\hat{\boldsymbol{y}}^*} = \sigma^2(\boldsymbol{X}'\boldsymbol{X})^{-1},$$
$$Var(\hat{\boldsymbol{y}}) = \boldsymbol{\Sigma}_{\hat{\boldsymbol{y}}} = \sigma^2[\boldsymbol{I} + (\boldsymbol{X}'\boldsymbol{X})^{-1}].$$

Substituting σ^2 with the estimated variance $\hat{\sigma}^2$ above defined in the last two relations, we have

$$\widehat{Var}(\hat{\boldsymbol{\beta}}) = \widehat{Var}(\hat{\boldsymbol{\beta}}|\boldsymbol{X}) = \hat{\boldsymbol{\Sigma}}_{\hat{\boldsymbol{\beta}}} = \hat{\sigma}^2(\boldsymbol{X}'\boldsymbol{X})^{-1},$$

$$\widehat{Var}(\hat{\boldsymbol{y}}^*) = \hat{\boldsymbol{\Sigma}}_{\hat{\boldsymbol{y}}^*} = \hat{\sigma}^2(\boldsymbol{X}'\boldsymbol{X})^{-1},$$

$$\widehat{Var}(\hat{\boldsymbol{y}}) = \hat{\boldsymbol{\Sigma}}_{\hat{\boldsymbol{y}}} = \hat{\sigma}^2[\boldsymbol{I} + (\boldsymbol{X}'\boldsymbol{X})^{-1}].$$

In addition, to evaluate the goodness of fit of the estimated regression model a commonly used index is the *coefficient of linear determination* R^2, that relies on the decomposition of the total deviance of $\boldsymbol{y}$.

Let $\boldsymbol{s}_y = \boldsymbol{y} - \bar{\boldsymbol{y}}$ be the difference between actual $\boldsymbol{y}$ and its mean $\bar{\boldsymbol{y}}$; then $\boldsymbol{s}_y$ is decomposed as

$$\boldsymbol{s}_y = \boldsymbol{y} - \bar{\boldsymbol{y}} = \boldsymbol{y} - \hat{\boldsymbol{y}} + \hat{\boldsymbol{y}} - \bar{\boldsymbol{y}} = \boldsymbol{s}_{\hat{e}} + \boldsymbol{s}_{\hat{y}}.$$

Consequently, the total deviance of $\boldsymbol{y}$, which is defined by $||\boldsymbol{s}_y||^2 = (\boldsymbol{y} - \bar{\boldsymbol{y}})'(\boldsymbol{y} - \bar{\boldsymbol{y}})$, can be decomposed as

$$||\boldsymbol{s}_y||^2 = \boldsymbol{s}_y'\boldsymbol{s}_y = \boldsymbol{s}_{\hat{e}}'\boldsymbol{s}_{\hat{e}} + \boldsymbol{s}_{\hat{y}}'\boldsymbol{s}_{\hat{y}} = ||\boldsymbol{s}_{\hat{e}}||^2 + ||\boldsymbol{s}_{\hat{y}}||^2,$$

where $||\boldsymbol{s}_{\hat{y}}||^2$ is that part of total deviance explained by the regression model (*regression deviance*) and $||\boldsymbol{s}_{\hat{e}}||^2$ is that part of total deviance that is due to random disturbances (*residual deviance*). Hence, the ratio between regression deviance and total deviance, known as coefficient of linear determination R^2

$$R^2 = \frac{\text{regression deviance}}{\text{total deviance}} = \frac{||\boldsymbol{s}_{\hat{y}}||^2}{||\boldsymbol{s}_y||^2} = 1 - \frac{||\boldsymbol{s}_{\hat{e}}||^2}{||\boldsymbol{s}_y||^2},$$

provides a measurement of the goodness of fit of the estimated linear regression model. Obviously, $0 \leq R^2 \leq 1$, with $R^2 = 0$ denoting the absence of linear dependence between $\boldsymbol{y}$ and $\boldsymbol{X}$ and $R^2 = 1$ denoting a perfect fit of observed data to the linear model.

Note that the numerical value of R^2 strictly depends on the number of explanatory variables included in the model. For this reason, to measure the goodness of fit of the regression model to the observed data, one usually should refer to the so-called adjusted R^2, which is defined by

$$R_{\text{adj}}^2 = 1 - \frac{n}{n-k}(1 - R^2).$$

The computation of the OLS estimator does not require any specific distributive assumption on the error component. However, to make inference on the model parameters we need to introduce a further hypothesis concerning the probabilistic distribution of the disturbances.

In more detail, if we introduce the assumption of normality

$$\boldsymbol{e} \sim N(\boldsymbol{0}, \sigma^2 \boldsymbol{I}_n),$$

we have
$$\boldsymbol{y} \sim N(\boldsymbol{X}\boldsymbol{\beta}, \sigma^2 \boldsymbol{I}).$$

In addition, the absence of correlation between the RVs y_i ($i = 1, \ldots, n$) implies independence.

The likelihood function of vector $\boldsymbol{y}$ is defined as

$$\mathcal{L}(\boldsymbol{\beta}, \sigma^2) = \prod_{i=1}^{n} f(y_i) = (2\pi\sigma^2)^{-n/2} \exp\left[-\frac{1}{2\sigma^2}(\boldsymbol{y} - \boldsymbol{X}\boldsymbol{\beta})'(\boldsymbol{y} - \boldsymbol{X}\boldsymbol{\beta})\right].$$

The estimators of $\boldsymbol{\beta}$ and σ^2 resulting from the maximization of $\mathcal{L}(\boldsymbol{\beta}, \sigma^2)$ are

$$\tilde{\boldsymbol{\beta}} = \hat{\boldsymbol{\beta}} = (\boldsymbol{X}'\boldsymbol{X})^{-1}\boldsymbol{X}'\boldsymbol{y},$$

$$\tilde{\sigma}^2 = \frac{||\tilde{\boldsymbol{e}}'\tilde{\boldsymbol{e}}||}{n} = \frac{||\hat{\boldsymbol{e}}'\hat{\boldsymbol{e}}||}{n} = \frac{1}{n}\sum_{i=1}^{n}\hat{e}_i^2 \neq \hat{\sigma}^2.$$

In practice, the OLS estimator and the maximum likelihood estimator of $\boldsymbol{\beta}$ are the same, whereas the estimators of variance σ^2 differ, being the OLS estimator $\hat{\sigma}^2$ unbiased and the maximum likelihood estimator $\tilde{\sigma}^2$ biased. Furthermore, it turns out that the maximum likelihood estimator $\tilde{\boldsymbol{\beta}}$ of $\boldsymbol{\beta}$ minimizes the variance in the class of unbiased estimators (Cramér-Rao limit), that is, $\tilde{\boldsymbol{\beta}}$ is *the best unbiased estimator (BUE)*.

From the above we derive the following properties:

$$\tilde{\boldsymbol{\beta}} = \hat{\boldsymbol{\beta}} \sim N(\boldsymbol{\beta}, \boldsymbol{\Sigma}_{\hat{\boldsymbol{\beta}}}),$$

$$\tilde{\boldsymbol{y}}^* = \hat{\boldsymbol{y}}^* \sim N(\boldsymbol{X}\boldsymbol{\beta}, \boldsymbol{\Sigma}_{\hat{\boldsymbol{y}}^*}),$$

$$\tilde{\boldsymbol{y}} = \hat{\boldsymbol{y}} \sim N(\boldsymbol{X}\boldsymbol{\beta}, \boldsymbol{\Sigma}_{\hat{\boldsymbol{y}}}),$$

$$\frac{n\tilde{\sigma}^2}{\sigma^2} = \frac{(n-k)\hat{\sigma}^2}{\sigma^2} = \frac{\tilde{\boldsymbol{e}}'\tilde{\boldsymbol{e}}}{\sigma^2} = \frac{\hat{\boldsymbol{e}}'\hat{\boldsymbol{e}}}{\sigma^2} = \frac{\sum_{i=1}^{n}\hat{e}_i^2}{\sigma^2} \sim \chi_{n-k}^2,$$

where $\boldsymbol{\Sigma}_{\hat{\boldsymbol{y}}^*} = \sigma^2(\boldsymbol{X}'\boldsymbol{X})^{-1}$ and $\boldsymbol{\Sigma}_{\hat{\boldsymbol{y}}} = \sigma^2[\boldsymbol{I} + (\boldsymbol{X}'\boldsymbol{X})^{-1}]$; furthermore, $\tilde{\boldsymbol{\beta}}$ and $\frac{n\tilde{\sigma}^2}{\sigma^2}$ are independent.

2.11.3   Hypothesis testing

The introduction of the normality assumption allows us to make inference on the model parameters. As concerns the hypothesis testing on regression coefficients $\boldsymbol{\beta}$, a completely general approach may be adopted to verify assumptions about specific linear combinations of regression coefficients.

Let $\boldsymbol{R}$ be a $q \times k$ matrix of rank $q \leq k$ and let $\boldsymbol{r}$ be a vector of dimension q, such that a system of q linear constraints is specified as follows

$$H_0 : \boldsymbol{R}\boldsymbol{\beta} = \boldsymbol{r}$$
$$H_1 : \boldsymbol{R}\boldsymbol{\beta} \neq \boldsymbol{r}.$$

From the above results, it turns out that

$$R\tilde{\beta} \sim N(R\beta, \sigma^2 R(X'X)^{-1}R')$$

and, under the null hypothesis $H_0 : R\beta = r$,

$$\frac{1}{\sigma^2}(R\tilde{\beta} - r)' \left[R(X'X)^{-1}R'\right]^{-1} (R\tilde{\beta} - r) \sim \chi_q^2.$$

To solve the hypothesis test based on $H_0 : R\beta = r$ an F-type statistic is used, usually called *Wald's statistic*,

$$F = \frac{\left\{\frac{1}{\sigma^2}(R\tilde{\beta} - r)' \left[R(X'X)^{-1}R'\right]^{-1} (R\tilde{\beta} - r)\right\}/q}{\left[(n-k)\hat{s}^2/\sigma^2\right]/(n-k)} =$$

$$= \frac{(R\tilde{\beta} - r)' \left[R(X'X)^{-1}R'\right]^{-1} (R\tilde{\beta} - r)}{\hat{s}^2 q}$$

$$\sim F_{q,n-k}. \tag{2.28}$$

Statistic F in Eq. (2.28) may be used to verify a linear constraint of type $R\beta = r$, with matrix R specified in a suitable way. For instance, if we put $q = 1$, $r = 0$ and

$$R' = (0, 0, \ldots, 1, \ldots, 0),$$

where number 1 occupies the $j+1$-th position referred to regression coefficient β_j ($j = 2, \ldots, k$), then the test statistic

$$F \sim F_{1,n-k} = t_{n-k}^2$$

solves the testing problem about the significance of regression coefficient β_j, that is,

$$H_0 : \beta_j = 0$$
$$H_1 : \beta_j \neq 0.$$

Another hypothesis test of particular interest concerns the global fit of the regression model and is specified constraining all regression coefficients to be equal to 0, that is,

$$H_0 : \beta_2 = \ldots = \beta_k = 0 \quad \text{no fit}$$
$$H_1 : \text{at least one } \beta_j \neq 0 \quad j = 2, \ldots, k.$$

To solve this problem we set $q = k - 1$, $r = 0$ and

$$R = \begin{bmatrix} 0 & 1 & 0 & \ldots & 0 \\ 0 & 0 & 1 & \ldots & 0 \\ \ldots & \ldots & \ldots & \ldots & \ldots \\ 0 & 0 & 0 & \ldots & 1 \end{bmatrix} = [\mathbf{0}_{k-1} : I_{k-1}].$$

The test statistic F becomes

$$
\begin{aligned}
F &= \frac{||\boldsymbol{s}_{\hat{y}}||^2/[\sigma^2(k-1)]}{||\hat{\boldsymbol{e}}||^2/[\sigma^2(n-k)]} \\
&= \frac{\text{regression deviance}/(k-1)}{\text{residual deviance}/(n-k)} \\
&= \frac{R^2}{1-R^2}\frac{n-k}{k-1} \\
&\sim F_{k-1,n-k}.
\end{aligned}
$$

making clear that the null hypothesis $H_0 : \beta_2 = \ldots = \beta_k = 0$ can be expressed both in terms of analysis of variance (Table 2.3) and in terms of relationship between the value assumed by R^2 and the significance of the test: the closer to 1 the value of R^2 is, the more significant the test is.

TABLE 2.3

Analysis of variance.

Source of variation	Sum of squares	Degrees of freedom	Avg. sum of squares	Statistic F
Regression	$||\boldsymbol{s}_{\hat{y}}||^2 = \boldsymbol{s}'_{\hat{y}}\boldsymbol{s}_{\hat{y}}$	$k-1$	$||\boldsymbol{s}_{\hat{y}}||^2/(k-1)$	$F = \frac{||\boldsymbol{s}_{\hat{y}}||^2/(k-1)}{||\hat{\boldsymbol{e}}||^2/(n-k)}$
Residuals	$||\hat{\boldsymbol{e}}||^2 = \hat{\boldsymbol{e}}'\hat{\boldsymbol{e}}$	$n-k$	$||\hat{\boldsymbol{e}}||^2/(n-k)$	$= \frac{R^2}{1-R^2}\frac{n-k}{k-1}$
Total	$||\boldsymbol{s}_y||^2 = \boldsymbol{s}'_y\boldsymbol{s}_y$	$n-1$		

2.11.4 Bayesian regression

In the Bayesian regression coefficients $\boldsymbol{\beta}$ and σ^2 are not constant but, being unknown entities, assume the nature of RVs with their own probability distribution. Therefore, we can write

$$
f(\boldsymbol{y},\boldsymbol{\beta},\sigma^2|\boldsymbol{X}) = \pi(\boldsymbol{\beta},\sigma^2)f(\boldsymbol{y}|\boldsymbol{X},\boldsymbol{\beta},\sigma^2) = \pi(\boldsymbol{\beta},\sigma^2)\mathcal{L}(\boldsymbol{\beta},\sigma^2),
$$

where $\pi(\boldsymbol{\beta},\sigma^2)$ is the prior distribution of the $k+1$ unknown parameters $\beta_1, \beta_2, \ldots, \beta_k$ and σ^2, while the likelihood becomes

$$
\begin{aligned}
\mathcal{L}(\boldsymbol{\beta},\sigma^2) &= f(\boldsymbol{\beta},\sigma^2|\boldsymbol{y},\boldsymbol{X}) = (2\pi\sigma^2)^{-n/2}\exp\left[-\frac{1}{2\sigma^2}(\boldsymbol{y}-\boldsymbol{X}\boldsymbol{\beta})'(\boldsymbol{y}-\boldsymbol{X}\boldsymbol{\beta})\right] \\
&= (2\pi\sigma^2)^{-n/2}\exp\left\{-\frac{1}{2\sigma^2}\left[(n-k)S^2 + (\boldsymbol{\beta}-\tilde{\boldsymbol{\beta}})'\boldsymbol{X}'\boldsymbol{X}(\boldsymbol{\beta}-\tilde{\boldsymbol{\beta}})\right]\right\}
\end{aligned}
$$

Taking into account what was illustrated in the Example 2.36 about the conjugate prior distribution of a normal RV, a possible specification of the prior distribution in the present case is

$$
\pi(\boldsymbol{\beta},\sigma^2) = \pi(\boldsymbol{\beta}|\sigma^2)\pi(\sigma^2) = \frac{e^{-\frac{1}{2\sigma^2}(\boldsymbol{\beta}-\boldsymbol{\beta}^*)'\boldsymbol{\Sigma}_\beta^{-1}(\boldsymbol{\beta}-\boldsymbol{\beta}^*)}}{(2\pi\sigma^2)^{1/2}|\boldsymbol{\Sigma}_\beta|^{1/2}} \cdot \frac{(\sigma^2)^{-\alpha-1}\delta^\alpha e^{-\frac{\delta}{\sigma^2}}}{\Gamma(\alpha)},
$$

where $\pi(\boldsymbol{\beta}|\sigma^2)$ is a normal distribution with mean $\boldsymbol{\beta}^*$ and variance-covariance matrix $\sigma^2\boldsymbol{\Sigma}_\beta$ and $\pi(\sigma^2)$ is an inverse Gamma with parameters α and δ. Then, the prior distribution $\pi(\boldsymbol{\beta},\sigma^2)$ is a normal inverse Gamma RV

$$(\boldsymbol{\beta},\sigma^2) \sim \mathrm{NInv\Gamma}(\boldsymbol{\beta}^*,\sigma^2\boldsymbol{\Sigma}_\beta;\alpha,\delta), \qquad (2.29)$$

which is a conjugate prior distribution of a RV belonging to the same family. Indeed, the posterior joint distribution is given by

$$\pi(\boldsymbol{\beta},\sigma^2|\boldsymbol{y}) = \frac{\pi(\boldsymbol{\beta},\sigma^2)\mathcal{L}(\boldsymbol{\beta},\sigma^2|\boldsymbol{y})}{f(\boldsymbol{y})} = \frac{\pi(\boldsymbol{\beta}|\sigma^2)\pi(\sigma^2)\mathcal{L}(\boldsymbol{\beta},\sigma^2|\boldsymbol{y})}{f(\boldsymbol{y})}$$

$$\propto \exp\left[-\frac{1}{2\sigma^2}(\boldsymbol{\beta}-\bar{\boldsymbol{\beta}})'\bar{\boldsymbol{\Sigma}}^{-1}(\boldsymbol{\beta}-\bar{\boldsymbol{\beta}})\right](\sigma^2)^{-\alpha^*-1}\exp\left(-\frac{\delta^*}{\sigma^2}\right),$$

where

$$\bar{\boldsymbol{\beta}} = (\boldsymbol{\Sigma}_\beta^{-1} + \boldsymbol{X}'\boldsymbol{X})^{-1}[\boldsymbol{\Sigma}_\beta^{-1}\boldsymbol{\beta}^* + (\boldsymbol{X}'\boldsymbol{X})\hat{\boldsymbol{\beta}}]^{-1},$$

$$\bar{\boldsymbol{\Sigma}}^{-1} = (\boldsymbol{\Sigma}_\beta^{-1} + \boldsymbol{X}'\boldsymbol{X})^{-1},$$

$$\alpha^* = n/2 - \alpha,$$

$$\delta^* = \frac{\delta}{2}\left\{(n-k)S^2 + (\boldsymbol{\beta}^*-\hat{\boldsymbol{\beta}})'\bar{\boldsymbol{\Sigma}}^{-1}(\boldsymbol{\beta}^*-\hat{\boldsymbol{\beta}})\right\}.$$

Therefore,

$$(\boldsymbol{\beta},\sigma^2|\boldsymbol{y}) \sim \mathrm{NInv\Gamma}(\bar{\boldsymbol{\beta}},\sigma^2\bar{\boldsymbol{\Sigma}};\alpha^*,\delta^*),$$

which belongs to the same family as the prior distribution (2.29) (normal inverse gamma distribution).

2.12 Structural equation model

A direct and relevant generalization of the multiple linear regression model above described is represented by the *simultaneous equation models*[8] [91, 134], which are characterized by a system of multiple regression equations where the response variable in one equation may appear as a predictor in another equation. In other words, variables in a simultaneous equation model may influence one another reciprocally, either directly or through other variables.

In more detail, simultaneous equation models are based on two types of observed variables:

---

[8]Authors note that simultaneous equation models are alternatively called structural equation models in the econometric setting. However, to avoid a misunderstanding with other settings, such as the psychometric one, here we use the expression "simultaneous equation model" limited to those models that involve only observed variables and the expression "structural equation model" for the more general class of models with multiple equations involving observed as well as unobservable (or latent) variables, as will be clear in what follows.

- endogenous variables y_i that are directly affected by other variables in the model and by their disturbances e_i,

- exogenous variables x_i that are not explained within the model and are fixed.

Let us denote by Y the $(t \times n)$ matrix of endogenous observed variables, X the $(t \times k)$ matrix of exogenous observed variables, and E the $(t \times n)$ matrix of unobservable errors; then a simultaneous equation model specifies as

$$Y\Gamma = XB + E, \tag{2.30}$$

where Γ and B denote the matrices of unknown regression coefficients relating, respectively, the endogenous variables to one another and the endogenous variables to the exogenous ones.

Under the assumptions $E(E) = 0$, $rank(E) \leq k$, and $(X'X)^{-1}$ existing and nonsingular, model (2.30) may be written in reduced form as

$$Y = XB\Gamma^{-1} + E\Gamma^{-1}, \tag{2.31}$$

which expresses endogenous variables in terms of exogenous variables and disturbances. If the parameters of a simultaneous equation model can be obtained from the estimated reduced form in Eq. (2.31) we say that the system is identified; if this cannot be done, we say that the system under consideration is unidentified or underidentified.

A useful generalization of the simultaneous equation models is represented by the wide class of *structural equation models* (SEMs; [23, 56, 67, 82]) representing a multivariate approach that combines aspects related to multiple and multivariate (linear) regression and aspects related to factor analysis, through a strong graphical support.

In addition to the presence of observed (endogenous and exogenous) variables that characterize the simultaneous equation models, SEMs are also based on a third type of variable: the unobservable or *latent variables*. Latent variables are (exogenous or endogenous) variables not directly observable and, then, are usually indirectly measured through a set of observable or manifest variables (or items or indicators).

SEMs encompass a wide range of multivariate statistical techniques (regression models, simultaneous equation models, factor analysis, latent class models, item response models) and were born to deal with continuous variables; however nowadays they have been generalized to also deal with categorical and discrete responses. From an historical perspective, SEMs trace back to three different traditions from specific fields of knowledge: path analysis from biostatistics [221, 55]; simultaneous equation modeling from econometrics [91, 134]; factor analysis from psychometrics [203, 6]. The main contribution to merge these three approaches comes from Joreskog [124] who developed the computer program LISREL (LInear Structural RELations), which for many years was the only software for structural equation modeling.

A SEM analysis is based on a sequence of steps. First, a theoretical model is developed, which detects exogenous and endogenous as well as observable and latent variables and defines the (causal) relationships among them (see Section 6.3.3 for more insights about the causal interpretation of SEMs). Second, a path diagram is usually built to display the relations among variables and, then, it is translated into a system of multiple equations. After having verified that the model is identifiable, its estimation and evaluation of fit follow. A revision of the initially specified model is also possible, whenever the goodness of fit is not satisfying.

A SEM is characterized by a system of multiple equations, one for each dependent endogenous variable, which discern between two sub-models: (i) a structural model or latent variable model to explain the relations among latent variables (as in a simultaneous equation model) and (ii) a measurement model to link the endogenous latent variables to the observed variables (as in a factor model).

For the sake of clarity, we introduce some basic notation:

- $\boldsymbol{\eta}_i$: vector of latent endogenous variables for individual i,

- $\boldsymbol{\xi}_i$: vector of latent exogenous variables for individual i,

- $\boldsymbol{y}_i$: vector of observed indicators of $\boldsymbol{\eta}_i$ (as usual in linear regression modeling),

- $\boldsymbol{x}_i$: vector of observed indicators of $\boldsymbol{\xi}_i$ (as usual in linear regression modeling),

- $\boldsymbol{\zeta}_i$: vector of errors in the structural model,

- $\boldsymbol{e}_i$, $\boldsymbol{\epsilon}_i$: vectors of errors (or unique factors) in the measurement model; $\boldsymbol{e}_i$ refers to $\boldsymbol{y}_i$ and $\boldsymbol{\epsilon}_i$ refers to $\boldsymbol{x}_i$,

- $\boldsymbol{\Phi}$: variance-covariance matrix of latent variables $\boldsymbol{\xi}_i$,

- $\boldsymbol{\Psi}$: variance-covariance matrix of errors $\boldsymbol{\zeta}_i$,

- $\boldsymbol{\Theta}_e$ and $\boldsymbol{\Theta}_{\epsilon}$: variance-covariance matrices of errors $\boldsymbol{e}_i$ and $\boldsymbol{\epsilon}_i$.

Moreover, as SEMs can be well represented by path diagrams, in Figure 2.8 we propose a system of symbols to use in graphs to denote the different types of variables and relationships involved; note that the proposed notation differs from that usually adopted in the literature about SEMs where ellipses denote latent variables and rectangles denote observed variables. An extension of these symbols that takes explicitly into account causal relationships will be presented in Figure 6.5.

Under the assumptions $E(\boldsymbol{\zeta}_i) = \boldsymbol{0}$, $cov(\boldsymbol{\xi}_i, \boldsymbol{\zeta}_i) = \boldsymbol{0}$, and $(\boldsymbol{I} - \mathbf{B})$ invertible matrix, the structural sub-model is defined as

$$\boldsymbol{\eta}_i = \boldsymbol{\alpha}_{\eta} + \mathbf{B}\boldsymbol{\eta}_i + \boldsymbol{\Gamma}\boldsymbol{\xi}_i + \boldsymbol{\zeta}_i, \tag{2.32}$$

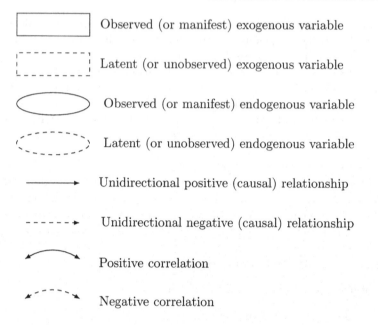

Observed (or manifest) exogenous variable

Latent (or unobserved) exogenous variable

Observed (or manifest) endogenous variable

Latent (or unobserved) endogenous variable

Unidirectional positive (causal) relationship

Unidirectional negative (causal) relationship

Positive correlation

Negative correlation

FIGURE 2.8
Symbols used in the path diagrams to distinguish between observed and latent variables, exogenous and endogenous variables, positive and negative relationships, unidirectional and reciprocal relationships (correlations).

with $\boldsymbol{\alpha}_\eta$ vector of intercepts and $\mathbf{B}$ and $\boldsymbol{\Gamma}$ matrices of regression coefficients. Alternatively, it may be expressed in a reduced form with each endogenous variable written as a function of only exogenous variables and errors, that is,

$$\boldsymbol{\eta}_i = (\boldsymbol{I} - \mathbf{B})^{-1}(\boldsymbol{\alpha}_\eta + \boldsymbol{\Gamma}\boldsymbol{\xi}_i + \boldsymbol{\zeta}_i). \tag{2.33}$$

For instance, let us consider two latent endogenous variables and one latent exogenous variable, whose relations are displayed by the path diagram in Figure 2.9.

The related structural model specifies as

$$\begin{bmatrix} \eta_{1i} \\ \eta_{2i} \end{bmatrix} = \begin{bmatrix} \alpha_{\eta_1} \\ \alpha_{\eta_2} \end{bmatrix} + \begin{bmatrix} 0 & 0 \\ \beta_{21} & 0 \end{bmatrix} \begin{bmatrix} \eta_{1i} \\ \eta_{2i} \end{bmatrix} + \begin{bmatrix} \gamma_{11} \\ \gamma_{21} \end{bmatrix} \begin{bmatrix} \xi_{1i} \end{bmatrix} + \begin{bmatrix} \zeta_{1i} \\ \zeta_{2i} \end{bmatrix}$$

with the following covariance structure

$$\boldsymbol{\Phi} = \begin{bmatrix} \Phi_{11} \end{bmatrix} = Var(\xi_1)$$

and

$$\boldsymbol{\Psi} = \begin{bmatrix} \Psi_{11} & 0 \\ 0 & \Psi_{22} \end{bmatrix} = \begin{bmatrix} Var(\zeta_1) & 0 \\ 0 & Var(\zeta_2) \end{bmatrix}.$$

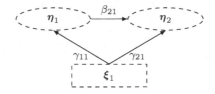

FIGURE 2.9
Example of path diagram (only structural model; variances and covariances omitted for sake of parsimony).

In addition to the sub-model in Eq. (2.32), under the assumptions that $E(e_i) = 0 = E(\epsilon_i)$, errors e_i and ϵ_i are uncorrelated one with the other and with respect to latent variables ξ_i and errors ζ_i, a measurement sub-model is specified to connect latent and observed variables, that is,

$$y_i = \alpha_y + \Lambda_y \eta_i + e_i$$
$$x_i = \alpha_x + \Lambda_x \xi_i + \epsilon_i, \tag{2.34}$$

with α_y and α_x vectors of intercepts and Λ_y and Λ_x matrices of factor loadings. The related reduced form follows

$$y_i = \alpha_y + \Lambda_y (I - B)^{-1} (\alpha_\eta + \Gamma \xi_i + \zeta_i) + \epsilon_i.$$

A simple example of a measurement model with one latent variable and three observed variables is displayed by the path diagram in Figure 2.10, to which the following equations refer:

$$\begin{bmatrix} x_{1i} \\ x_{2i} \\ x_{3i} \end{bmatrix} = \begin{bmatrix} 0 \\ \alpha_{x2} \\ \alpha_{x3} \end{bmatrix} + \begin{bmatrix} 1 \\ \lambda_{x21} \\ \lambda_{x31} \end{bmatrix} \begin{bmatrix} \xi_{1i} \end{bmatrix} + \begin{bmatrix} \epsilon_{1i} \\ \epsilon_{2i} \\ \epsilon_{3i} \end{bmatrix}$$

with

$$\Theta_\epsilon = \begin{bmatrix} Var(\epsilon_1) & 0 & 0 \\ 0 & Var(\epsilon_2) & 0 \\ 0 & 0 & Var(\epsilon_3) \end{bmatrix}.$$

More in general, a complete SEM will assume a structure similar to that displayed in Figure 2.11.

As concerns the model identification phase, as above outlined, if the parameters of a SEM can be obtained from the estimated reduced form we say that the system is identified; if this cannot be done, we say that the system under consideration is unidentified or underidentified. This means that there will be two or more values of the parameters that are equally consistent with the data. Unfortunately, there does not exist a general rule that covers all possible situations to establish whether a model is identifiable.

Symptoms of identification problems are usually represented by high standard errors for one or more estimated parameters, a singular information matrix, estimates of one or more parameters that are unreasonable or impossible

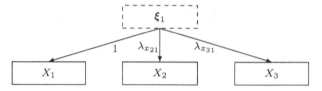

FIGURE 2.10
Example of path diagram (only measurement model; variances and covariances omitted for sake of parsimony).

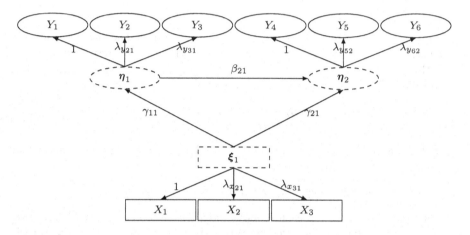

FIGURE 2.11
Example of path diagram (structural and measurement sub-models; variances and covariances omitted for sake of parsimony).

(e.g., negative variances), and high correlation between estimated coefficients, whereas possible causes of the identifiability problems are given by a high number of regression coefficients with respect to the number of means, variances, and covariances of the observed variables, the presence of too many reciprocal effects, errors in the latent variables scales. In the most common situations, the identification question is positively solved if the model parameters can be written as unique functions of means, variances, and covariances of the observed variables. More in general, the solution to make identifiable a SEM consists of adding information to the model specification according to the theoretical knowledge of the researcher. In practice, this requires the normalization of the scale of latent variables (e.g., put the first factor loading of each latent variable equal to 1), the omission of one or more variables (i.e., constraining to be equal to 0 one or more regression coefficients), and setting linear constraints on the variance and covariance components.

Finally, here we do not dwell on the aspects related to model estimation, providing only some brief hints. Given that the ordinary least squares approach applied in the SEM frame will generally get biased and inconsistent estimators, one can refer to two main types of alternative estimators: the full information estimators, which consider the full system of equations at the same time, and the limited information estimators, which are based on an equation-by-equation approach.

The most commonly used full information estimator is the maximum likelihood one, which is consistent, asymptotically unbiased, asymptotically efficient, and asymptotically normally distributed. As in linear regression modeling, the maximum likelihood approach is based on the assumption of multivariate normality (Example 2.18) of the observed variables and it is robust to the violation of this distributive assumption. The main drawback of the full information approach is due to the propagation of a specification error in one equation of the system through the other equations (even if these are correctly specified).

Among the limited information estimators, the most commonly used are the instrument variable estimator and the two stage least squares estimator. In the first case, the estimation process starts selecting a first equation to estimate. Then, the process is based on some instrumental variables, which are chosen among the observed variables of the SEM that are uncorrelated with the errors of the same equation and have a significant association with the problematic independent variables. Alternatively, in the two stage least squares approach, the latent variables in the SEM are expressed as a function of the observed covariates and eliminated from the model; then, a least squares estimation is implemented that is based on the instrumental variables approach.

A special case of SEM that is particularly simple to estimate is the *recursive model* that is obtained when the following two conditions hold: (i) there are no reciprocal directed paths among the endogenous variables and (ii) errors are independent of one another. In other words, in a recursive model there is a unidirectional dependency among the endogenous variables such that, for given values of the exogenous variables, values for the endogenous variables can be determined sequentially rather than jointly; moreover, explicative variables in a certain structural equation are always independent of the errors of that equation. As a consequence of these properties, the ordinary least squares approach may be appropriately applied in this specific type of SEM (for more details, see Section 6.3.3).

3

Utility theory

CONTENTS

3.1 Introduction

In this chapter we aim at exploring in depth what was introduced in Section 1.4 about decisions under certainty and risk, through the illustration of the fundamental aspects related to the *normative decision theory* (Section 1.2). Normative decision theory concerns how decisions *should be taken by ideally rational individuals* in order to maximize their own well-being. Therefore, it does not describe actually observable behaviors, but it deals with behavioral rules to which rational individuals must conform.

We first introduce the problem of taking decisions when the consequences (outcomes, rewards) of actions are known with certainty: in such a context, axioms of rational behavior and the concept of value function are illustrated.

Next, we focus on decisions under risk, when probabilities of outcomes are externally defined or when the probability distribution of outcomes is subjectively elicited. In the first case, we mainly refer to the expected utility theory formulated by von Neumann and Morgenstern [213], whereas the subjective expected utility theory of Savage [191] and Anscombe and Aumann [7] represents the core of decisions in the second case. Under both approaches the rational behavioral axioms are illustrated and the theorems of representation and uniqueness of the utility function are enunciated and demonstrated.

It is worth noting that the conceptual frame adopted in this chapter is the *Bayesian decision theory* (Section 1.5); therefore the possibility to update the (objective or subjective) prior knowledge about the states of nature through sample surveys is ignored (it will be taken into account later, in Chapter 5).

The remainder of the chapter is devoted to paradoxes of rational behavioral axioms and empirical failures of expected utility theory. In the end, some generalizations of utility theory are illustrated that rely on a more flexible axiomatic frame.

3.2 Binary relations and preferences

Before proceeding with the illustration of rational behavioral axioms, it is worth being reminded of some basic notions about binary relations.

Let us consider a finite set C of elements $\{c_1, \ldots c_k\}$ and a binary relation R. Then, for any pair of elements $(c_i, c_j) \in C$ $(i, j = 1, \ldots, k)$, either $c_i \, R \, c_j$ or $c_i \, \bar{R} \, c_j$. Moreover, any binary relation R can satisfy the following properties:

Definition 3.1. Transitivity. *R is transitive if, for any $c_i, c_j, c_h \in C$ such that $c_i \, R \, c_j$ and $c_j \, R \, c_h$, then $c_i \, R \, c_h$;*

Definition 3.2. Asymmetry. *R is asymmetric if, for any $c_i, c_j \in C$, $c_i \, R \, c_j \implies c_j \, \bar{R} \, c_i$;*

Definition 3.3. Symmetry. *R is symmetric if, for any $c_i, c_j \in C$, $c_i \, R \, c_j \implies c_j \, R \, c_i$;*

Definition 3.4. Reflexivity. *R is reflexive if, for any $c_i \in C$, $c_i \, R \, c_i$;*

Definition 3.5. Completeness. *R is complete if, for any $c_i, c_j \in C$, either $c_i \, R \, c_j$ or $c_j \, R \, c_i$ or both of them are true. Similarly, either $c_i \, \bar{R} \, c_j$ or $c_j \, \bar{R} \, c_i$ or both of them are true;*

Definition 3.6. Negative transitivity. *R is transitive in a negative sense if, for any $c_i, c_j, c_k \in C$ such that $c_i \, \bar{R} \, c_j$ and $c_j \, \bar{R} \, c_h$, then $c_i \, \bar{R} \, c_h$;*

Definition 3.7. Antisymmetry. *R is antisymmetric if, for any $c_i, c_j \in C$, $c_i \, R \, c_j$ and $c_j \, R \, c_i \implies c_i = c_j$.*

In the context of decision theory, some binary relations are introduced such that the decision maker can express his/her preference towards any pair of consequences (or outcomes) c_i, c_j:

Definition 3.8. Strict preference $\succ$: $c_i \succ c_j$ *means that consequence c_i is strictly preferred to consequence c_j;*

Definition 3.9. Weak preference $\succeq$: $c_i \succeq c_j$ *means that consequence c_i is at least as preferred as (or weakly preferred to) outcome c_j;*

Definition 3.10. Indifference $\sim$: $c_i \sim c_j$ *means that consequences c_i and c_j are equally preferred;*

Definition 3.11. Exchange $\leftarrow$: $c_i \leftarrow c_j$ *denotes that c_j is given away in favor of c_i; note that preference relations can be defined also on exchanges. In what follows we will denote by $\succ_e$, $\succeq_e$, and $\sim_e$ the relations of strict preference, weak preference, and indifference on exchanges.*

The absence of a relation is denoted by $\not\succ$, $\not\succeq$, $\not\sim$, $\not\leftarrow$.

3.3 Decisions under certainty: Value theory

Under certainty, the state of nature is assumed to be known and, then, every possible action a_i is associated with a specific consequence c_i. In such a context, the following rational behavioral axioms are introduced:

Axiom 3.1. Completeness: *preference relation is complete, that is,* $\forall c_i, c_j \in C$, *either $c_i \succeq c_j$ or $c_j \succeq c_i$ or both of them are true.*

Axiom 3.2. Transitivity: *preference relation is transitive, that is,* $\forall c_i, c_j, c_h \in C$,
$$c_i \succeq c_j \cap c_j \succeq c_h \implies c_i \succeq c_h.$$

Axiom 3.3. Coherence between indifference and weak preference: $\forall c_i, c_j \in C$,
$$c_i \sim c_j \iff (c_i \succeq c_j) \cap (c_j \succeq c_i).$$

Axiom 3.4. Coherence between weak preference and strict preference: $\forall c_i, c_j \in C$,
$$c_i \succ c_j \iff c_j \not\succeq c_i.$$

The mentioned four axioms have relevant logic implications. The first axiom states that a decision maker cannot be uncertain about his/her preferences: he/she is always able to provide an ordering of consequences. At most, in virtue of the third axiom he/she can be indifferent between two alternatives, but he/she cannot say "I don't know". In addition, transitivity and coherence axioms imply that a decision maker does not express inconsistent preferences.

Theorem 3.1. *From Axioms 3.1-3.4 it follows that*

- *strict preference relation $\succ$ is transitive and antisymmetric;*

- *indifference relation $\sim$ is transitive, reflexive, and symmetric;*

- $\forall c_i, c_j, c_h \in C, (c_i \sim c_j \cap c_j \succ c_h) \implies c_i \succ c_h;$

- $\forall c_i, c_j \in C,$ *one and only one of the following relations holds:* $c_i \succ c_j,$ $c_i \sim c_j, c_j \succ c_i.$

Proof.

- $\succ$ is transitive.
  Let us assume that a decision maker expresses a cyclic preference, that is, $c_i \succ c_j, c_j \succ c_h, c_h \succ c_i$. Then, the following results hold:

    1. $c_j \succ c_h \iff c_h \not\succeq c_j$, from Axiom 3.4;
    2. $c_j \succeq c_h$, from Axiom 3.1;
    3. $c_j \succeq c_h, c_h \succeq c_i \implies c_j \succeq c_i$, from Axiom 3.2;
    4. $c_i \succ c_j \iff c_j \not\succeq c_i$, from Axiom 3.4.

  However, result 4 contradicts result 3. Hence, our initial assumption $c_i \succ c_j, c_j \succ c_h, c_h \succ c_i$ is false, so the transitivity of strict preference relation is true.

- $\succ$ is antisymmetric.
  Let us assume that $\succ$ is symmetric; that is, there exist c_i and c_j elements of C such that $c_i \succ c_j$ and $c_j \succ c_i$. Then, the following results hold:

    1. $c_i \succ c_j \implies c_j \not\succ c_i$, from Axiom 3.4;
    2. $c_j \succ c_i \implies c_i \not\succ c_j$, from Axiom 3.4.

  However, results 1 and 2 contradict Axiom 3.1 (completeness). Hence, the initial assumption of symmetry of $\succ$ is false, so $\succ$ is antisymmetric.

- $\sim$ is transitive.
  Let us assume that $c_i \sim c_j$ and $c_j \sim c_h$. By virtue of Axiom 3.3, the first relation implies $c_i \succeq c_j \cap c_j \succeq c_i$ and the second relation implies $c_j \succeq c_h \cap c_h \succeq c_j$. Then, by virtue of Axiom 3.2, $c_i \succeq c_h \cap c_h \succeq c_i$ implying (Axiom 3.3) the transitivity of $\sim$. Similarly, one can prove that $\sim$ is reflexive and antisymmetric.

- $\forall c_i, c_j, c_h \in C, (c_i \sim c_j \cap c_j \succ c_h) \implies c_i \succ c_h.$
  Let us assume $c_i \sim c_j, c_j \succ c_h, c_i \not\succ c_h$. Then, the following results hold:

    1. $c_i \not\succ c_h \implies c_h \succeq c_i$, from Axiom 3.4;
    2. $c_i \sim c_j \implies c_i \succeq c_j$, from Axiom 3.3.

Results 1 and 2 above imply $c_h \succeq c_j$ from Axiom 3.2. However, $c_j \succ c_h$ implies $c_h \not\succeq c_j$ that contradicts the previous result. Hence, the initial assumption $c_i \sim c_j, c_j \succ c_h, c_i \not\succ c_h$ is false, so it follows that $(c_i \sim c_j \cap c_j \succ c_h) \implies c_i \succ c_h$.

- $\forall c_i, c_j \in C$, one and only one of the following relations holds: $c_i \succ c_j$, $c_i \sim c_j$, $c_j \succ c_i$.
  For all pairs of consequences c_i, c_j one and only one of the following alternatives holds:

  1. $c_i \succeq c_j \cap c_j \not\succeq c_i$, which implies $c_i \succ c_j$ from Axiom 3.4;
  2. $c_i \succeq c_j \cap c_j \succeq c_i$, which implies $c_i \sim c_j$ from Axiom 3.3;
  3. $c_i \not\succeq c_j \cap c_j \succeq c_i$, which implies $c_j \succ c_i$ from Axiom 3.4;
  4. $c_i \not\succeq c_j \cap c_j \not\succeq c_i$, which contradicts Axiom 3.1 and, then, it is impossible.

$\qquad\qquad\qquad\qquad\qquad\qquad\qquad\qquad\qquad\qquad\qquad\qquad\qquad$ □

As a relevant consequence of Theorem 3.1, the existence and uniqueness of a *value function* defined on an ordinal scale equivalent to the preference structure of the decision maker may be proved.

Theorem 3.2. Representation theorem (value function on ordinal scale). *Given a finite set of consequences $C = (c_1, c_2, \ldots, c_k)$ and a relation of weak preference $\succeq$ satisfying Axioms 3.1-3.4, there exists, and can be elicitated, a real-value function $V(\cdot)$ on an ordinal scale such that*

$$c_i \succeq c_j \iff V(c_i) \geq V(c_j).$$

Proof.
Let $V(c_i)$ denote the number of elements $c_i \in C$ such that $c_i \succeq c_j$.

- $c_i \succeq c_j \implies V(c_i) \geq V(c_j)$.
  Let c_h be an element of C such that $c_j \succeq c_h$. If $c_i \succeq c_j$, then $c_i \succeq c_h$ for Axiom 3.2. Therefore, every element accounted by $V(c_j)$ is also taken into account in $V(c_i)$, that is, $V(c_i) \geq V(c_j)$.

- $c_i \succ c_j \implies V(c_i) > V(c_j)$.
  From the above point it follows that $c_i \succ c_j \implies c_i \succeq c_j \implies V(c_i) \geq V(c_j)$; in addition, $c_i \succ c_j \implies c_j \not\succeq c_i$. Therefore, there exists at least one element c_i that is taken into account in $V(c_i)$ (as $c_i \succeq c_i$), but not in $V(c_j)$. Formally, $c_i \succ c_j \implies V(c_i) \geq V(c_j) + 1 > V(c_j)$.

- $V(c_i) \geq V(c_j) \implies c_i \succeq c_j$.
  Let us assume that there exist $c_i, c_j \in C$ such that $V(c_i) \geq V(c_j)$ and $c_i \not\succeq c_j$. However, $c_i \not\succeq c_j$ implies that $c_j \succ c_i$ and, from the above point,

this implies that $V(c_j) > V(c_i)$ and, then, $V(c_i) \not\succeq V(c_j)$. This last result contradicts the initial assumption. Hence, the initial assumption is false, so $V(c_i) \geq V(c_j) \implies c_i \succeq c_j$. $\qquad\qquad\qquad\qquad\qquad\qquad\qquad\qquad\qquad\qquad\qquad$ □

Theorem 3.3. Uniqueness theorem (value function on ordinal scale).
Given a finite set of consequences $C = (c_1, c_2, \ldots, c_k)$ and a relation of weak preference $\succeq$ satisfying Axioms 3.1-3.4, there exist two value functions $V(\cdot)$ and $W(\cdot)$ such that

$$c_i \succeq c_j \iff V(c_i) \geq V(c_j)$$
$$c_i \succeq c_j \iff W(c_i) \geq W(c_j)$$

if and only if $V(\cdot) = h(W(\cdot))$, where $h(\cdot)$ is a monotone increasing function.

In other words, the uniqueness theorem states that a real-value function on an ordinal scale is unique, up to a monotone increasing transformation.

Proof.
Let us assume $W(\cdot)$ a real-value function on an ordinal scale satisfying the weak preference relation and $h(\cdot)$ a strictly increasing function. Therefore, relation $c_i \succeq c_j$ holds if and only if

- $W(c_i) \geq W(c_j)$, as $W(\cdot)$ is a function with ordinal values;

- $h(W(c_i)) \geq h(W(c_j))$, as $h(\cdot)$ is a strictly increasing function;

- $V(c_i) \geq V(c_j)$, as $V(\cdot)$ is a function with ordinal values.

In other words, $V(\cdot)$ and $W(\cdot)$ are both real-value functions on an ordinal scale and are coherent with the same preference structure.

Now, let us see that this result implies that there exists an increasing monotone function $h(\cdot)$ such that $V(\cdot) = h(W(\cdot))$. As illustrated in Figure 3.1, point A with coordinates $(W(c_j), V(c_j))$ splits the system of Cartesian axes into four quadrants. If relation $c_i \succeq c_j$ holds, then point B of coordinates $(W(c_i), V(c_i))$ surely belongs to the I quadrant; otherwise assumption that $V(\cdot)$ and $W(\cdot)$ are coherent with the same preference structure would be violated. Hence, any curve that links points A and B is described by a strictly increasing function $h(\cdot)$ such that $V(\cdot) = h(W(\cdot))$. $\qquad\qquad\qquad$ □

The real-value function built on an ordinal scale provides a ranking of outcomes c_i according to the preference structure of the decision maker; however it does not provide any information about the *distance* between consecutive outcomes. In other words, a function on an ordinal scale does not provide the answer to the question "How much is consequence c_i preferred to consequence c_j?". For this aim, we need to measure the value of consequences on an interval scale. For this reason we introduce the exchange relation between outcomes,

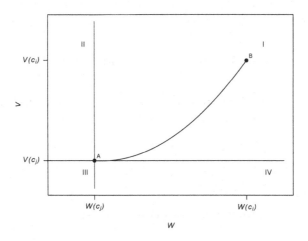

FIGURE 3.1
Graphical illustration of Theorem 3.4.

denoted by $c_i \leftarrow c_j$ (see Section 3.2) and a new set of rational behavioral axioms:

Axiom 3.5. Weak ordering: *Both relation $\succeq$ and $\succeq_e$ satisfy Axioms 3.1-3.4.*

This axiom confirms that, in the presence of exchanges between consequences, uncertainty about preferences is not admitted and preferences cannot contradict each other.

Axiom 3.6. Coherence between $\succeq$ and $\succeq_e$: *$\forall c_i, c_j \in C$*

$$c_i \succeq c_j \iff (c_i \leftarrow c_j) \succeq_e (c_h \leftarrow c_h) \forall c_h \in C.$$

Axiom 3.7. Internal coherence of exchange relation:

- $(c_i \leftarrow c_j) \succeq_e (c_h \leftarrow c_l) \iff (c_l \leftarrow c_h) \succeq_e (c_j \leftarrow c_i)$, *that is, a preference relation between exchanges reverses when the exchange reverses, too;*

- $(c_i \leftarrow c_j) \succeq_e (c_h \leftarrow c_l) \cap (c_j \leftarrow c_m) \succeq_e (c_l \leftarrow c_n) \iff (c_i \leftarrow c_m) \succeq_e (c_h \leftarrow c_n)$, *that is, a preference relation remains constant both in the presence of direct exchanges and when the exchange is mediated through a third element.*

Axiom 3.8. Solvibility:

- $\forall c_i, c_j, c_h \in C, \ \exists y \in C : (y \leftarrow c_i) \sim_e (c_j \leftarrow c_h)$

- $\forall c_i, c_j \in C, \ \exists y \in C : (c_i \leftarrow y) \sim_e (y \leftarrow c_j)$

This axiom states that there always exists an outcome $y \in C$ such that the relation of indifference between other elements in C is satisfied, apart from the decision maker's preferences.

The following axiom is based on the definition of strictly limited standard sequence.

Definition 3.12. Strictly limited standard sequence. *Let $c_1, \ldots, c_k$ and y be possible outcomes. Sequence $c_1, \ldots, c_k$ is strictly limited and standard if*

$$y \succ c_k; \quad (c_k \leftarrow c_{k-1}) \sim_e (c_2 \leftarrow c_1).$$

Note that the second relation implies $(c_2 \leftarrow c_1) \sim_e (c_3 \leftarrow c_2) \sim_e \ldots \sim_e (c_j \leftarrow c_{j-1}) \sim_e \ldots \sim_e (c_k \leftarrow c_{k-1})$.

Axiom 3.9. Archimedean property: *Every strictly limited standard sequence is finite.*

In practice, the archimedean axiom requires that an outcome cannot be incommensurably better (or worse) than another outcome.

As a consequence of Axioms 3.5-3.9, the existence and uniqueness of a value function measurable on an interval scale equivalent to the preference structure of the decision maker may be proved.

Theorem 3.4. Representation theorem (value function on interval scale). *Given a finite set of consequences $C = (c_1, c_2, \ldots, c_k)$ and relations of weak preference $\succeq$ and $\succeq_e$ satisfying Axioms 3.5-3.9, there exists, and can be elicitated, a real-value function $V(\cdot)$ on an interval scale such that*

$$c_i \succeq c_j \iff V(c_i) \geq V(c_j);$$
$$(c_i \leftarrow c_j) \succeq_e (c_h \leftarrow c_l) \iff V(c_i) - V(c_j) \geq V(c_h) - V(c_l).$$

Proof.
Let us assume that some positive money rewards $0 < c_1 \leq \ldots \leq c_i \leq \ldots \leq c_k$ are offered to the decision maker and that $c_j \succeq c_i$ if and only if $c_i \leq c_j$. In addition, we assume $V(0) = 0$ with $0 = c_0$ and $V(c_1) = 1$.

By virtue of Axiom 3.8 (solvability), there always exists a money reward c_2 such that

$$(c_2 \leftarrow c_1) \sim_e (c_1 \leftarrow c_0).$$

Such a relation implies

$$V(c_2) - V(c_1) = V(c_1) - V(c_0)$$
$$V(c_2) - 1 = 1 - 0 \implies V(c_2) = 2.$$

Similarly, a sequence of money rewards $c_3, c_4, \ldots$ can be identified such that

$$(c_1 \leftarrow c_0) \sim_e (c_2 \leftarrow c_1) \sim_e (c_3 \leftarrow c_2) \sim_e \ldots \sim_e (c_i \leftarrow c_{i-1}) \sim_e \ldots$$

and, for all $c_i \in C$, $V(c_i) = i$ $(i = 1, \ldots, k)$:

$$V(c_0) = 0, V(c_1) = 1, V(c_2) = 2, \ldots, V(c_i) = i, \ldots$$

In addition, for a generic reward c, value $V(c)$ may be determined splitting the preference intervals $(c_i \leftarrow c_{i-1})$. More in detail, if $c = c_i$ then $V(c) = V(c_i)$, as above shown. If $c \neq c_i$, then, by virtue of the archimedean property (Axiom 3.9), there exists a generic i such that

$$c_{i+1} \succeq c \succeq c_i,$$

which implies $i + 1 \geq V(c) \geq i$. Now, the decision maker is asked for reward $c_{i+1/2}$ such that

$$(c_{i+1} \leftarrow c_{i+1/2}) \sim_e (c_{i+1/2} \leftarrow c_i),$$

which implies $V(c_{i+1/2}) = i + 1/2$. As $c_{i+1} > c_{i+1/2} > c_i$, either $c_{i+1} \leq c \leq c_{i+1/2}$ or $c_{i+1/2} \leq c \leq c_i$ is surely true. The splitting procedure may be repeated until a measure for $V(c)$ is obtained as accurately as possible.

In the end, points $(c_i, V(c_i))$ define a function that provides a measurement on an interval scale of the preferences of the decision maker with respect to elements $c_i \in C$ (money rewards or any other type of outcomes), independently of the value of c_i. □

Similarly to the value function defined on an ordinal scale, also the value function defined on an interval scale is unique, up to a positive linear transformation.

Theorem 3.5. Uniqueness theorem (value function on interval scale).
Given a finite set of consequences $C = (c_1, c_2, \ldots, c_k)$ and relations of weak preference $\succeq$ and $\succeq_e$ satisfying Axioms 3.5-3.9, there exist two value functions $V(\cdot)$ and $W(\cdot)$ measured on an interval scale satisfying the following relations

$$c_i \succeq c_j \iff V(c_i) \geq V(c_j);$$
$$(c_i \leftarrow c_j) \succeq_e (c_h \leftarrow c_l) \iff V(c_i) - V(c_j) \geq V(c_h) - V(c_l);$$
$$c_i \succeq c_j \iff W(c_i) \geq W(c_j);$$
$$(c_i \leftarrow c_j) \succeq_e (c_h \leftarrow c_l) \iff W(c_i) - W(c_j) \geq W(c_h) - W(c_l)$$

if and only if $V(\cdot) = a + bW(\cdot)$ with $b > 0$.

Proof.
Let us assume that $V(\cdot) = a + bW(\cdot)$ with $b > 0$. Then,

$$c_i \succeq c_j \iff W(c_i) \geq W(c_j)$$
$$\iff a + bW(c_i) \geq a + bW(c_j), \quad \text{as } b > 0$$
$$\iff V(c_i) \geq V(c_j).$$

Moreover,

$$
\begin{aligned}
(c_i \leftarrow c_j) \succeq_e (c_h \leftarrow c_l) &\iff W(c_i) - W(c_j) \geq W(c_h) - W(c_l) \\
&\iff b[W(c_i) - W(c_j)] \geq b[W(c_h) - W(c_l)], \text{ as } b > 0 \\
&\iff b[W(c_i) - W(c_j)] + (a - a) \geq b[W(c_h) - W(c_l)] \\
&\quad + (a - a) \\
&\iff V(c_i) - V(c_j) \geq V(c_h) - V(c_l).
\end{aligned}
$$

Hence, $V(\cdot) = a + bW(\cdot)$ is a value function on an interval scale coherent with the same preference structure represented by $W(\cdot)$. $\square$

To conclude this section it is worth noting that, under certainty, the decision rule of a rational decision maker consists of choosing that action and, then, that consequence to which the maximum of the value function is associated.

3.4   Decisions under risk: Utility theory

Decision theory under certainty is useful to introduce some rational behavioral axioms; however its practical utility is limited to the situations where the decision maker knows with certainty the state of nature associated with each possible action. More generally, the decision maker has to select the optimal action in risky contexts, that is, when a probability distribution is associated with the states of nature. In such a setting, a number of consequences corresponds to each action, as is illustrated in Table 3.1 (which is distinguished from the decision table in Table 1.1 for the presence of the probabilities of the states of nature), instead of having a single consequence for each action, as happens under certainty.

TABLE 3.1

Decision table in the presence of prior probabilities of the states of nature, $\pi(\theta_j)$ $(j = 1, \ldots, k)$.

	Prior probabilities			
Action	$\pi(\theta_1)$	$\pi(\theta_2)$	$\ldots$	$\pi(\theta_k)$
a_1	c_{11}	c_{12}	$\cdots$	c_{1k}
a_2	c_{21}	c_{22}	$\cdots$	c_{2k}
$\vdots$	$\cdots$	$\cdots$	$\vdots$	$\cdots$
a_m	c_{m1}	c_{m2}	$\cdots$	c_{mk}

In the decision theory under risk the concept of *lottery* or *gambling* is used to represent in a synthetic way the set of consequences associated with a certain action and the probabilities of the states of nature. In more detail, choosing action a_i $(i = 1, \ldots, m)$ corresponds to playing a *simple lottery* l_i that pays c_{ij} with probability $p_j = \pi(\theta_j)$ $(j = 1, \ldots, k)$:

$$l_i = \langle p_1, c_{i1}; p_2, c_{i2}; \ldots; p_j, c_{ij}; \ldots; p_k, c_{ik} \rangle.$$

In the case of just two outcomes $(k = 2)$ lottery l_i simplifies to

$$l_i = \langle 0, c_{i1}; 0, c_{i2}; \ldots; p, c_{ih}; \ldots; (1 - p), c_{ij}; \ldots; 0, c_{ik} \rangle$$

and is briefly denoted as $l_i = \langle c_{ih} \; p \; c_{ij} \rangle$; in what follows subscript i will be omitted when superfluous. When the output of a lottery consists of participating in another lottery, a *compound lottery* is defined as follows

$$l_i = \langle q_1, l_{i1}; q_2, l_{i2}; \ldots; q_j, l_{ij}; \ldots; q_r, l_{ir} \rangle,$$

where $l_{i1}, \ldots, l_{ir}$ represent simple lotteries and $q_1, \ldots, q_r$ the related probabilities. Hence, the output of a lottery may consist either of paying a final reward (consequence) or participating to another lottery. Here we assume that the number of compound lotteries until the final reward is finite.

It has also to be noted that, under certainty, any consequence c_{ij} may be interpreted as a *degenerate lottery*, that is,

$$c_{ij} = \langle 0, c_{i1}; 0, c_{i2}; \ldots; 1, c_{ij}; \ldots; 0, c_{ik} \rangle.$$

From an historical perspective (see, for instance, the work of Huygens [114]), one of the first approaches to decisions in risky situations was based on the *mathematical expectation criterion*, according to which a rational decision maker should prefer the action that provides the maximum (monetary) expected reward. For instance, let us consider the following choice between two alternative lotteries:

- lottery $l_1 = \langle 100 \; 0.50 \; -60 \rangle$ gives a fifty-fifty chance of either winning 100 euros or losing 60 euros;

- lottery $l_2 = \langle 1{,}000 \; 0.50 \; -500 \rangle$ gives a fifty-fifty chance of either winning 1,000 euros or losing 500 euros.

According to the mathematical expectation criterion, lottery l_2 should be preferred to lottery l_1 by any rational decision maker, as

$$\mathrm{E}(l_1) = 0.50 \cdot 100 - 0.50 \cdot 60 = 20 < \mathrm{E}(l_2) = 0.50 \cdot 1000 - 0.50 \cdot 500 = 250.$$

However, it is not unusual to observe individuals that prefer lottery l_1 to l_2, as the potential monetary loss related with l_1 is significantly smaller than the loss related with l_2.

Another notorious example of the fallacy of the mathematical expectation approach is represented by the St. Petersburg paradox, studied in-depth by Bernoulli in 1738 (see [18] for the English translation of the original work). Presented with a game paying an infinitely positive expected reward, no decision maker is available to spend an arbitrarily large amount of money to play, contrary to what the mathematical expectation criterion suggests. On the other hand, an individual will usually pay a really limited sum of money to play a game with a small probability of large winnings.

In other words, individuals' choices are driven by the *moral expectation* [18] that they attribute to an action, which may differ from its mathematical expectation. It is worth noting that the mathematical expectation criterion detects the optimal action for *any decision maker*, because the expected monetary reward of an action is univocally defined, whereas the moral expectation of an action is subject-specific. In general, a risk averse decision maker will attribute a high moral value to actions with small loss, independent of the entity of winning, whereas risk seeking is characterized by a high moral value for actions with large winnings. For instance, as far as the choice between lotteries l_1 and l_2 let us assume the following moral values for a certain individual:

$$mv(1000) = 1.00$$
$$mv(100) = 0.20$$
$$mv(-60) = -0.10$$
$$mv(-500) = -1.00,$$

with $mv(\cdot)$ denoting the moral value. Hence, comparing the expected moral values of l_1 and l_2, that is,

$$EMV(l_1) = 0.50 \cdot mv(100) + 0.50 \cdot mv(-60) = 0.50 \cdot 0.20 - 0.50 \cdot 0.10 = 0.05$$

and

$$EMV(l_2) = 0.50 \cdot mv(1000) + 0.50 \cdot mv(500) = 0.50 \cdot 1.00 - 0.50 \cdot 1.00 = 0.00,$$

the decision maker will prefer l_1 to l_2 as $EMV(l_1) = 0.05 > EMV(l_2) = 0.00$.

The concept of moral value introduced by Bernoulli in 1738 corresponds to what von Neumann and Morgenstern [213] in 1944 name *utility*. Starting from some basic axioms of rational behavior, the two authors show the existence and uniqueness of a subject-specific function, named the utility function, representing the preferences of a decision maker in a specific context.

3.4.1 von Neumann and Morgenstern's theory

In the expected utility theory by von Neumann and Morgenstern [213], the choice criterion is based on the quantitative evaluation of the utility of consequences. The rational decision maker acts as if he/she computed the expected utility of each action, that is, the average utility of consequences weighted with respect to the related probabilities. In von Neumann and Morgenstern's theory, probabilities of the states of nature are assumed to be known (decisions under risk). The final choice consists of selecting that action that provides the maximum expected utility.

The logical step from the value function measured on an interval scale to the utility function requires the introduction of further rational behavioral axioms.

Axiom 3.10. Weak ordering*: Preferences on a finite set of lotteries $L = (l_1, \ldots, l_n)$ satisfy Axioms 3.1-3.9.*

Axiom 3.11. Continuity*: $\forall l_1, l_2, l_3 \in L$, if $l_1 \succeq l_2 \succeq l_3 \implies \exists p$, with $0 \leq p \leq 1$, such that $l_2 \sim \langle l_1 \, p \, l_3 \rangle$.*

Axiom 3.12. Monotonicity*: Given two lotteries $l_1, l_2 \in L$ and such that $l_1 \succeq l_2$ and two probabilities p_1, p_2, then*

$$\langle l_1 \, p_1 \, l_2 \rangle \succeq \langle l_1 \, p_2 \, l_2 \rangle \iff p_1 \geq p_2.$$

In other words, the decision maker will choose the lottery with the highest probability of obtaining the preferred result.

Axiom 3.13. Independence*: Given $c_i, c_j, c_h \in C$ such that $c_i \sim c_j$, then $\langle c_i \, p \, c_h \rangle \sim \langle c_j \, p \, c_h \rangle$.*

In other words, the independence axiom says that the ordering of preferences between two outcomes is independent of the presence of a third outcome. This axiom is particularly interesting, because its empirical violation is often observed and, as a main consequence, generalizations of expected utility theory are based on weak versions of this axiom. Three special cases of the independence axiom follow:

Axiom 3.14. Compound lotteries reduction*: Let $l = \langle q_1, l_1; q_2, l_2; \ldots; q_j, l_j; \ldots; q_r, l_r \rangle$ be a compound lottery with $l_j = \langle p_{j1}, c_1; p_{j2}, c_2; \ldots; p_{jk}, c_k \rangle$. In addition, let $l' = \langle p_1, c_1; p_2, c_2; \ldots; p_k, c_k \rangle$ be a simple lottery with $p_i = q_1 p_{j1} + q_2 p_{j2} + \ldots + q_k p_{jk}$, for $i = 1, \ldots, k$. Hence, $l \sim l'$.*

The compound lotteries reduction axiom states that preferences depend only on final outcomes (i.e., $c_1, \ldots, c_k$) and related probabilities (i.e., $p_1, \ldots, p_k$): the number of steps to attain these results and the specific mechanism to compute the probabilities are irrelevant. Further, any pleasure in participating in a lottery game or, on the other hand, any ethical hesitation do not play any role in the decisional process. More in general, this axiom complies with the general principle of invariance according to which different representations of the same decisional problem lead to the same preference structure.

Axiom 3.15. Substitution or deletion: *Let $c_i, c_j \in C$ such that $c_i \sim c_j$; moreover, l and l' are two (simple or compound) lotteries defined as follows: $l = \langle \ldots; q, c_i; \ldots \rangle$ and $l' = \langle \ldots; q, c_j; \ldots \rangle$. Hence, $l \sim l'$.*

Axiom 3.16. Interrelation: *Let $l, l' \in L$ such that $l \succeq l'$. Hence, $l \succeq \langle l \, p \, l' \rangle \succeq l'$.*

The axioms above illustrated imply two further conditions that should hold in a rational decisional context: *dynamic coherence* and *consequentiality*. A decision maker is coherent from a dynamic point of view when, all the other elements being constant, the same choice at different time points leads to the same decision. For instance, let us assume that a compound lottery is proposed in time t_0, whose outputs are 0 with probability 0.90 and the choice between two simple lotteries A and B in time t_1 with probability 0.10. Let us also consider two simple lotteries D and E that are proposed in t_0 such that $D = A$ and $E = B$. Hence, if the decision maker prefers D to E in t_0, he/she should prefer A to B in t_1.

The concept of consequentiality [150, 96] is related to dynamic coherence. Consequentiality of decisions holds when, at any time point, the decision maker only focuses on the present choice, ignoring the past choices and the already achieved outcomes. In other words, a rational decisional process must not be affected by past events, but only by future consequences. If this condition holds, the optimal choices of a multi-step decisional problem may be detected going back to the decision tree. In the business decisional context the decision to go on with a certain project because of the huge sum of money already invested and despite some preliminary negative results is completely irrational in light of the consequentiality condition. Indeed, if the context conditions are mutated such that the final consequences of the project are worse than initially expected it could be rational to give up the project, independent of past investments.

Compliance with the rational behavioral axioms above introduced allows us to prove the existence and the uniqueness of a utility function that resembles the preferences of the decision maker.

Theorem 3.6. Representation theorem (existence of utility function). *Given a finite set of consequences $C = (c_1, c_2, \ldots, c_k)$ and relations of weak preference $\succeq$ and $\succeq_e$ satisfying Axioms 3.1-3.16, there exists, and can be elicitated, a utility function $U(\cdot)$ on C such that*

    1. Ordering property:

$$\forall c_i, c_j \in C, c_i \succeq c_j \iff u(c_i) \geq u(c_j);$$

    2. Linearity property:
    Given $l_1 = \langle p_1, c_1; p_2, c_2; \ldots; p_k, c_k \rangle$ and $l_2 = \langle q_1, c_1; q_2, c_2; \ldots; q_k, c_k \rangle$,

with $p_1, p_2, \ldots, p_k$ and $q_1, q_2, \ldots, q_k$ *probabilities related with consequences in* $\mathcal{C}$, *the following relations hold:*

$$
\begin{aligned}
l_1 \succeq l_2 &\iff u(l_1) \geq u(l_2) \\
&\iff u(\langle p_1, c_1; p_2, c_2; \ldots; p_k, c_k \rangle) \geq u(\langle q_1, c_1; q_2, c_2; \ldots; q_k, c_k \rangle) \\
&\iff p_1 u(c_1) + p_2 u(c_2) + \ldots + p_k u(c_k) \geq q_1 u(c_1) + q_2 u(c_2) + \ldots \\
&\quad + q_k u(c_k) \\
&\iff \sum_{i=1}^{k} p_i u(c_i) \geq \sum_{i=1}^{k} q_i u(c_i) \\
&\iff EU(l_1) \geq EU(l_2),
\end{aligned}
$$

where $EU(l_1) = \sum_{i=1}^{k} p_i u(c_i)$ *and* $EU(l_2) = \sum_{i=1}^{k} q_i u(c_i)$ *denote the expected utility of lotteries* l_1 *and* l_2, *respectively.*

Theorem 3.6 states that the preferences of a rational decision maker can be represented through a utility function and the optimal action is that one with the maximum expected utility. Note that the utility function is measured on an interval scale; a more general approach that takes into account the possibility of measuring utility on a ratio scale is treated by Peterson [170].

Proof.
Let us consider a simple lottery $\langle p_1, c_1; p_2, c_2; \ldots; p_k, c_k \rangle$. By virtue of Axiom 3.11 (continuity) there exists a simple lottery $\langle c_1 \ u_i \ c_k \rangle$ such that $c_i \sim \langle c_1 \ u_i \ c_k \rangle$, with $i = 1, \ldots, k$ and $u_i = u(c_i) = p_i$. Hence:

- Ordering property.
  By virtue of Axioms 3.10 (weak ordering), 3.11 (continuity), and 3.12 (monotonicity)

$$
\begin{aligned}
c_i &\succeq c_j \\
\langle c_1 \ u_i \ c_k \rangle &\succeq \langle c_1 \ u_j \ c_k \rangle \\
u_i &\geq u_j.
\end{aligned}
$$

- Linearity property.
  Let $u(c_1) = u_1 = 0$ and $u(c_k) = u_k = 1$. By virtue of the transitivity of the indifference relation (Theorem 3.1) and Axiom 3.15 (substitution or deletion) the following relations hold:

$$
\begin{aligned}
l = &\langle p_1, c_1; p_2, c_2; \ldots; p_k, c_k \rangle \\
\sim &\langle p_1, \langle c_1 \ u_1 \ c_k \rangle; p_2, c_2; \ldots; p_k, c_k \rangle \\
\sim &\langle p_1, \langle c_1 \ u_1 \ c_k \rangle; p_2, \langle c_1 \ u_2 \ c_k \rangle; \ldots; p_k, c_k \rangle \\
\sim &\langle p_1, \langle c_1 \ u_1 \ c_k \rangle; p_2, \langle c_1 \ u_2 \ c_k \rangle; \ldots; p_k, \langle c_1 \ u_k \ c_k \rangle \rangle.
\end{aligned}
$$

Moreover, Axiom 3.14 implies

$$l = \langle p_1, c_1; p_2, c_2; \ldots; p_k, c_k \rangle$$
$$\sim \langle (p_1 u_1 + p_2 u_2 + \ldots + p_k u_k), c_1; 0, c_2; \ldots; 0, c_{k-1};$$
$$[p_1(1 - u_1) + p_2(1 - u_2) + \ldots + p_k(1 - u_k)], c_k \rangle$$
$$= \langle c_1 \sum_{i=1}^{k} p_i u_i \; c_k \rangle.$$

In other words, the simple lottery l is indifferent to a lottery that pays c_1 with probability $\sum_{i=1}^{k} p_i u_i$.

Similarly, given the simple lottery $l' = \langle q_1, c_1; q_2, c_2; \ldots; q_k, c_k \rangle$, the following relation holds

$$l' \sim \langle c_1 \sum_{i=1}^{k} q_i u_i \; c_k \rangle.$$

In the end, by virtue of Axioms 3.10 (weak ordering) and 3.12 (monotonicity) it follows

$$l \succeq l' \iff \langle c_1 \sum_{i=1}^{k} p_i u_i \; c_k \rangle \succeq \langle c_1 \sum_{i=1}^{k} q_i u_i \; c_k \rangle$$
$$\iff \sum_{i=1}^{k} p_i u_i \geq \sum_{i=1}^{k} q_i u_i.$$

$\square$

Theorem 3.7. Uniqueness theorem (uniqueness of utility function). *Given a finite set of consequences $C = (c_1, c_2, \ldots, c_k)$ and a utility function $U(\cdot)$ on C, then function $W(\cdot)$ such that $w(\cdot) = \alpha u(\cdot) + \beta$, with $\alpha > 0$, is a utility function representing the same preferences as $U(\cdot)$. Moreover, if $U(\cdot)$ and $W(\cdot)$ are two utility functions defined on C representing the same preference structure, then there exist $\alpha > 0$ and β such that $w(\cdot) = \alpha u(\cdot) + \beta$.*

In other words, the theorem states that the utility function of a decision maker is unique, up to a positive linear transformation.

Proof.

- Let $U(\cdot)$ be a utility function on C and let $w(\cdot) = \alpha u(\cdot) + \beta$ with $\alpha > 0$.

$$\sum_{i=1}^{k} p_i w(c_i) \geq \sum_{i=1}^{k} q_i w(c_i) \iff \alpha \left[\sum_{i=1}^{k} p_i u(c_i) \right] + \beta \geq \alpha \left[\sum_{i=1}^{k} q_i u(c_i) \right] + \beta$$

$$\iff \sum_{i=1}^{k} p_i u(c_i) \geq \sum_{i=1}^{k} q_i u(c_i).$$

Hence, the expected value of $W(\cdot)$ provides the same ordering of lotteries as the expected value of $U(\cdot)$: we conclude that $W(\cdot)$ is a utility function, too.

- Let $U(\cdot)$ and $W(\cdot)$ be two utility functions defined on C and representing the same structure of preferences.
Let us assume that $w(\cdot) \neq \alpha u(\cdot) + \beta$ (non linearity assumption). Moreover, let (u_i, w_i) be the coordinates of point $c_i \in C$, for all $i = 1, \ldots, k$ (Figure 3.2), with $u_i = u(c_i)$ and $w_i = w(c_i)$. Under assumption of non-linearity there exist three points $c_1, c_2, c_3 \in C$ whose coordinates $(u_1, w_1), (u_2, w_2), (u_3, w_3)$ do not lie on the same straight line. Let us assume that $u_1 < u_2 < u_3$ and let us consider lottery $\langle c_3\ p\ c_1 \rangle$ with $p = (u_2 - u_1)/(u_3 - u_1)$. With respect to function $U(\cdot)$ the expected utility of $\langle c_3\ p\ c_1 \rangle$ is

$$u[pc_3 + (1-p)c_1] = pu(c_3) + (1-p)u(c_1)$$
$$= \frac{u_2 - u_1}{u_3 - u_1} u_3 + \frac{u_3 - u_2}{u_3 - u_1} u_1$$
$$= u_2.$$

Hence, under utility function $U(\cdot)$, $c_2 \sim \langle c_3\ p\ c_1 \rangle$. However, the non linearity assumption implies

$$w_2 \neq w_2^*,$$

with $w_2^* = pw_3 + (1-p)w_1$; that is, under utility function $W(\cdot)$, relation $c_2 \sim \langle c_3\ p\ c_1 \rangle$ is false. This last result contradicts the initial assumption that $U(\cdot)$ and $W(\cdot)$ represent the same preferences. Hence, the non linearity assumption is false, so we conclude that $W(\cdot)$ is a linear transformation of $U(\cdot)$.

$\square$

In practice, the above theorems about the existence and uniqueness of the utility function imply that the optimal decision for the rational decision maker is represented by action a^* that maximizes the expected utility:

$$a^* = \arg \left\{ \max_{i=1}^{m} \mathrm{E}_\theta \left[u(a_i, \theta_j) \right] \right\} = \arg \left[\max_{i=1}^{m} \sum_{j=1}^{k} u(a_i, \theta_j) \pi(\theta_j) \right], \qquad (3.1)$$

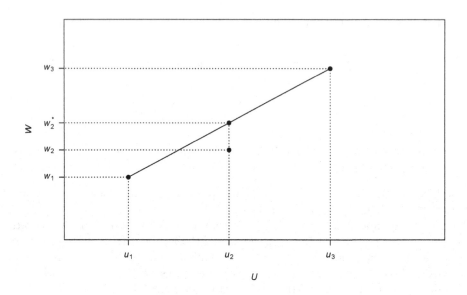

FIGURE 3.2
Graphical illustration of Theorem 3.7.

where $u(a_i, \theta_j)$ denotes the utility of action a_i under state of nature θ_j and $\pi(\theta_j)$ is the prior probability that θ_j happens. The decision table shown in Table 3.1 may be here reformulated substituting consequences c_{ij} with utilities $u_{ij} = u(a_i, \theta_j)$ (Table 3.2).

TABLE 3.2
Decision table in the presence of prior probabilities of the states of nature, $\pi(\theta_j)$, and with consequences expressed in terms of utilities, $u_{ij} = u(a_i, \theta_j)$ $(i = 1, \ldots, m;$ $j = 1, \ldots, k)$.

| | Prior probabilities | | | |
Action	$\pi(\theta_1)$	$\pi(\theta_2)$	...	$\pi(\theta_k)$
a_1	u_{11}	u_{12}	...	u_{1k}
a_2	u_{21}	u_{22}	...	u_{2k}
$\vdots$	...	...	$\vdots$	...
a_m	u_{m1}	u_{m2}	...	u_{mk}

3.4.2   Savage's theory

The expected utility theory by von Neumann and Morgenstern [213] suffers by the main limit that probabilities of the states of nature are assumed to be given. This assumption may be acceptable in some specific settings, such as games of chance (e.g., card games, roulette, dice games) and contexts in which reliable statistical data are available, that is all those situations in which there is a general shared consensus about the value of probabilities of the states of nature.

More realistically, probabilities as well as utilities are usually unknown and the decision maker has to elicit both of them. The first relevant contribution in such a direction is due to Ramsey [179] (see also [162]), even if his theory is limited to numerically measurable and additive results. A more extensive approach is the *subjective expected utility theory*, which was developed by Savage in 1954 [191] (for a detailed illustration see also [36, 78, 162, 170]) and by Anscombe and Aumann in 1963 [7]. In what follows Savage's theory is illustrated; for details about the formulation of Anscombe and Aumann the reader may refer to [78], Chap. 14, and [162], Chap. 6.

The subjective expected utility theory aims at proving - at the same time - the existence for each decision maker of a utility function, which describes how much outcomes are desirable (as in von Neumann and Mongenstern's theory), as well as a probability function, which provides a measurement of the subjective plausibility of the states of nature. Neither the utility function nor the probability function are known, but both of them derive from subjective preferences over actions and outcomes.

Savage's theory is based on some rational behavioral axioms that involve the following primitive concepts:

- $\Theta = \{\theta_1, \theta_2, \ldots\}$: states of nature, which may be split into subsets, such as $\Theta_0 \subseteq \Theta$, called events;

- $\mathcal{A} = \{a_1, a_2, \ldots\}$: actions;

- $\mathcal{C} = \{c_1, c_2, \ldots\}$: outcomes (or consequences) of taking an action under a specific state of nature

$$c_i = a_i(\theta_j) \quad i = 1, 2, \ldots; j = 1, 2, \ldots.$$

Note that an action is a function that maps states of nature to outcomes, that is, $a_i : \Theta \to \mathcal{C}$.

Axiom 3.17. $\succeq$ *on* $\mathcal{A}$ *is a weak ordering relation.*

This axiom resembles Axiom 3.5 above stated, including the concepts of transitivity and completeness of preference relations over actions.

Axiom 3.18. *Given actions* $a_1, a_1', a_2, a_2' \in \mathcal{A}$ *and event* $\Theta_0 \subseteq \Theta$ *such that*

- $a_1(\theta_j) = a_1'(\theta_j)$ *and* $a_2(\theta_j) = a_2'(\theta_j)$ $\quad \forall \theta_j \in \Theta_0;$

- $a_1(\theta_j) = a_2(\theta_j)$ and $a_1'(\theta_j) = a_2'(\theta_j)$ $\forall \theta_j \in \bar{\Theta}_0$.

Hence, $a_1 \succ a_2 \iff a_1' \succ a_2'$.

This axiom resembles Axiom 3.13 by von Neumann and Morgenstern's theory. In practice, it states that the preference relation between two actions should depend only on the outcomes of the two actions when they differ. For instance, let us consider the decisional situation in Table 3.3, where completely general (i.e., not necessarily of numerical type) outcomes are present. Under $\Theta_0 \subseteq \Theta$ actions a_1 and a_1' provide the same outcome ($\lhd$) as well as actions a_2 and a_2' ($\diamond$). On the contrary, under $\bar{\Theta}_0 \subseteq \Theta$ a_1 and a_2 pay the same reward ($\circ$) as well as a_1' and a_2' ($\star$). It turns out that the preference relation between a_1 and a_2 is the same as the preference relation between a_1' and a_2', as in the subspace Θ_0 $a_1 = a_1'$ and $a_2 = a_2'$, being of no relevance outcomes payed under $\bar{\Theta}_0$.

TABLE 3.3
Illustration of Axiom
3.18.

	Events	
Actions	Θ_0	$\bar{\Theta}_0$
a_1	$\lhd$	$\circ$
a_2	$\diamond$	$\circ$
a_1'	$\lhd$	$\star$
a_2'	$\diamond$	$\star$

The next axiom rests on the concept of *null state*, which is introduced to remedy the absence of the concept of impossible event (whose definition requires the concept of probability that will turn out as a consequence of the rational behavioral axioms). An event $\Theta_0 \subseteq \Theta$ is null if $a_i \sim a_h$ under Θ_0 for all pairs of actions $(a_i, a_h) \in \mathcal{A}$. In other words, a state is null when it is perceived as unlikely to be observed. Another concept of some relevance for the next axiom is that of *static action* [170], which denotes an action whose outcome is independent of the state of nature over a subspace Θ_0.

Axiom 3.19. *Let $\Theta_0 \subseteq \Theta$ be a non-null event and let $a_1(\cdot)$ and $a_2(\cdot)$ be static actions, that is, $a_1(\theta_j) = c_1$ and $a_2(\theta_j) = c_2$ for all $\theta_j \in \Theta_0$. Then, $a_1 \succeq a_2$ given $\Theta_0 \iff c_1 \succeq c_2$.*

This axiom asserts that the knowledge of a non-null event cannot modify the preference ordering among actions; in addition, a preference relation over actions complies with a preference relation over the related outcomes, assuming that these are independent of the true state of nature. From a more general perspective, Axiom 3.19 states that, in order to assure the objectivity of the decision maker's preferences, the comparison among outcomes in terms of their desirability cannot be affected by the measurement instrument, which is represented by states of nature and actions [78].

Axiom 3.20. *Given actions* $a_1, a_1', a_2, a_2' \in \mathcal{A}$, *outcomes* $c_1, c_1', c_2, c_2' \in \mathcal{C}$, *and events* $\Theta_0, \Theta_1 \subseteq \Theta$ *such that*

- $a_1(\theta_j) = c_1$ *and* $a_2(\theta_j) = c_1'$ $\forall \theta_j \in \Theta_0$;

- $a_1(\theta_j) = c_2$ *and* $a_2(\theta_j) = c_2'$ $\forall \theta_j \in \bar{\Theta}_0$;

- $a_1'(\theta_j) = c_1$ *and* $a_2'(\theta_j) = c_1'$ $\forall \theta_j \in \Theta_1$;

- $a_1'(\theta_j) = c_2$ *and* $a_2'(\theta_j) = c_2'$ $\forall \theta_j \in \bar{\Theta}_1$.

Then, $a_1 \succ a_1' \iff a_2 \succ a_2'$.

This axiom is particularly relevant in the Savage's theory setting, because it introduces the concept of *qualitative probability* providing a tool to define a ranking of events in terms of subjective belief about their happening. For the sake of clarity, let us consider the simple decisional problem illustrated in Table 3.4.

TABLE 3.4
Illustration of Axiom 3.20.

	Events			
Actions	Θ_0	$\bar{\Theta}_0$	Θ_1	$\bar{\Theta}_1$
a_1	100	0	–	–
a_1'	–	–	100	0
a_2	50	10	–	–
a_2'	–	–	50	10

Now, let us observe that if the decision maker states $a_1 \succ a_1'$, it is reasonable to conclude that, in the opinion of the decision maker, event Θ_0 is *more likely than* event Θ_1, as the possible outcomes related to the two actions are the same. On the other hand, if the same decision maker stated $a_2' \succ a_2$, then he/she should also conclude that event Θ_1 is more likely than event Θ_0. Axiom 3.20 avoids such types of incoherent orderings, requiring the independence of the measurement instrument for the ranking of states of nature.

Axioms 3.19 and 3.20 may be considered complementary one other: both of them require that the subjective ordering of outcomes (Axiom 3.19) and states of nature (Axiom 3.20) are independent of the measurement tools, that is, states of nature and actions in the former case and outcomes in the latter case.

Axiom 3.21. *There exist* $c_i, c_h \in \mathcal{C}$ *such that* $c_i \succ c_h$.

Axiom 3.21 states that a decision maker is always able to express at least some strict preference over outcomes. If he/she were indifferent about all pairs of outcomes, no subjective probability would be elicited.

A decision maker complying with Axioms 3.17-3.21 is able to rank states of nature from the least likely to the most likely, but he/she is not able to assign them a quantitative meaning. The next axiom allows us to pass from qualitative probability to quantitative probability. In addition, this axiom also makes it possible to establish the uniqueness of the quantitative probability representation.

Axiom 3.22. *For all actions $a_1, a_2 \in \mathcal{A}$ such that $a_1 \succ a_2$, there exists a finite partition of Θ, denoted by $\Theta_1, \Theta_2, \ldots, \Theta_k$, such that*

- $a_1'(\theta_j) \neq a_1(\theta_j)$ for all $\theta_j \in \Theta_1$ and $a_1'(\theta_j) = a_1(\theta_j)$ for all $\theta_j \notin \Theta_1$ $(j = 1, 2, \ldots)$, or

- $a_2'(\theta_j) \neq a_2(\theta_j)$ for all $\theta_j \in \Theta_1$ and $a_2'(\theta_j) = a_2(\theta_j)$ for all $\theta_j \notin \Theta_1$ $(j = 1, 2, \ldots)$.

This axiom resembles the archimedean property (Axiom 3.9), which guarantees that there does not exist outcomes infinitely better or worse than others. In practice, Axiom 3.22 states a sort of continuity condition: if two actions provide the same outcomes throughout all the space of events up to an arbitrarily small sub-space, then they are close enough to imply the same preference relation with respect to a third action.

The last axiom is a technical condition that makes it possible to represent the decision maker's preferences in the presence of infinite outcomes $\mathcal{C} = (c_1, c_2, \ldots)$, the opposite of the von Neumann and Morgestern's theory that explicitly requires that $\mathcal{C}$ is a finite set (see Theorem 3.6).

Axiom 3.23. *Given actions $a_1, a_2 \in \mathcal{A}$ and event $\Theta_0 \subseteq \Theta$ such that $a_1 \succeq a_2(\theta_j)$ for all $\theta_j \in \Theta_0$, then $a_1 \succeq a_2$.*

In other words, Axiom 3.23 asserts that if an action is preferred to any specific outcome of another action, then the former action is preferred to the latter one.

From the above axioms the following theorem results, which states the existence and uniqueness of a quantitative subjective probability function as well as an utility function for a decision maker.

Theorem 3.8. Representation and uniqueness theorem (existence and uniqueness of probability and utility functions). *Given a set of states of nature $\Theta = \{\theta_1, \theta_2, \ldots\}$, a set of consequences $\mathcal{C} = (c_1, c_2, \ldots)$, a set of actions $\mathcal{A} = \{a_1, a_2, \ldots\}$, and a relation of weak preference $\succeq$ satisfying Axioms 3.17-3.23, then:*

> *1. there exists an unique and finitely additive probability measure $\pi(\cdot)$ on Θ such that*

$$\Theta_0 \succeq \Theta_1 \iff \pi(\Theta_0) \geq \pi(\Theta_1),$$

with $\Theta_0, \Theta_1 \subseteq \Theta$;

2. there exists a bounded utility function $U(\cdot) : \mathcal{C} \to \mathbb{R}$ such that

$$a_i \succeq a_h \iff \sum_j u(a_i(\theta_j))\pi(\theta_j) \geq \sum_j u(a_h(\theta_j))\pi(\theta_j)$$

$$\iff \sum_j u(c_i)\pi(\theta_j) \geq \sum_j u(c_h)\pi(\theta_j);$$

3. $U(\cdot)$ is unique up to a positive linear transformation.

Proof of Theorem 3.8 is omitted: for details, see [191] and [135].

Note that point 2 above may be generalized to the case of an uncountable space of outcomes as follows:

$$a_i \succeq a_h \iff \int_\Theta u(a_i(\theta))\pi(\theta)d\theta \geq \int_\Theta u(a_h(\theta))\pi(\theta)d\theta.$$

To conclude, it is worth noting that the optimal decision rule deriving from Savage's theory resembles that suggested by the von Neumann and Morgenstern's theory, that is, the maximization of the expected utility. The main difference lies in the type of probabilities of the states of nature: subjective according to the Savage's setting and objective (i.e., externally defined) according to the von Neumann and Morgenstern's setting.

3.5 Empirical failures of rational behavioral axioms

The empirical violation of the rational behavioral axioms is the most relevant weakness of the expected utility theory. This violation undermines the descriptive validity of the expected utility theory but not its normative validity. Anyway, the manifestation of actual behaviors that contradict the axiomatic background justifies the development of some generalizations about the expected utility theory, which are typically based on weaker versions of some rational behavioral axioms (Section 3.6). A review of the main criticisms of von Neumann and Morgestern's axiomatic approach is available in Peterson[170] and Parmigiani and Inoue [162], whereas Gilboa [78] (Chap. 12) provides an accurate analysis of the weaknesses of Savage's approach.

3.5.1 Violation of transitivity

Axiom 3.2 states the coherence in the preference ordering of outcomes, such that if action a_1 is preferred to action a_2, and action a_2 is preferred to action a_3, then a rational decision maker should also prefer a_1 over a_3. In practice this axiom is not always satisfied. This is the case with *cyclic preferences*, that is, when $a_1 \succ a_2 \succ a_3 \succ a_1$. A typical situation is when the appeal of an action is evaluated with respect to several criteria. For instance, the preferences over three possible partners (e.g., Sam, Tom, John), which are characterized by different levels of beauty, intelligence, and wealth, may result in a cyclic ordering, whenever each alternative is perceived as superior to the others over two out of three characteristics (e.g., Sam is more beautiful and more intelligent than Tom; Tom is more intelligent and wealthier than John; John is more beautiful and wealthier than Sam); see the experiment by May [155] for an illustration of cyclic preferences. More generally, the evaluation of preferences in the presence of multiple attributes of outcomes is an object of interest in the multi-attribute utility theory [131, 117, 57].

Another example of violation of transitivity is the phenomenon of *preference reversal*, often observed in the monetary lotteries setting [88, 142, 143]. Let $l_1 = \langle 30\$ \ 0.90 \ 0 \rangle$ and $l_2 = \langle 100\$ \ 0.30 \ 0 \rangle$ be two lotteries and let us assume that the decision maker prefers l_1 to l_2, that is, $l_1 \succ l_2$. In addition, let $c_1^* = 25\$$ and $c_2^* = 27\$$ be the prices that the decision maker would be willing to pay for sure for l_1 and l_2, respectively, such that $c_1^* \sim l_1$ and $c_2^* \sim l_2$. The selling prices c_1^* and c_2^* represent the so called *sure-things* or *certainty equivalents* of l_1 and l_2.

Definition 3.13. Sure-thing or certainty equivalent. *The sure-thing or certainty equivalent c_l^* [191] of a lottery l is that outcome which makes the decision maker indifferent between having it with certainty and playing lottery l, that is,*

$$c_l^* \sim l$$

or, equivalently,

$$u(c_l^*) = \sum_{i=1}^{k} u(c_i)\pi(c_i) = EU(l) \iff c_l^* = EU^{-1}(l).$$

As far as the above example with $c_1^* = 25$ and $c_2^* = 27$, under the transitivity axiom one would expect

$$c_1^* \sim l_1 \succ l_2 \sim c_2^* \implies c_1^* \succ c_2^*.$$

However, under the common sense assumption that more money is better than less, $c_2^* \succ c_1^*$ will be usually observed, with an evident contradiction of the transitivity axiom.

A further situation that leads to violate transitivity is due to the difficulty in perceiving small changes in outcomes, as happens with the perception of temperature. For instance, given an environmental temperature equal to $c = 20$ degrees centigrade, any individual is indifferent between taking the action of staying in a room with 20 degrees centigrade and staying in a room with $20 + 0.01$ degrees centigrade, and, similarly, he/she will be indifferent between 20.01 and 20.02 degrees centigrade. Along the same reasoning one observes

$$20 \sim 20.01 \quad 20.01 \sim 20.02 \quad 20.02 \sim 20.03 \ldots 29.01 \sim 29.02 \ldots 99.99 \sim 100.00$$

and, for the transitivity axiom, one would expect the absurd conclusion of indifference between 20 and 100 degrees centigrade.

3.5.2   Certainty effect

The expected utility theory implies the principle of linearity of probabilities: a decision maker is expected to assign the same relevance to the probabilities of the events, independently of their value. However, empirical evidence often shows that individuals tend to perceive as more appealing sure outcomes with respect to probable outcomes. Such a type of phenomenon is known as *certainty effect* [126, 209]. A classical example of certainty effect is provided by the *Allais' paradox* [4]. Let us consider two different types of situations:

- situation 1: the decision maker is asked to choose between the sure result $l_1 = c = 100$ (mln of US dollar) and playing the lottery $l_1' = \langle 500, 0.10; 100, 0.89; 0, 0.01 \rangle$;

- situation 2: the decision maker is asked to choose between lottery $l_2 = \langle 100\ 0.11\ 0 \rangle$ (mln of US dollar) and lottery $l_2' = \langle 500\ 0.10\ 0 \rangle$.

In general, preference of l_1 over l_1' is often observed together with preference of l_2' over l_2. Indeed, the certainty of the result is favored in the first situation, whereas in the second situation the two probabilities are perceived as really similar (0.11 vs. 0.10); thus the decision maker focuses on the amounts the two possible gains differ in a sensible way (100 vs. 500). Each of these preferences is perfectly rational *per sé*, but both of them violate the expected utility principle. Indeed, $l_1 \succ l_1'$ implies

$$u(100) > 0.10u(500) + 0.89u(100) + 0.01u(0)$$
$$0.11u(100) > 0.10u(500),$$

whereas $l_2' \succ l_2$ implies the opposite

$$0.10u(500) + 0.90u(0) > 0.11u(100) + 0.89u(0)$$
$$0.11u(100) < 0.10u(500).$$

Such type of phenomenon is explained [126] by the major pleasure related to sure results with respect to probable results. As a main consequence, the certainty effect affects the shape of the utility function, as illustrated in Chapter 4.

The certainty effect also implies a violation of the independence Axiom 3.13 (see also Parmigiani and Inoue [162]). Let us consider the lottery

$$q = \langle 500\ 0.11\ 0 \rangle$$

and let $l_1 \succ q$. Then, for the independence axiom the following preference relation holds for any lottery q^*

$$l_1 \succ q \iff \langle l_1\ 0.11\ q^* \rangle \succ \langle q\ 0.11\ q^* \rangle.$$

In detail, for $q^* = l_1$

$$l_1 \succ q \iff l_1 \succ \langle q \; 0.11 \; l_1 \rangle.$$

However, $\langle q \; 0.11 \; l_1 \rangle$ is a compound lottery such that

$$\langle q \; 0.11 \; l_1 \rangle \sim \langle 500, \; 0.10; \; 100, \; 0.89; 0, \; 0.01 \rangle = l_1'.$$

Then, $l_1 \succ q$ implies $l_1 \succ l_1'$. On the other hand, for $q^* = 0$, according to the same lines as above one obtains $l_1 \succ q$ if and only if $l_2 \succ l_2'$. To conclude, the independence axiom requires $l_1 \succ l_1'$ together with $l_2 \succ l_2'$, whereas the empirical evidence often shows $l_1 \succ l_1'$ and $l_2' \succ l_2$ at the same time.

An effect specular to the certainty effect leading to the violation of the expected utility principle is known as *reflection effect*, which denotes an instinctive aversion to sure losses.

3.5.3 Pseudo-certainty effect and isolation effect

Not only the certainty of an outcome, but also the pseudo-certainty, that is, the perception of an outcome *as if* it would be certain while it is actually uncertain, leads to a biased perception of probabilities and, then, a violation of the expected utility principle. Let us consider the following two situations [126]:

- situation 1: a two-steps game is proposed

$$\langle 0 \; 0.75 \; l \rangle,$$

  with l compound lottery paying $l_1 = 3000\$$ with certainty and $l_1' = \langle 4000\$ \; 0.80 \; 0 \rangle$. A decision maker is asked to choose between l_1 and l_1' before the game starts;

- situation 2: the decision maker is asked to choose between lottery $l_2 = \langle 3000\$ \; 0.25 \; 0 \rangle$ and lottery $l_2' = \langle 4000\$ \; 0.20 \; 0 \rangle$.

The two above situations denote identical choice problems, although presented under different forms, as is easily verified by applying the compound lotteries reduction Axiom (3.14) to situation 1, which immediately results in a game that pays \$3000 with probability $0.25 \times 1 = 0.25$ (0 otherwise) against a game that pays \$4000 with probability $0.25 \times 0.80 = 0.20$ (0 otherwise). Hence, according to the expected utility principle, preference of l_1 over l_1' is consistent with preference of l_2 over l_2' (or vice-versa). However, some experiments [126] often report preference of l_1 over l_1' together with preference of l_2' over l_2. Such empirical failure of the expected utility principle is explained through the so called *isolation effect*, denoting the tendency of individuals to disentangle complex problems, rather than deal with them together, and to focus on single aspects. As a main consequence, the isolation of consecutive sub-games may lead to inconsistent preferences due to the biased perception of the probabilities. For instance, in situation 1 the outcome of \$3000 payed by lottery l_1 is often perceived as if it would be certain, even if the related probability is just 0.20 (*pseudo-certainty effect*).

The experiment above described is particularly interesting, as it outlines the violation of Axioms (3.13) and (3.14) through the simultaneous presence of multiple effects due to the isolation of sub-games, the perception of some outcomes as pseudo-certain, and, more in general, the sensibility of individuals to the form under which a choice is presented, known also as *framing effect*.

3.5.4 Framing effect

According to the independence axiom, the specific form under which a choice is presented should not affect the decision about the optimal action. Other than the example provided in the previous section, other interesting experiments have been proposed in the literature that outline the relevance of the frame under which the decision maker is asked to decide.

Insurance frame Let us consider the following choices:

- insurance frame: the decision maker is asked to choose between exposing himself/herself to the relatively small risk of losing a given amount of money, $l_1 = \langle \$5000 \ 0.001 \ 0 \rangle$, and the possibility of paying a premium, say $l_1' = \$5$, to insure against such a risk;

- lottery frame: the decision maker is asked to choose between playing lottery $l_2 = \langle \$5000 \ 0.001 \ 0 \rangle$ and the sure loss $l_2' = \$5$.

According to the expected utility principle, preference of l_1 over l_1' should be observed together with preference of l_2 over l_2' (or vice-versa). However, several studies (see, for instance, [104] and [198]) showed inconsistent preferences, with individuals tending to prefer the payment of the premium in the insurance frame and to play the gamble in the lottery frame. Although the two above decision problems are equivalent from a substantial point of view, the payment of an insurance premium (first frame) is psychologically perceived as different from choosing a sure loss (second frame). In the insurance frame the individual is aware of exposing himself/herself to a risk and the insurance premium is the instrument to avoid such a risk. On the contrary, the payment of a sure loss is not perceived as an instrument to avoid participation in an unwanted game.

Health frame The sensitivity of individuals to the form under which a problem is presented is particularly evident in the health frame, where speaking in terms of survival rate rather than mortality rate often leads to a different perception of the decisional problem. Let us assume that two treatments are available, which are related with different survival rates:

- treatment A presents a survival rate of 90% after the surgery, 68% after one year, and 34% after five years;

- treatment B provides a survival rate equal to 100% after the surgery, 77% after one year, and 22% after five years.

Alternatively, the outcome of treatments A and B may be expressed in terms of mortality rates:

- treatment A presents a mortality rate of 10% after the surgery, 32% after one year, and 66% after five years;

- treatment B provides a mortality rate equal to 0% after the surgery, 23% after one year, and 78% after five years.

In an experiment conducted by Tversky and Kahneman [209], treatment B was preferred by 18% of patients in the first frame and by 44% of patients in the second frame. This shows how individuals are more sensitive to a reduction of the present risk of death (from 10% to 0%), rather than to an increase of the survival rate from 90% to 100%.

More generally, the two examples above described confirm the greater sensitivity of individuals towards losses rather than gains. The topic will be take up in Chapter 4, when the effect of sensitivity on the shape of the utility function is illustrated.

A special case of framing effect is represented by the *inertia effect* [104], according to which individuals tend to maintain their current position (status quo) rather than risk an uncertain gain, unless the alternative is definitely advantageous. An example is here provided:

- situation 1: the decision maker may decide to accept the following lottery $\langle \$100\ 0.50\ -\$100 \rangle$, without charge;

- situation 2: the decision maker may decide to transfer the lottery $\langle \$100\ 0.50\ -\$100 \rangle$ to someone else, without charge.

The inertia effect induces the decision maker to not accept the lottery in situation 1 and, at the same time, to not transfer the same lottery to someone else in situation 2.

3.5.5   Extreme probability effect

The certainty and pseudo-certainty effects are part of a more general phenomenon consisting of the biased perception of extreme probability levels (i.e., probabilities close to 0 and 1): the relevance of small probabilities is usually underweighted, whereas the importance of high probabilities is often overweighted. To clarify this point let us consider the following situation:

- situation 1: the decision maker is asked what lottery he/she prefers between $l_1 = \langle 6000\ 0.45\ 0 \rangle$ and $l_1' = \langle 3000\ 0.90\ 0 \rangle$;

- situation 2: the decision maker is asked what lottery he/she prefers between $l_2 = \langle 6000\ 0.001\ 0 \rangle$ and $l'_2 = \langle 3000\ 0.002\ 0 \rangle$.

In an experiment proposed by Kahneman and Tversky [126] (see also [104]) the 86% of preferences in situation 1 fall on l_1 against the 73% of preferences in situation 2 that fall on l_2. This violation of the expected utility principle is mainly due to the different perception of probability levels: in situation 2 individuals have difficulty in perceiving the difference between 0.001 and 0.002, such that some of them tend to select l_2 as it pays a greater amount of money with respect to l'_2, regardless of the probability levels. On the other hand, in situation 1 individuals tend to prefer l'_1 as it assures a gain with a higher probability (0.90 vs 0.45). The extreme probability effect reflects the general difficulty of individuals in combining information from different sources (e.g., probabilities and outcomes) and the tendency to focus on one element at a time.

3.5.6 Aversion to uncertainty

A specific example of biased perception of probabilities is the aversion to ambiguous situations, also known as Ellsberg's paradox [58]. Ellsberg observed that individuals tend to prefer decisional contexts where the probability distribution of states of nature is known to situations where the probabilities are "ambiguous" (i.e., are not known with certainty). For instance, let us assume that two urns are given: urn A has 50 red balls and 50 black balls, whereas urn B has 100 balls either red or black. The decision maker is placed in front of the following situations:

- situation 1: the decision maker is asked if he/she prefers to bet on a *red* ball from urn A or from urn B;

- situation 2: the decision maker is asked if he/she prefers to bet on a *black* ball from urn A or from urn B.

Many decision makers often prefer to bet on urn A in both situations, even if this pair of preferences leads to violate Axiom (3.18) in Savage's theory and, hence, the expected utility principle. Indeed, betting on urn A in situation 1 implies a subjective probability of drawing a red ball in urn B less than 0.50; on the other hand, betting on urn A also in situation 2 implies a subjective probability of drawing a black ball in urn B less than 0.50, but this contradicts the previous statement (p(red ball in urn B) + p(black ball in urn B) < 1). Such a type of phenomenon is explained through an instinctive aversion for ambiguous situations; then preferences tend to shift towards sure situations. In the above example individuals prefer to bet on urn A because they know the probability of selecting a red ball as well as a black ball, whereas no information is available about the composition of urn B.

3.6   Alternative utility theories

Empirical failures of rational behavioral axioms outline the inadequacy of expected utility theory to describe the actual behavior of individuals in a completely satisfactory way. As a main consequence, alternative normative theories have been developed to better comply with the empirical evidence. All of them share the principle that the decisional process is based not only on the expected utility (i.e., sum of utilities of single results weighted with probabilities), but other elements are involved, such as the certainty of results, the frame of the decisional problem, and so on. To realize a satisfactory compliance with the actual behavior, these theories adopt a more general axiomatic base with respect to the expected utility theory. As far as this point, Starmer [206] distinguishes between *conventional theories* and *non-conventional theories*. Conventional theories, which are also commonly known as *generalized utility theories*, require the existence of a single function representing the individual preferences that satisfies Axioms (3.10) (weak ordering), (3.11) (continuity), and (3.12) (monotonicity), whereas the violation of Axiom (3.13) (independence) is admitted and weak formulations of this property are assumed. Note that the failure of the independence principle implies the non-linearity of probabilities. In such a frame, the expected utility theory represents a special case of generalized utility theories. On the other hand, non-conventional theories admit not only weak versions of the independence axiom, but also different preference functions in relation to the domain of results as well as violations of ordering, continuity, and monotonicity [64, 65]. The most relevant example of non-conventional theory is represented by the *prospect theory* [126] (see Section 3.6.2 for details).

A nice instrument for the comprehension of the main differences among alternative theories is the *triangle diagram* [26, 153, 154]. The triangle diagram is a graphical instrument that displays in a two-dimensional space lotteries of type $l = \langle c_1, p_1; c_2, p_2; c_3, p_3 \rangle$, with $c_3 \succ c_2 \succ c_1$. In more detail, given a right-angled triangle (Figure 3.3), the lower edge denotes p_1 (i.e., probability of the worst outcome c_1) and the left edge denotes p_3 (i.e., probability of the best outcome c_3), whereas p_2 is implicitly obtained as $p_2 = 1 - p_1 - p_3$.

A specific lottery is identified by a point whose location in the diagram depends on the values of p_1 and p_3. In general, l is located inside the triangle (e.g., point A in Figure 3.3); however it will be placed on the left edge if $p_1 = 0$ (Figure 3.4, top left panel), on the lower edge if $p_3 = 0$ (Figure 3.4, top right panel), on the hypothenuse if $p_2 = 0$ (Figure 3.4, bottom left panel), and on a corner of the triangle in the presence of a sure result (see, for instance, Figure 3.4, bottom-right panel, where $p_1 = p_3 = 0$ and $p_2 = 1$).

Moving along the hypothenuse from the lower right corner to the upper left corner denotes an increase of preferences, as probability p_1 of the worst outcome reduces and, at the same time, probability p_3 of the best outcome

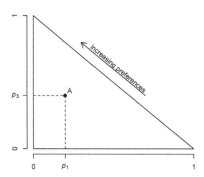

FIGURE 3.3
Triangle diagram: example.

increases. Lotteries that are equally preferred are connected by a curve, known as an *indifference curve*: if lotteries l and l' lie on the same indifference curve then $l \sim l'$, whereas if l lies on a higher indifference curve then $l \succ l'$ (and vice-versa).

The shape of the indifference curves provides a lot of information about the underlying utility theory. For instance, in the case of the expected utility theory,

- the continuity axiom guarantees the absence of open spaces in the indifference curves;

- the completeness axiom guarantees that any pair of points in the triangle are either on the same curve or on two different curves

- the transitivity axiom guarantees that curves do not cross inside the triangle;

- the independence axiom guarantees that indifference curves are parallel straight lines (Figure 3.5, top left panel).

In addition, the slope of the indifference curve measures how much p_3 has to be compensated for a unit increase in p_1: a higher slope suggests an increase in the risk aversion of the decision maker (see Chapter 4 for details about the attitude towards risk). Since in the expected utility theory the indifference curves are parallel, their slope is constant.

In the case of the violation of independence axiom, indifference curves are not parallel any more. For instance, in the *weighted utility theory* ([192] and references therein) a weaker version of the independence Axiom (3.13) holds.

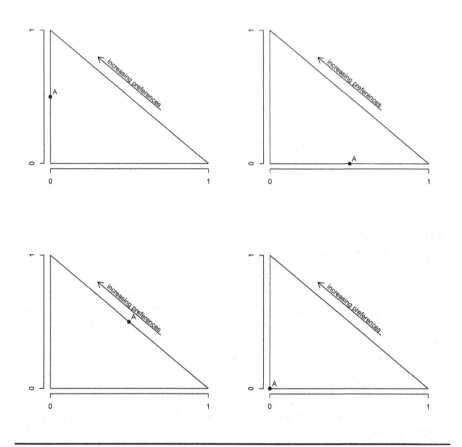

FIGURE 3.4
Special cases of triangle diagrams: $p_1 = 0$ (top left), $p_3 = 0$ (top right), $p_2 = 0$ (bottom left), $p_1 = p_3 = 0$ and $p_2 = 1$ (bottom right).

Axiom 3.24. Weak substitution axiom*: Given $c_i, c_j \in C$ such that $c_i \sim c_j$, then $\forall p_1 \in (0, 1)$, there exists $p_2 \in (0, 1)$ such that $\langle c_i \, p_1 \, c_h \rangle \sim \langle c_j \, p_2 \, c_h \rangle$, with $p_1 \neq p_2$ and $\forall c_h \in C$.*

In practice, the weak substitution axiom allows for the so called fanning phenomenon, consisting of indifference curves that meet at a point outside the triangle (Figure 3.5, top right panel). As a main consequence, the indifference curves become steeper (fanning out) or less steep (fanning in) as one moves in the direction of increasing preferences. Thus, a varying slope and, hence, a varying level of aversion towards risk is allowed, differently from the expected utility theory (see Section 4.2 for a detailed definition of risk aversion).

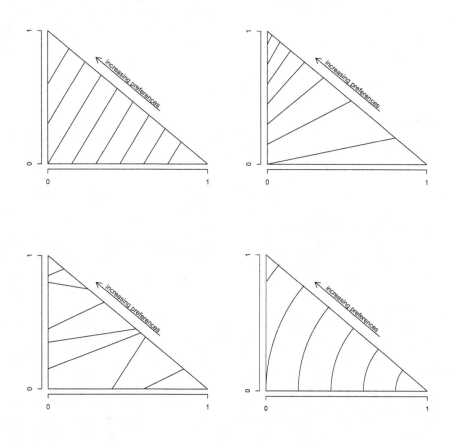

FIGURE 3.5
Indifference curves under expected utility theory (top left), weighted utility theory (top right), implicit weighted utility theory (bottom left), and rank-dependent utility theory (bottom right).

Further flexibility is allowed by the *implicit weighted utility theory* ([192] and references therein), which is based on an even weaker form of the independence axiom: not only p_2 differs from p_1 with p_2 depending on p_1 (as in Axiom (3.24)), but with p_2 also depending on a third outcome c_h.

Axiom 3.25. Very weak substitution axiom: *Given* $c_i, c_j, c_h \in C$ *such that* $c_i \sim c_j$, *then* $\forall p_1 \in (0, 1)$, *there exists* $p_2 \in (0, 1)$ *such that* $\langle c_i \, p_1 \, c_h \rangle \sim \langle c_j \, p_2 \, c_h \rangle$, *with* $p_1 \neq p_2$.

Under the very weak substitution axiom indifference curves do not cross over at all, as shown in Figure 3.5 (bottom left panel).

Alternative utility theories that further generalize weighted utility theory and implicit weighted utility theory rely on weighting probability p_i of outcome c_i to properly account for extreme levels of probabilities. More precisely, under the expected utility theory the value assigned to a lottery l is (see Theorem 3.6)

$$EU(l) = \sum_{i=1}^{k} p_i u(c_i),$$

which is non-linear in the outcomes but linear in the probabilities, whereas in the *non-linear utility* theories the expected utility function is replaced by a more general functional form characterized by non-linearity in outcomes as well as in probabilities:

$$\sum_{i=1}^{k} \omega(p_i) u(c_i), \tag{3.2}$$

with $\omega(\cdot)$ generic transformation of probabilities p_i.

The most relevant examples of non-linear utility theories are provided by the rank-dependent utility theory (Section 3.6.1) and by the prospect theory (Section 3.6.2). In practice, the indifference curves are neither linear nor parallel, as shown in Figure 3.5 (bottom right) that refers to the rank-dependent utility theory.

Before proceeding, it is worth noting that the literature agrees that the best theory from a prescriptive point of view does not exist. Indeed, alternative theories designed to explain violations of expected utility are themselves violated in other ways. Expected utility theory performs reasonable well as long as the lotteries are not too close to the edges of the triangle and have the same support [97]. On the contrary, alternative theories outperform expected utility theory when extreme probabilities or large rewards are involved [59]. Overall, prospect theory (Section 3.6.2) is the most coherent with the empirical evidence. More generally, every theory represents a trade-off between parsimony and accuracy and the goodness of a theory also depends on the use of it. If the aim is to help people to make rational and coherent decisions, expected utility theory represents a good and parsimonious instrument; on the other hand, if one aims at explaining the actual behavior of people as accurately as possible prospect theory and its variants have to be preferred. An interesting direction for future research consists of integrating a stochastic error term into models for decision making in order to account for the heterogeneity in observed behaviors [192].

3.6.1 Rank-dependent utility theory

In the presence of weighted probabilities, value (3.2) assigned to a lottery will not generally satisfy the monotonicity assumption, up to specific constraints. The violation of monotonicity is quite problematic, as it means that stochastically dominated lotteries may be preferred to dominating lotteries. To avoid this issue, in the rank-dependent utility theory [177] weighting function $\omega(\cdot)$ is defined as a suitable transformation on cumulative probabilities that accounts for the ranking of outcomes, that is,

$$\omega(p_i) = G\left(\sum_{j=1}^{i} p_j\right) - G\left(\sum_{j=1}^{i-1} p_j\right),$$

with $G\left(\sum_{j=1}^{i} p_j\right)$ subjective weight attached to the probability of getting c_i or worse, and $G\left(\sum_{j=1}^{i-1} p_j\right)$ subjective weight attached to the probability of getting something worse than c_i. The shape of function $G(\cdot)$ depends on the behavior of the decision maker and it reflects his/her attitude towards risk (see Section 4.2 for details). For instance, if $G(\cdot)$ is convex, then the indifference curves are concave (i.e., risk aversion), implying an overweighting of the lower-ranked consequences relative to higher-ranked consequences; on the contrary, if $G(\cdot)$ is concave, then the indifference curves are convex (i.e., risk seeking), implying an underweighting of the lower-ranked consequences relative to higher-ranked consequences.

There exist several axiomatizations coherent with the rank-dependent utility theory that rely on somewhat different weak forms of the independence Axiom (3.13), such as

Axiom 3.26. Co-monotonic independence axiom *[216]. Preferences over lotteries are unaffected by substitution of common consequences so long as these substitutions have no effect on the rank order outcomes in either lottery.*

In other words, while the independence Axiom (3.13) states that the preference ordering between c_i and c_j is completely independent of a third outcome c_h, for any $c_h \in C$, this may not be true under the rank-dependent utility approach, as changing $c_h \in C$ with another element of C may modify the ranking of outcomes and, hence, the weights.

3.6.2 Prospect theory and cumulative prospect theory

Prospect theory was introduced by Kahneman and Tversky [126] and it became quite popular for its capacity of satisfactorily describing actual behaviors of people. In the prospect theory the decisional process is decomposed in two phases: the editing phase and the evaluation phase.

In the *editing phase* competing lotteries are preliminarily analyzed in order to obtain a simple representation and avoid incoherent choices. In more detail, the following operations are carried out:

- dominance heuristic: the problem of violation of monotonicity due to the presence of the weighting function $\omega(\cdot)$ is faced by removing dominated lotteries;

- coding: outcomes are coded as gains and losses relative to a reference point, which usually is the current asset position (status quo) of the decision maker;

- combination: lotteries are simplified by combining probabilities associated with identical consequences (e.g., $\langle 200, \ 0.25; \ 200, 0.25; \ 0, \ 0.50 \rangle$ is reduced to $\langle 200 \ 0.50 \ 0 \rangle$);

- cancellation: elimination of elements common to the lotteries under consideration (e.g., $\langle 200 \ 0.5 \ l_1 \rangle$ vs $\langle 200 \ 0.5 \ l_2 \rangle$ may be reduced to l_1 vs l_2);

- segregation: the sure component of a lottery is disentangled by the risky component (e.g., lottery $\langle 300 \ 0.8 \ 200 \rangle$ is decomposed in the sure gain of 200 and a risky lottery $\langle 100 \ 0.8 \ 0 \rangle$);

- simplification: probabilities and/or outcomes are rounded (e.g., $\langle 101 \ 0.79 \ 0 \rangle$ becomes $\langle 100 \ 0.80 \ 0 \rangle$).

In the *evaluation phase*, the decision maker selects that action with the maximum value. The overall value of a lottery, provided by (3.2), depends on two elements: the utility $u(c_i)$ of outcomes, as in the expected utility theory, and the weight $\omega(p_i)$ assigned to each outcome ($i = 1, \ldots, k$). The exact shape of $u(\cdot)$ and $\omega(\cdot)$ is an empirical issue; however Kahneman and Tversky provide some hints, driven by some experiments.

Empirical failures of expected utility theory illustrated in Section 3.5 suggest that individuals assign a subjective relevance to probabilities of results so that the utility of each outcome is multiplied not by its probability but by a subjective decisional weight. Note that the subjective weight $\omega(p_i)$ is not a probability and it can be defined both for objective probabilities (used in von Neumann and Morgenstern's theory) and for subjective probabilities (used in Savage's theory). Probability weighting function accounting for the empirical violations of expected utility theory should satisfy the following properties:

- $\omega(p_i)$ increases with p_i and is not defined for probabilities very close to 0 and 1 (i.e., it has discontinuity points in the extremes);

- subadditivity: for small values of p_i, $\omega(r p_i) > r\omega(p_i)$ for all $0 < r < 1$;

- overweighting of small probabilities: for small values of p_i, $\omega(p_i) > p_i$;

- subcertainty: moderate and large probabilities are underweighted, that is, $\omega(p_i) + \omega(1 - p_1) < 1$ for $0 < p_i < 1$.

Based on the above properties, probability weighting function may look as in Figure 3.6 (left). The straight line holds when $\omega(p_i) = p_i$, that is, when the linearity property of probabilities is satisfied.

As concerns the utility function, experiments of Kahneman and Tversky suggest a shape of $u(\cdot)$ as shown in Figure 3.6 (right). In particular, $u(\cdot)$ is kinked at the reference point, with different shapes for losses and gains. Thus, under the prospect theory the decision maker does not have a unique utility function, as under the generalized utility theories, but different functions related to the lottery domains, as under the non-conventional theories. In addition, $u(\cdot)$ is concave for gains and convex for losses and is steeper in the

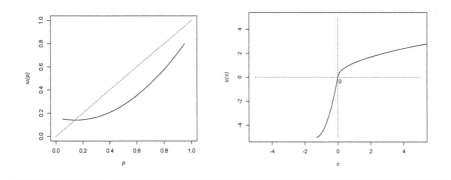

FIGURE 3.6
Prospect theory: probability weighting function (left) and utility function (right).

losses domain; more details about the interpretation of $u(\cdot)$ are provided in Section 4.2.

Even if in the prospect theory direct violations of monotonicity are avoided through the deletion of dominated lotteries in the editing phase, the risk of indirect violation of monotonicity remains (this may rise in the presence of pairwise comparisons [206] or, more trivially, if the decision maker is not sufficiently careful during the editing phase). To avoid violations of monotonicity, the fundamentals of rank-dependent utility theory and prospect theory are combined in a new class of non-linear utility theories, known as *rank- and sign-dependent utility theories*, among which is the *cumulative prospect theory* [210]. Similarly to the rank-dependent utility theory, in the cumulative prospect theory the transformation involves cumulative probabilities instead of the probabilities themselves. More precisely, different decisional weights are computed separately for gains and losses, so that the same level of probability p_i is associated with a different weight according to whether c_i is a gain or a loss. This leads to overweight only for extreme events, rather than all events (i.e., extreme and non-extreme), that occur with small probability. In practice, value (3.2) of a lottery is decomposed into the sum of two components

$$\sum_{i=-h}^{0} \omega^{-}(p_i)u(c_i) + \sum_{i=0}^{k} \omega^{+}(p_i)u(c_i), \tag{3.3}$$

with transformation involving cumulative probabilities in the domain of losses and decumulative probabilities in the domain of gains, that is,

$$\omega^-(p_i) = G\left(\sum_{j=-h}^{i} p_j\right) - G\left(\sum_{j=-h}^{i-1} p_j\right), \quad -h \le i \le 0,$$

$$\omega^+(p_i) = G\left(\sum_{j=i}^{k} p_j\right) - G\left(\sum_{j=i+1}^{k} p_j\right), \quad 0 \le i \le k.$$

4

Utility function elicitation

CONTENTS

4.1 Introduction

In Chapter 3 we discussed the existence and uniqueness, up to linear positive transformations, of a utility function that reflects the preferences of a rational individual. In this chapter, we deal with the practical issue of the utility function elicitation. The expression *utility function elicitation* denotes the process of drawing out the preference structure from a decision maker and maps it into a real values function. Similarly in the Bayesian context, one speaks about *probability function elicitation* to denote the process of formulating individual knowledge and beliefs about uncertain quantities into a probability distribution. These two issues share common aspects and involve similar challenges. In this chapter we focus on the utility function elicitation, while we refer the reader to the literature about Bayesian statistics for probability function elicitation.

We start by introducing the definition of risk aversion, risk seeking, and risk neutrality and we illustrate the relation between the shape of a utility function (i.e., convex, concave, S-shaped, inverted S-shaped) and the attitude towards risk. A measure to assess changes in the attitude towards risk as well as to compare individuals is also described.

Next, the classical elicitation paradigm is illustrated, which relies on complete information for the utility elicitation. In such a setting, two main types of approach are distinguished: the standard gamble methods and the paired gamble methods. In both cases, the decision maker has to compare two lotteries and to express his/her preference. In the standard gamble approach the comparison is made easier because one lottery degenerates in a sure outcome, whereas in the paired gamble approach both lotteries usually consist of two uncertain results. Among the paired gamble methods, the tradeoff one is especially appealing because it is robust to the violation of the expected utility principle. Classical approaches can also be combined in multi-step procedures, which are used under the generalized utility theories when both the utility function and the weighting probability function have to be elicited.

Alternatively to elicitation approaches that are based on complete information, nowadays automated procedures are developing that rely on partial preference information. We first provide details on the Bayesian elicitation approach, which is based on the principle that the utility function may be treated as a random variable having a prior distribution. Then, the concept of expected expected (double expected) utility is illustrated. A further paradigm is also illustrated, based on the idea of clustering in homogenous groups a set of fully elicited utility functions. Then, decision makers are classified in such groups according to just a few observable characteristics.

The remainder of the chapter is devoted to the problem of elicitation when the decision maker is a collective body consisting of a multitude of individuals. In the end, a case study is illustrated concerning the elicitation of the utility function of a banking foundation.

4.2   Attitude towards risk

Knowledge about the decision maker's attitude towards risk (aversion, propensity, indifference) is relevant to choose the optimal action.

Definition 4.1. *Given a lottery l, an individual is risk averse if he/she strictly prefers the expected value of l for sure to participating in the lottery l,*

$$E(l) \succ l.$$

Similarly, an individual is risk seeking if he/she prefers participating in lottery l to receiving the expected value of l for sure,

$$l \succ E(l).$$

Finally, an individual is risk neutral if the indifference relation holds between lottery l and its expected value $E(l)$, that is,

$$l \sim E(l).$$

An important piece of information about the decision maker's attitude towards risk is provided by the shape of his/her utility function. Generally speaking, a utility function may be concave, convex, or linear, as shown in Figure 4.1 .

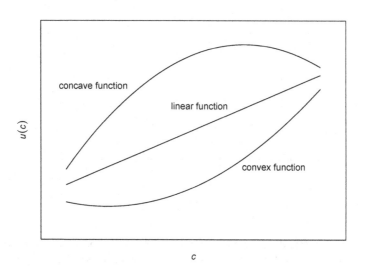

FIGURE 4.1
Shape of utility function: concave, linear, and convex.

Let us consider a lottery $l = \langle c_1 \; p \; c_2 \rangle$, where c_1 and c_2 are considered numerical outcomes[1] and $c_2 > c_1$, and a risk averse decision maker; then relation $E(l) \succ l$ holds. From the linearity property of the expected utility theory (Theorem (3.6)) it turns out that

$$u[pc_1 + (1-p)c_2] > pu(c_1) + (1-p)u(c_2)$$
$$\iff u[E(l)] > EU(l).$$

In other words, the utility of the expected value of l ($u[E(l)]$) is greater than the expected utility of l ($EU(l)$). From a geometrical perspective (Figure 4.2), this means that a decision maker is risk averse if and only if his/her utility function is concave or if the segment connecting any pair of points of $u(\cdot)$ is everywhere below $u(\cdot)$.

---

[1]Note that, in this chapter outcomes $c_1, c_2, \dots$ of lotteries are numerical quantities.

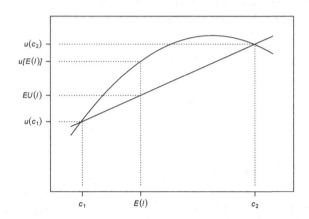

FIGURE 4.2
Concave utility function and risk aversion attitude.

The geometrical interpretation of the curvature of utility function for risk seeking and risk indifference develops along the same lines. In the case of a convex utility function (Figure 4.3), the segment connecting c_1 and c_2 lies everywhere above $u(\cdot)$:

$$u[pc_1 + (1 - p)c_2] < pu(c_1) + (1 - p)u(c_2)$$
$$\Longleftrightarrow \ u[E(l)] < EU(l).$$

This means that, in the presence of a convex utility function, the utility of the expected value of a lottery for sure is smaller than the expected utility of the same lottery: hence, a decision maker is risk seeking if and only if the utility function is convex. Finally, a linear utility function complies with a neutral attitude towards risk, as

$$u[pc_1 + (1 - p)c_2] = pu(c_1) + (1 - p)u(c_2)$$
$$\Longleftrightarrow \ u[E(l)] = EU(l).$$

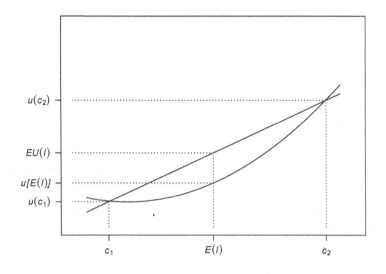

FIGURE 4.3
Convex utility function and risk seeking attitude.

Another interesting piece of information from the graphical representation of a utility function is given by the *risk premium* (*insurance premium*; Figure 4.4). The risk premium is defined as the difference between the expected value of a lottery $E(l)$ and its certainty equivalent c_l^* (defined as in eq. (3.13)),

$$RP(l) = E(l) - c_l^*. \tag{4.1}$$

In the presence of a concave utility function, when negative outcomes are involved as happens in insurance contracts, the risk premium represents how much more the decision maker is willing to pay to relieve himself/herself of the risk of a possible but relevant loss. On the contrary, when positive outcomes are involved as happens in the gamble context, the risk premium represents how much less the decision maker is available to receive to avoid the lottery. In the presence of a linear utility function, the risk premium is null.

The above considerations about the one-to-one relationship between curvature of utility function and attitude towards risk hold under the expected utility theory and derive from the linearity property of probabilities. However, this relationship no longer exists under non-linear utility theories, as was formally proved by Chateauneuf and Cohen [35] and empirically verified by Abdellaoui, Bleichrodt and L'Haridon [2] in the prospect theory setting.

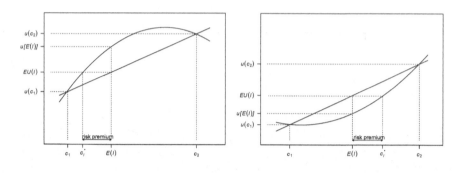

FIGURE 4.4
Risk premium for a concave utility function (left) and for a convex utility function (right).

In particular, when the underweighting of probabilities is strong enough, then a risk aversion attitude may coexist with a linear or a convex utility function. Similarly, overweighting of probabilities may exert an upward impact on the elicited utilities and a concave utility may be observed for a risk seeking individual. Hence, under the prospect theory (and its cumulative generalization) the shape of the utility function is affected by the probability weighting function, whose specific form depends on the attitude towards risk. Several experiments [20, 169, 210] lead to propose an inverse-S shaped probability weighting function, which is coherent with overweighting of small probabilities and underweighting of intermediate and large probabilities.

More generally, in the literature were conducted several experiments aimed at identifying some regularities in the shape of the utility functions. Generally speaking, the most common shape consists of a convex trait for losses followed by a concave trait for gains (convex-concave trend; Figure 4.5, top left panel), but there exists empirical evidence in favor of different trends.

According to some experiments of Friedman and Savage [72], a concave-convex shaped utility function is able to explain the tendency of low income individuals both to pay insurance premiums and, at the same time, to buy lottery tickets. This statement is made clearer by Figure 4.5 (top right panel), where point c_0 denotes the status quo of a decision maker, c_1 the relevant loss which he/she may incur without insurance, and c_2 and c_3 the consistent reward and the small loss associated with a gamble. This type of utility function was generalized by Friedman and Savage [72] to also account for the risk attitude of high income individuals, adding a concave trait after the convex one, such that a concave-convex-concave shaped function results, as is shown in Figure 4.5 (bottom left panel). Such type of utility function is able to accommodate

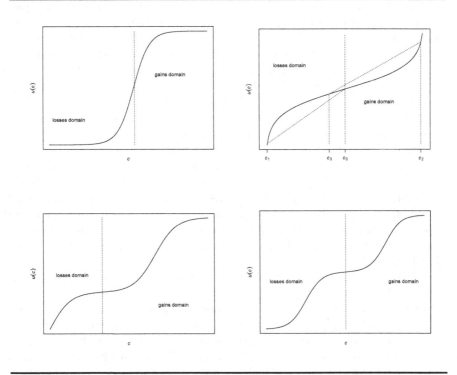

FIGURE 4.5
Examples of mixed shaped utility functions: convex-concave or S shaped (top left), concave-convex or inverse-S shaped (top right), Friedman and Savage's function (bottom left), Markowitz's function (bottom right); vertical dotted line denotes the status quo.

the transition among socio-economic classes. Indeed, an increase of income associated with an uncertain situation that allows an improvement of the relative position within the present social class but does not allow movement towards the next higher class usually generates a decreasing marginal utility (i.e., concave shape). On the other hand, an increase of income that allows access to a superior socio-economic status gives rise to an increasing marginal utility (i.e., convex shape). Similarly, a decision maker whose present position is already high tends to be risk averse to avoid the relegation in an inferior socio-economic class.

The proposal of Friedman and Savage cannot explain some part of the empirical evidence mainly related to medium income decision makers (e.g., their aversion to fair lotteries, having a fifty-fifty probability to win c and to lose $-c$, $c > 0$). As far as this point, Markowitz [152] proposes a four traits utility function centered around the status quo of the decision maker

(Figure 4.5, bottom right panel). In the loss domain, a convex-concave trend explains why individuals tend to prefer to pay a small sure amount to avoid a lottery involving a relevant loss (concave trait), but the risk attitude reverses when the insurance premium gets too high (convex trait). On the contrary, in the gain domain a substantial risk seeking is observed when low rewards are involved (convex trait), followed by risk aversion in the presence of relevant rewards and, then, relevant sure counterparts (concave trait). The resulting utility function (convex-concave-convex-concave) has three inflection points: the intermediate one represents the status quo of the decision maker and the location of the other two depends on the sensitivity of the decision maker to gains and losses. Individuals with a high sensitivity tend to maintain an attitude of risk seeking in the gain domain and of risk aversion in the loss domain: hence, the two inflection points are located far from the status quo. In addition, as usually individuals are more sensitive towards losses than gains, the utility function of Markowitz is steeper in the losses domain rather than in the gain domain, that is $u(c) < |u(-c)|$ ($c > 0$), implying an aversion to fair lotteries.

The central role of the status quo as well as the different slope of the utility function according to the domain are resumed by Kahneman and Tversky [126] in the prospect theory setting, as outlined in Section 3.6.2. As illustrated in Figure 3.6 (right), decision makers tend to be risk averse when positive outcomes are involved, whereas an attitude of risk seeking is usually encountered with negative outcomes. Moreover, the more outcomes involved in the decision that are far from the decision maker's status quo, the more the sensitivity towards gains and losses reduces. Hence, utility function is concave in the gains domain and is convex in the losses domain. Indeed, when one departs from the status quo to the gain domain, individual wealth increases and the utility of wealth increases too, but following a marginally decreasing trend (concave curve). Similarly, when one departs from the status quo to the losses domain, individual wealth reduces and the disutility of wealth reduces too, but again with a marginally decreasing trend (convex curve). Moreover, as losses are usually perceived as more relevant than gains with the same absolute value ($u(c) < |u(-c)|$), the utility trait in the losses domain is steeper than the utility trait in the gains domain. The difference in the slope of utility function may be measured through the ratio [63, 215]

$$R(0) = \frac{u'_+(0)}{u'_-(0)},$$

where $u'_+(0)$ and $u'_-(0)$ denote the first right derivative and the first left derivative of $u(\cdot)$ evaluated in the reference point (i.e., status quo). Ratio $R(0)$ is less than 1 when the curve is steeper in the losses domain than in the gains domain (i.e., $u'_+(0) < u'_-(0)$). For further studies about the relation between the shape of the utility function and the sensitivity that lead to conclusions similar to those of Kahneman and Tversky see also [61] and [1].

4.3 A measure of risk aversion

An useful index to measure the attitude towards risk and to compare individuals is the Arrow-Pratt index [175], which is defined as:

$$r(c) = -\frac{u''(c)}{u'(c)}, \quad \forall c \in \mathcal{C},$$

where $u'(c)$ and $u''(c)$ are the first derivative and the second derivative, respectively, of $U(\cdot)$ computed in c ($c \in \mathcal{C}$).

As $u(\cdot)$ is a non-decreasing function, the first derivative $u'(\cdot)$ is positive throughout and the sign of $r(c)$ is driven by the sign of the second derivative $u''(\cdot)$:

- if $u''(c) < 0$, $u(\cdot)$ is concave (risk aversion attitude) and $r(c) > 0$;

- if $u''(c) > 0$, $u(\cdot)$ is convex (risk seeking attitude) and $r(c) < 0$;

- if $u''(c) = 0$, $u(\cdot)$ is linear (risk neutrality attitude) and $r(c) = 0$.

In other words, the sign of $r(c)$ provides information about the attitude towards risk of the decision maker.

In addition, the Arrow-Pratt index is an appropriate measure of concavity/convexity of a utility function (i.e., $r(c)$ allows us to answer the question "How concave/convex is the utility function?"). Indeed, as a utility function is defined up to a positive linear transformation, neither the comparison between the second derivatives nor the comparison between the graphical displays of the utility functions allow us to satisfactorily assess the different intensity of the risk attitude in different individuals (or the same individual in different time points). On the contrary, as the Arrow-Pratt index $r(\cdot)$ does not change when $u(\cdot)$ is positively scaled, it is appropriate to compare utility functions. Given two concave utility functions U_A and U_B and related Arrow-Pratt indices $r_A(\cdot)$ and $r_B(\cdot)$, U_A is *locally more risk averse* than U_B if and only if $r_A(c) \geq r_B(c)$ and is *globally more risk averse* if relation $r_A(c) \geq r_B(c)$ holds for all $c \in \mathcal{C}$. Equivalently, U_A is globally more risk averse than U_B if and only if $u_A(c) = f(u_B(c))$ for all $c \in \mathcal{C}$, where $f(\cdot)$ is a concave function. In terms of risk premium, it turns out that $r_A(c) \geq r_B(c)$ implies that $RP_A \geq RP_B$, where RP is the risk premium defined in eq. (4.1). Hence, if individual A is uniformly more risk averse than individual B, the risk premium associated with A, RP_A, is greater than the risk premium associated with B, RP_B.

Finally, the Arrow-Pratt index $r(\cdot)$ is specially useful to assess how the decision maker's attitude changes when his/her status quo modifies. In more detail, a utility function is said to be *decreasingly risk averse* if and only if index $r(c)$ is non-increasing in c. Equivalently, one says that a decision maker is decreasingly risk averse if the risk premium of any lottery is non-increasing in c (or the certainty equivalent is non-decreasing): in other words, the risk aversion of an individual tends to decrease when wealth increases. As outlined in the previous section (Section 4.2), several experiments lead to empirically verify that the sensitivity to gains tends to reduce when wealth increases and, hence, individuals tend to be decreasingly risk averse [43]. It is worth noting also that when the amount of money involved in a lottery is small compared to the status quo of an individual, a constant risk aversion is often empirically observed (see, among others, [208] and [73]). Formally, a utility function is said to be *constantly risk averse* if and only if $r(c)$ is constant in c. For more details see also Parmigiani and Inoue [162] and Pratt, Raiffa, and Schlaifer [176] (Appendix 3).

4.4   Classical elicitation paradigm

One of the more challenging aspects connected with the practical use of the utility theory concerns the elicitation of the utility function and the elicitation of the subjective probability distribution of the states of nature. As above outlined, problems related to elicitation of utility function and elicitation of probabilities are similar; therefore in what follows we will explicitly deal with the elicitation of utility function and refer the reader to the wide literature about subjective Bayesian statistics for the elicitation of probabilities.

A first approach to elicitation aims to achieve a complete description of preferences of the decision maker. For this aim, an experimental procedure is usually implemented based on subsequent implicit or direct queries. Implicit queries consist of observing the individual's behavior in a supervised environment (e.g., time spent on a web page and links followed, patterns and strategies adopted to solve a problem) so that preference relations can be revealed by observed choices and actions. Instead, with direct queries the decision maker is faced with an explicit choice.

Elicitation of utility function through a *complete preference information* approach presents several shortcomings. First, queries involving numbers and probabilities are hard to answer by individuals who do not have a basic background in statistics and probability. In practice, the implementation of these elicitation approaches requires a preliminary training and the presence of an expert (e.g., the analyst or the researcher that is interested in the study) that helps in answering the queries, translating qualitative judgements in quantities. Second, framing effects and difficulty in perceiving very small and very high probabilities may induce biased answers. The general conclusion is that the assessed utility function may not reflect the decision maker's true preferences. Third, assessing utility function through the empirical determination of some (usually few) points is useless in complex decisional problems that require utility functions expressed in analytical form. Recently, approaches based on partial information about preference relations found a remarkable development (for details see Section 4.6). The main advantage of these approaches is that they are easier to implement and, hence, are especially appropriate for adaptive elicitation processes, mainly when, in the frame of alternative utility theories, the elicitation of the utility function as well as of the probability weighting function are involved.

Despite the above cited shortcomings, complete preference elicitation is commonly used in the practice of decision making and it represents the basis for more complex approaches. Elicitation methods are characterized by the nature of queries proposed to the decision maker. The simplest case consists of order comparisons between pairs of alternatives: the individual is asked whether he/she prefers c_i to c_j or c_j to c_i or is indifferent. A variant consists of asking a preference ranking for a set of alternatives $c_1, c_2, \ldots, c_k$. Among methods based on order comparisons, there are the *analog (or direct) scaling* approach and the *magnitude estimation* approach. In both cases individuals compare a set of outcomes with a standard one that is definitely worse (analog scaling) or better (magnitude estimation). To facilitate the response process a graphical support may also be used, where the standard value is located along a straight line continuum and the decision maker places each outcome on the left or on the right of the standard one. The distance between the evaluated outcome and the standard denotes "how much" the former is worse or better than the latter one. Such types of query are really simple, but they are also not very informative.

Alternative direct elicitation approaches are the gamble methods, based on queries involving lotteries, usually with two outcomes. Gamble elicitation methods are divided into two main classes [60]:

- *standard gamble methods* require a comparison between a sure outcome and a lottery

$$c_i \ ? \ \langle c_{ij} \ p \ c_{ih} \rangle,$$

with ? standing for $\succ$, $\sim$, or $\not\succ$ and values of c_i, c_{ij}, c_{ih}, p given;

- *paired gamble methods* require a comparison between two lotteries

$$\langle c_{ij} \ p \ c_{ih} \rangle \ ? \ \langle c_{il} \ q \ c_{im} \rangle, \tag{4.2}$$

with ? standing for $\succ$, $\sim$, or $\not\succ$ and the other elements given.

In both cases, one or two reference outcomes are usually identified with arbitrary utilities: for instance, the smallest outcome c_1 is associated to a null utility $(u(c_1) = 0)$ and the highest outcome c_k $(c_k > c_1)$ is associated with a utility equal to 1 $(u(c_k) = 1)$. Then, a utility in the range $(0, 1)$ is assigned to the remaining outcomes $c_2, c_3, \ldots, c_{k-1}$, according to individuals' answers. For instance, let $u(500) = 0$ and $u(1000) = 1$ utilities assigned to \$500 and \$1000, respectively; then the utility assigned to \$600 is surely in the range $(0, 1)$. To determine the exact value of the utility for a certain individual, he/she may be asked to detect the value of p that makes him/her indifferent between \$600 for sure and participating in a lottery that pays \$500 with probability p and \$1000 with probability $1 - p$

$$600 \sim \langle 500 \; p \; 1000 \rangle.$$

On the basis of the expected utility principle, it turns out that

$$u(600) = pu(500) + (1 - p)u(1000) \implies u(600) = p \cdot 0 + (1 - p) \cdot 1 = 1 - p.$$

The procedure is repeated a number of times so that, in the end, several points with related utilities are available and can be plotted on a Cartesian axes system. These points can be interpolated (e.g., using spline functions) to obtain the individual's utility function.

Generally, the decision problem involves non-monetary outcomes: in such cases it is preliminarily necessary to formulate qualitative outcomes in quantitative terms, using suitable measurement units. For instance, one can adopt the number of years of life in good health or the health costs, in the health setting; the expected number of visitors for an exhibition or the number and the assessed value of exhibited works, in the cultural heritage setting; the occupational rate a year after graduation, in the educational field; the number of published papers, citations, and patents, in the scientific research setting. In the business field, decisions of management should be driven, other than by monetary returns, also by the ecological, social, human, and political impact of the operative choices.

4.4.1 Standard gamble methods

In what follows the most common standard gamble methods are briefly illustrated.

Preference comparisons method The decision maker is asked about a sequence of n preference relations (i.e., $\succ$, $\sim$, or $\not\succ$) between a lottery $l_i = \langle c_{ij} \; p \; c_{ih} \rangle$ and a sure outcome c_i,

$$c_i \; ? \; \langle c_{ij} \; p \; c_{ih} \rangle, \quad i = 1, \dots, k, \tag{4.3}$$

with c_{ij}, c_{ih}, c_i, p given. Alternatively, an iterative procedure is implemented, adjusting one or more terms of the lottery after each comparison. The process stops when an indifference relation is obtained. This approach is usually adopted to preliminarily investigate an individual's risk attitude and to evaluate the coherence with utility functions elicited through other methods.

Lottery equivalent (LE) method In the LE method, values of $c_i = c_1$, $c_{ih} = c_0$ and p in Eq. (4.3) are given and the decision maker is asked about the value of $c_{ij} = c_2 = ?$, with $c_0 < c_1 < c_2$, that makes him/her indifferent with respect to receiving c_1 for sure. Then, the comparison is

$$c_1 \sim \langle c_2 \; p \; c_0 \rangle, \quad \text{with } c_2 =?.$$

From the expected utility principle, the following relation results

$$u(c_1) = pu(c_2) + (1 - p)u(c_0).$$

As c_0 is the worst outcome, one may fix $u(c_0) = 0$; moreover, an arbitrary positive value may be assigned to the utility of the sure outcome, as $u(c_1) = u^*$. Hence,

$$u^* = pu(c_2) \implies u(c_2) = \frac{1}{p}u^*.$$

In an iterative procedure, the second step consists of choosing c_3, with $c_1 < c_2 < c_3$, such that the following indifference relation holds

$$c_2 \sim \langle c_3 \; p \; c_1 \rangle, \quad \text{with } c_3 =?,$$

from which

$$u(c_2) = pu(c_3) + (1 - p)u(c_1) \implies \frac{1}{p}u^* = pu(c_3) + (1 - p)u^*$$

$$\implies u(c_3) = \frac{1 - p + p^2}{p^2}u^*.$$

In general, the utility of the $(i + 1)$-st value elicited by the decision maker is a function of u^* and p and is obtained through the comparison

$$c_i \sim \langle c_{i+1} \; p \; c_{i-1} \rangle, \quad \text{with } c_{i+1} =? \text{ and } i = 1, \ldots, k.$$

As values of u^* and p are arbitrary, a common choice consists of $u^* = 1/k$ and $p = 0.5$. In such a case, it is immediately verified that

$$u(c_0)=0, u(c_1)=1/k, u(c_2) = 2/k, u(c_3) = 3/k, \ldots, u(c_i) = i/k, \ldots, u(c_k) = 1.$$

This variant of the LE method is named the *equisection method*, as it causes a partition of the utility scale in segments of the same length.

Similar to the other standard gamble methods, the LE approach suffers for some shortcomings. First, the sequentiality of the decision maker's choices (i.e., the value of c_i depends on the value assigned to c_{i-1}) may induce the propagation of initial response errors. Second, in practice a certain tendency is observed to maintain constant the difference between consecutive values (i.e., $c_i - c_{i-1} = K$, with K constant value), which implies an overestimation of linear utility functions. Moreover, no correction for the certainty effect (Section 3.5.2) is provided: empirical evidence from the LE method outlines an overestimation of risk aversion under the gains domain and of risk seeking under the losses domain. Finally, the LE method remains valid only under the expected utility theory and it is not robust to non-linear weighting of probabilities.

Certainty equivalent (CE) method The CE method is by far the most commonly used elicitation approach because of its simplicity. Given lottery $l_i = \langle c_{ij}\ p\ c_{ih} \rangle$, the decision maker has to elicit the certainty equivalent, that is, the outcome c_i that makes him/her indifferent between receiving c_i for sure and participating in the lottery l_i. There exist two main variants of this approach: the *fractile CE method* and the *chaining CE method*.

In the fractile CE method, outcomes of lottery l_i are fixed ($c_{ij} = c_k$ and $c_{ih} = 0$ for all $i = 1, \ldots, k$, with $c_0 < c_k$) and a sequence of probabilities $p_1, p_2, \ldots, p_i, \ldots, p_k$ is provided. The decision maker chooses c_i ($c_0 < c_i < c_k$) so that

$$c_i \sim \langle c_k\ p_i\ c_0 \rangle, \quad \text{with } c_i = ?;\ i = 1, \ldots, k.$$

According to the expected utility principle, when $u(c_0) = 0$ and $u(c_k) = 1$ it turns out that

$$u(c_i) = p_i.$$

The chaining CE method represents the chained variant of the fractile CE approach. After having elicited the certainty equivalent for each lottery $\langle c_k\ p_i\ c_0 \rangle$ ($i = 1, \ldots, k$), the decision maker chooses the certainty equivalents of the lotteries

$$c_2 \sim \langle c_k\ p_i\ c_1 \rangle$$
$$c_3 \sim \langle c_1\ p_i\ c_2 \rangle$$
$$\vdots \quad \vdots$$
$$c_h \sim \langle c_{h-1}\ p_i\ c_0 \rangle$$
$$\cdots \quad \cdots$$

Both fractile and chaining CE methods suffer for some bias when values of p_i are close to 0 and 1. To remedy this problem, the *midpoint chaining CE method* is usually implemented, where only one probability $p_i = 0.50$ is used. Similar to what was described above, the first comparison leads to eliciting c_1 ($c_0 < c_1 < c_k$) so that

$$c_1 \sim \langle c_k\ 0.5\ c_0 \rangle,$$

which implies $u(c_1) = 0.5$. The subsequent comparisons lead to eliciting $c_2, c_3, \ldots$ such that

$$c_2 \sim \langle c_1\ 0.5\ c_0 \rangle \text{ with } (c_0 < c_2 < c_1) \implies u(c_2) = 0.5 \cdot 0.5 = 0.25,$$
$$c_3 \sim \langle c_k\ 0.5\ c_1 \rangle \text{ with } (c_1 < c_3 < c_k) \implies u(c_3) = 0.5 + 0.5 \cdot 0.5 = 0.75,$$
$$c_4 \sim \langle c_2\ 0.5\ c_0 \rangle \text{ with } (c_0 < c_4 < c_2) \implies u(c_4) = 0.5 \cdot 0.25 = 0.125,$$
$$c_5 \sim \langle c_1\ 0.5\ c_2 \rangle \text{ with } (c_2 < c_5 < c_1) \implies u(c_5) = 0.5 \cdot 0.5 + 0.5 \cdot 0.25$$
$$= 0.375,$$

$$\cdots \quad \cdots \quad \cdots$$

In practice, the midpoint of the interval between values $u(c_{i-1})$ and $u(c_i)$ is detected and the corresponding certainty equivalent is elicited. The total number of steps depend on k. For $k = 4$, the iterative process requires three steps ($u(c_0) = 0$ and $u(c_4) = 1$):

$$c_1 \sim \langle c_4 \; 0.5 \; c_0 \rangle \text{ with } (c_0 < c_1 < c_4) \implies u(c_1) = 0.5,$$
$$c_2 \sim \langle c_1 \; 0.5 \; c_0 \rangle \text{ with } (c_0 < c_2 < c_1) \implies u(c_2) = 0.25,$$
$$c_3 \sim \langle c_4 \; 0.5 \; c_1 \rangle \text{ with } (c_1 < c_3 < c_4) \implies u(c_3) = 0.75.$$

In addition to the shortcomings above described with reference to the LE method and that hold also for the CE method, it is here worth noting that the two variants, the fractile and the midpoint chaining ones, tend to provide different results.

Probability equivalent (PE) method The PE method requires that the decision maker elicits a probability p_i such that he/she is indifferent between a sure outcome and a lottery. Similarly to the CE method, two variants of the PE method may be adopted.

Under the *extreme lotteries approach*, a sequence of certainty equivalents $c_1, c_2, \ldots, c_k$ is a priori defined and the decision maker is asked about corresponding values of p_i such that

$$c_i \sim \langle c_k \; p_i \; c_0 \rangle, \quad \text{with } p_i = ?; \; i = 1, \ldots, k,$$

and given c_0 and c_k. As usual under the expected utility theory, if $u(c_0) = 0$ and $u(c_k) = 1$, then $u(c_i) = p_i$. Under this approach subsequent responses are not chained and, then, the propagation of initial errors is not a problem at all. However, a possible shortcoming is represented by the size of interval (c_0, c_k). Indeed, the decision maker may encounter difficulties in eliciting the probability level, when the distance between the two extremes of the interval is too wide.

On the other hand, under the *adjacent lotteries approach* chained sequences of lotteries are used, where the extreme values c_0 and c_k are substituted with adjacent certainty equivalents, that is,

$$c_i \sim \langle c_{i+1} \; p_i \; c_{i-1} \rangle, \quad \text{with } p_i = ?; \; i = 1, \ldots, k.$$

Therefore, at step $i = 1$ the decision maker elicits p_1 that makes him/her indifferent between

$$c_1 \sim \langle c_2 \; p_1 \; c_0 \rangle,$$

for given c_0, c_1, c_2. At step $i = 2$, p_2 is elicited so that

$$c_2 \sim \langle c_3 \; p_2 \; c_1 \rangle,$$

and so on.

The PE method suffers from the same problems as the other standard gamble methods. In addition, it is cognitively demanding as individuals without a statistical background have difficulties in eliciting probabilities. To partially solve this problem, some precautions may be adopted [71]: for instance, instead of formulating a numerical value for p_i, the decision maker may be asked about shading a circle sector in such a way that the size of the corresponding angle reflects the value of p_i.

In the literature there exist several experimental studies [63, 104, 105, 214] that provide extensive evidence of systematic discrepancies between utility functions elicited through the standard gamble methods, mainly CE and PE. These studies confirm that the elicitation method affects the utility function, favoring some specific psychological attitudes and response strategies. In addition to the presence of random errors in the response process that lead to a certain variability in the elicited values, the major difficulties encountered in answering queries of the PE method favor a sort of "reconstruction" of queries from individuals. Lotteries involving only gains or only losses are thought of as mixed lotteries with outcomes coded in terms of gains and losses with respect to the status quo. For instance, a comparison between a certain \$100 and lottery $\langle \$200\ p_i\ \$0 \rangle$ is implicitly recoded as a comparison between a certain \$0 (i.e., no gain and no loss) and participating in the lottery $\langle \$100\ p_i\ -\$100 \rangle$. The main consequence is that under the PE approach individuals tend to be more risk seeking in the losses domain and more risk adverse in the gains domain than under the CE approach. In addition, complex experiments based on alternating PE and CE methods outline a certain tendency to change the attitude towards risk (from risk seeking to risk adverse, or vice-versa) according to whether the CE method is followed by the PE method or the PE method is followed by the CE one.

4.4.2   Paired gamble methods

To overcome the drawbacks that characterize the standard gamble methods, mainly the certainty effect, the comparison between a sure outcome and a lottery may be substituted by a comparison between two (non-degenerated) lotteries, as happens in the paired gamble methods. Similar to standard gamble methods, there exist several variants of paired gamble methods depending on elements of Eq. (4.2) that are given and those that have to be elicited by the decision maker.

Preference comparisons method The decision maker elicits a sequence of preference relations (i.e., $\succ$, $\sim$, or $\nsucc$) between pairs of lotteries, as in Eq. (4.2):

$$\langle c_{ij}\ 0.50\ c_{ih} \rangle\ ?\ \langle c_{il}\ 0.50\ c_{im} \rangle,$$

with c_{ij}, c_{ih}, c_{il}, and c_{im} known and $p = q = 0.5$ to simplify the elicitation process.

Probability equivalent (paired) method The decision maker has to choose the value of p that makes him/her indifferent between lottery $\langle c_{ij} \; p \; c_{ih} \rangle$ and lottery $\langle c_{il} \; q \; c_{im} \rangle$, with extremes of lotteries known and q usually set equal to 0.50 or equal to p. A simplified version of this approach [156] consists of proposing a sequence of n comparisons with $c_{ih} = c_{im} = 0$, $c_{ij} = R$, with R the best plausible outcome, and q fixed (e.g., $q = 0.50$ or $q = 0.75$), for $i = 1, \ldots, k$:

$$\langle R \; p_i \; 0 \rangle \sim \langle c_i \; q \; 0 \rangle.$$

Note that, as R is the best plausible outcome, c_i is always set by the analyst such that $c_i < R$ and the decision maker has to elicit a value for p_i less than q, to avoid that lottery $\langle R \; p_i \; 0 \rangle$ stochastically dominates lottery $\langle c_i \; q \; 0 \rangle$. Given these constraints, the utility of c_i under the expected utility theory is given by

$$p_i u(R) + (1 - p_i)u(0) = qu(c_i) + (1 - q)u(0)$$
$$\implies p_i = qu(c_i)$$
$$\implies u(c_i) = p_i/q,$$

where, without loss of generality, $u(R) = 1$ and $u(0) = 0$.

Lottery equivalent (paired) method - Tradeoff method Differently from the probability equivalent approach, probabilities p and q of lotteries as well as three out of four outcomes $c_{ij}, c_{ih}, c_{il}, c_{im}$ are chosen by an analyst and the decision maker has to elicit the remaining outcome. The most known lottery equivalent paired approach is represented by the *tradeoff (TO) method*, which was introduced by Wakker and Deneffe in 1996 [214] and extended by Fennema and Van Assen in 1999 [61].

The TO method is based on a sequence of comparisons between chained pairs of lotteries. The starting comparison is

$$\langle c_1 \; p \; R \rangle \sim \langle c_0 \; p \; r \rangle, \quad \text{with } c_1 = ?,$$

where p, c_0, R, and r ($r < R$) are freely set by the analyst and c_1 is elicited by the decision maker so that an indifference relation holds between the two lotteries. We recommend that values for R and r are close enough, such that the sequence of elicited values $c_1, \ldots, c_k$ is not too wide and the related utilities are sufficiently accurate. From this first comparison no direct measurement of utility can be elicited; however the following difference results

$$pu(c_1) + (1 - p)u(R) = pu(c_0) + (1 - p)u(r)$$
$$p[u(c_1) - u(c_0)] = (1 - p)[u(R) - u(r)].$$

The second comparison results from the previous one by substituting c_1 to c_0, all the other elements being constant:

$$\langle c_2 \, p \, R \rangle \sim \langle c_1 \, p \, r \rangle, \quad \text{with } c_2 =?,$$

from which the following difference results:

$$p[u(c_2) - u(c_1)] = (1 - p)[u(R) - u(r)].$$

Hence, it turns out that

$$p[u(c_2) - u(c_1)] = p[u(c_1) - u(c_0)] \implies u(c_2) = 2u(c_1).$$

In general, at the i-th step of the procedure we obtain

$$u(c_i) = iu(c_1),$$

where $u(c_1)$ is arbitrarily chosen. Note that the shape of the utility function is immediately inferred by the sequence of differences between pairs of consecutive elicited values. Hence, a linear utility function results if

$$c_1 - c_0 = c_2 - c_1 = c_3 - c_2 = \ldots = c_k - c_{k-1};$$

a concave utility function results if

$$c_1 - c_0 < c_2 - c_1 < c_3 - c_2 < \ldots < c_k - c_{k-1};$$

and a convex utility function is obtained by monotonic increasing differences

$$c_1 - c_0 > c_2 - c_1 > c_3 - c_2 > \ldots > c_k - c_{k-1}.$$

Mixed shapes are possible whenever subsequent differences are partly increasing and partly decreasing.

There exist two variants of the TO method: the outward TO and the inward TO. In the *outward TO method* outcomes elicited by the decision maker increase in their absolute value, without any implicit superior limit. For the sake of clarity, let us consider a lottery where the decision maker may lose \$$c_1$ with probability p and gain \$2000 otherwise versus a lottery where one may lose \$25 with probability p and gain \$1500 otherwise:

$$l_1 = \langle c_1 \, p \, 2000 \rangle \sim l_2 = \langle -25 \, p \, 1500 \rangle.$$

The decision maker has to elicit the value of c_1 that makes the above indifference relation true. Any loss greater than \$25 is eligible, to avoid a lottery dominating the other one. Let us assume that the decision maker chooses $c_1 = -\$60$. Then, the subsequent comparison is

$$l_1 = \langle c_2 \, p \, 2000 \rangle \sim l_2 = \langle -60 \, p \, 1500 \rangle. \tag{4.4}$$

Again, the absolute value of c_2 must be greater than 60, say $c_2 = -\$150$. Then, the comparison follows as

$$l_1 = \langle c_3\ p\ 2000 \rangle \sim l_2 = \langle -150\ p\ 1500 \rangle.$$

The procedure is repeated as many times as one wants: in the end, a sequence of monotonic increasing outcomes $c_1, c_2, c_3, \ldots, c_k$ is elicited with related utilities. The reasoning is the same if positive (rather than negative) outcomes are involved.

On the other hand, in the *inward TO method* queries are formulated such that outcomes are monotonic decreasing (in absolute value). Taking the above example, we can start with asking the decision maker about c_1 such that

$$l_1 = \langle -150\ p\ 2000 \rangle \sim l_2 = \langle c_1\ p\ 1500 \rangle. \tag{4.5}$$

Now, c_1 must be smaller (in absolute value) than \$150 to compensate the smaller gain (\$1500 vs. \$2000) of lottery l_2 with respect to lottery l_1. Let us assume that $c_1 = -110$. Then, the subsequent comparison is

$$l_1 = \langle -110\ p\ 2000 \rangle \sim l_2 = \langle c_2\ p\ 1500 \rangle.$$

The decision maker must now provide a value of c_2 smaller (in absolute value) than 110. In general, a sequence of decreasing outcomes is elicited that approximate to \$0 (the natural inferior limit).

It is worth noting that outward and inward variants of the TO method do not necessarily provide the same utility function. In theory, indifference relations under both methods should be satisfied for the same elicited values. For instance, if under outward TO eq. (4.4) is satisfied for $c_2 = -\$150$, one expects that under inward TO eq. (4.5) holds for $c_1 = -\$60$. However, empirical evidence [61] often provides conflicting elicited values (e.g., in the above example $c_1 = -\$110 \neq -\$60$). More generally, utility functions elicited through the inward TO approach tend to bend more sharply than utility functions elicited through the outward TO. Again, the diminishing sensitivity and the framing effect provide a plausible explanation for this phenomenon. In the above example, under the outward TO an increase of 33% of rewards (from \$1500 to \$2000) is compensated by an increase of 150% of losses (from \$60 to \$150); under the inward TO the perspective of the decision maker changes: a decrease of 25% of rewards (from \$2000 to \$1500) is compensated by a decrease of 27% of losses (from \$150 to \$110).

In the above discussion nothing has been said about the value of p. Indeed, probability p has no specific role in the application of the TO method and this makes particularly appealing this elicitation approach. One can substitute p with any value as well as with non-linear transformations $\omega(p)$: the final value of utility is always $u(c_i) = iu(c_1)$. Therefore, the TO method remains valid when probabilities are non-linearly weighted (as in the prospect theory) or even when they are unknown, so that it is applicable both to decisions under risk and under uncertainty.

To summarize, the TO method is the only non-parametric elicitation method that is robust to the non-linear probability weighting and, then, to the violation of the expected utility theory. It does not require any specific knowledge in probability and it also does not suffer from the certainty effect. On the other hand, it is a little cognitively demanding as it involves multiple comparisons between pairs of lotteries. In addition, discrepancies between the outward and the inward variants are empirically observed.

Some experimental studies (e.g.,[214]) outline how utility functions elicited through the standard gamble CE and PE approaches have a concavity systematically sharper with respect to the TO method, because of the certainty effect. As far as this point, Bleichrodt, Pinto and Wakker [21] show how utilities elicited under the expected utility theory through CE and PE approaches may be quantitatively corrected to account for the prospect theory assumptions (i.e., non-linear probability weighting and losses aversion): differences between CE- and PE- corrected utilities and TO utilities are definitely small and not statistically significant.

4.4.3   Other classical elicitation methods

In the health setting, some variants of the above described methods are commonly used: the *person tradeoff* and the *time tradeoff* method. The person tradeoff method is applied when decisions with a social impact on groups of individuals are involved, mainly in the presence of limited financial resources. The elicitation process relies on queries of the type: "How many people suffering from illness xx have to be treated to obtain a social benefit equal to treating 100 people suffering from illness yy?".

Differently, the time tradeoff is used for decisions involving single individuals facing complex choices regarding their own health. In such a context queries are addressed to the ill individual to elicit his/her utility function and, consequently, to detect the more appropriate health treatment. As in a standard gamble method, where the individual is placed in front of the choice between playing a lottery or an intermediate outcome for sure, in the time tradeoff method the patient is asked the number of years c that make him/her indifferent between living for t years in ill health and living for $c < t$ years in good health. For instance, the patient may be placed in front of the choice between no treatment and a therapy, with no treatment implying a number of t years of life with pain (i.e., poor health), whereas the therapy is associated with uncertain results, such as no effect (i.e., t years of poor health), no pain (i.e., c years of good health), death after the therapy. Alternatively, the patient may be asked to choose between two alternative therapies, providing intermediate states of health between good health and death: therapy 1 associated with t years with a quality of life of level q_1 and therapy 2 associated with c years with a quality of life of level q_2, with $q_2 > q_1$ and $c < t$. In such a situation, the patient incurs a loss of health in the time range $[0, t]$ equal to $t[1 - u(q_1)]$, if he/she chooses therapy 1, and equal to $c[1 - u(q_2)]$ if he/she chooses therapy 2, where 1 is the utility of good health, $u(q_1)$ the utility of q_1, and $u(q_2)$ is the utility of q_2. Patient has to elicit c until his/her attitude towards the two therapies is neutral. At this point, the utility of therapy 2 is obtained as a function of the utility of therapy 1:

$$t[1 - u(q_1)] = c[1 - u(q_2)] \implies u(q_1) = 1 - \frac{c}{t}[1 - u(q_2)],$$

where c/t is an adjustment factor denoting the individual tradeoff between length of life and quality of life.

It can be shown [218, 162] that the time tradeoff method and the certainty equivalent standard gamble method are equivalent (i.e., provide the same utility function) when three conditions are satisfied. First, length of life and quality of life are independent. Second, the tradeoff between length of life and quality of life is proportional; that is, if a patient is indifferent between c years in good health and t years in poor health he/she is also indifferent between kc years in good health and kt years in good health, for $k > 0$. Third, the marginal utility of years of life is constant; that is, the utility of living next year is the same as the utility of living each subsequent year.

4.5   Multi-step approaches

Classical elicitation methods described above, with the only exception being the TO approach, are valid only under the expected utility theory and are not robust when individuals distort or misperceive the probabilities. More generally, we need methods that allow us to simultaneously elicit both the utilities of a set of outcomes and the corresponding probability weights.

One of the main contributions to the simultaneous elicitation of utility function and probability weighting function is due to Gonzales and Wu [84] who propose an application of the CE standard gamble method through a multi-step approach.

First, the CE method is used to elicit a sequence of certainty equivalents, denoted as c_{ijh},

$$c_{ijh} \sim \langle c_i \, p_h \, c_j \rangle, \quad c_i < c_j, i, j = 1, \ldots, k, h = 1, \ldots, l,$$

with outcomes c_i and c_j and probabilities p_h given.

Second, an iterative non-parametric algorithm is implemented to estimate utilities $u(\cdot)$ and weighted probabilities $\omega(\cdot)$. A model is formulated to explain the utility of the certainty equivalents as a function of the expected utility of the lotteries, as in eq. (3.2),

$$u(c_{ijh}) = \omega(p_h)u(c_i) + [1 - \omega(p_h)]u(c_j),$$

with $u(\cdot)$ and $\omega(\cdot)$ unknown parameters.

Using an alternating least squares approach, at each iteration either the weights $\omega(p_h)$ are held fixed and utilities $u(c_i)$ and $u(c_j)$ are estimated, or the utilities $u(c_i)$ and $u(c_j)$ are held fixed and weights $\omega(p_h)$ are estimated. Utilities of certainty equivalents $u(c_{ijh})$ are obtained through a linear interpolation of estimates of $u(c_i)$ and $u(c_j)$.

Bleichrodt and Pinto [20] suggest a two-step approach, which can be applied under the rank-dependent utility theory and is based on the TO method and on the linear interpolation. First, a standard sequence of outcomes $c_1, \ldots, c_k$ is elicited through the TO method. Second, an outcome c^* is elicited such that an indifference relation holds between two lotteries

$$\langle c_i \, p \, c_j \rangle \quad \text{and} \quad \langle c_h \, p \, c^* \rangle,$$

with $c_h \geq c_i \geq c_j$ and $p \leq 0.5$, or between

$$\langle c_i \, p \, c_j \rangle \quad \text{and} \quad \langle c^* \, p \, c_l \rangle,$$

with $c_i \geq c_j \geq c_l$ and $p > 0.5$; $i, j, l = 1, \ldots, k$. The following deterministic relations hold between probability weights $\omega(p)$ and utilities $u(c^*)$ (see [20] for details)

$$\omega(p) = \frac{u(c_j) - u(c^*)}{[u(c_j) - u(c^*)] + [u(c_l) - u(c_i)]} \quad \text{if } c_l \geq c_i \geq c_j$$

and

$$\omega(p) = \frac{u(c_j) - u(c_l)}{[u(c^*) - u(c_i)] + [u(c_j) - u(c_l)]} \quad \text{if } c_i \geq c_j \geq c_l.$$

Then, given $u(c^*)$ probability weights are assessed without any further issue. Instead, as concerns $u(c^*)$, as c^* is not an element of the standard sequence $c_1, \ldots, c_k$, but $c_1 < c^* < c_k$, utilities $u(c^*)$ may be computed through a linear interpolation of adjacent values of the standard sequence c_i and c_{i+1}, with $i = 1, \ldots, k-1$ and $c_i < c^* < c_{i+1}$.

Different from the Bleichrodt and Pinto approach, the two-step proposal of Abdellaoui [1] consists of (*i*) applying the TO method to elicit a standard sequence of outcomes $c_1, \ldots, c_k$, followed by (*ii*) the PE standard gamble method to obtain a standard sequence of probabilities $p_1, \ldots, p_{k-1}$. In more detail, in the second step the sequence $c_1, \ldots, c_k$ is used as an input for the PE approach, such that the individual has to elicit that value of p_i that makes him/her indifferent between lottery $\langle c_k \, p_i \, c_0 \rangle$ and outcome c_i for sure, with $c_0 < c_i < c_k$ and $i = 1, \ldots, k-1$. As a shortcoming of this approach, elicited values may be biased because of the change in the measurement scale from the TO method, which refers to monetary or other types of outcome, to the PE method, which refers to probabilities.

Some variants of the above approach based on three-step procedures may be implemented that aim to improve the utility values elicited at the first step. For instance, a possibility [19] consists of a suitable combination of TO,

CE and PE methods. First, a TO approach is applied in accordance with the two-step proposal of Abdellaoui [1] to elicit the weights of a certain number of probabilities. Second, these weighted probabilities are used as an input in the chaining CE method to assess the utility function. Third, outcomes elicited through the CE method are used as an input in the PE method to improve the assessment of the individual's probability weighting function. Another example of a three-step procedure [3] starts with eliciting a probability whose weight is 0.5 ($\omega(p) = 0.5$) in accordance with the two-step proposal of Abdellaoui [1]. Then, a third step is added, where the midpoint chaining CE standard gamble method is applied with probability $p = \omega^{-1}(0.5)$ (instead of $p = 0.5$) to assess again the utility values. Note that in both the mentioned three-step approaches the CE method does not suffer from the non-robustness of the elicited utilities, because it takes weighted probabilities as input.

4.6 Partial preference information paradigm

A classical paradigm for utility elicitation aims at attaining complete information about the decision maker's preferences. However cognitive, computational, and financial costs often make the elicitation process demanding. Many decision makers are not experts so they find queries involving numbers and probabilities cognitively hard to answer without preliminary training. Moreover, the classical paradigm presents bottlenecks to designing automated decisions (i.e., decisions supported by automated systems): individuals' preferences are unknown and vary from individual to individual and, within the same individual, across time. In practice, complete elicitation of the utility function is generally a burden. An alternative paradigm to elicit the utility function is based on *partial information* about preferences, as happens with the *Bayesian elicitation* approach.

From a Bayesian perspective, if the utility function is not fully known it is treated as a random variable drawn from a prior distribution. So, even if the decision maker's preferences are not fully known, probabilistic information about them is available and enters into the decisional process.

Under the complete preference information paradigm, each action has a certain utility, given by the expected utility of its outcomes, where the expected value is computed with respect to the states of nature θ_j, $j = 1, \ldots, k$ (see also Theorem 3.6):

$$u(a_i) = EU(a_i) = E_\theta[u(a_i, \theta_j)] = \sum_{j=1}^{k} u(a_i, \theta_j)\pi(\theta_j) = \sum_{j=1}^{k} u(c_{ij})\pi(\theta_j).$$

As outlined in Chapter 3, the optimal decision is the one that maximizes $u(a_i)$, as in eq. (3.1).

Under the partial preference information paradigm, $u(a_i)$ is not fully known, but a prior probability distribution $\pi(u)$ is available, which is defined over the space of all possible utility functions. As usual in the Bayesian framework, the choice of the prior probability distribution may be problematic. We suggest a preferance for probabilistic models that are computationally manageable and flexible enough to adapt to any type of prior belief, such as mixtures of normal or Beta distributions.

In such a setting, the optimal decision is the one that maximizes $E_U[u(a_i)]$, that is, the expected utility of outcomes given a prior probability distribution, defined as

$$
\begin{aligned}
E_U[u(a_i)] &= E_U E_\theta[u(a_i, \theta_j)] \\
&= \int_{u \in U} \pi(u) E_\theta[u(a_i, \theta_j)] du \\
&= \int_{u \in U} \pi(u) \sum_{j=1}^{k} u(a_i, \theta_j) \pi(\theta_j) du \\
&= \int_{u \in U} \pi(u) \sum_{j=1}^{k} u(c_{ij}) \pi(\theta_j) du \\
&= \int_{u \in U} \pi(u) u(a_i) du.
\end{aligned}
$$

$E_U[u(a_i)]$ is also named the *expected expected utility* with double expected value computed on the states of nature, as usual, and on the utility functions.

Nowadays, Bayesian elicitation is particularly appealing because it provides a useful instrument to implement automated elicitation processes. According to *adaptive utility elicitation* [34], the probabilistic utility distribution $\pi(u)$ is updated to $\pi(u|\boldsymbol{x})$ in accordance with the Bayes' rule, with $\boldsymbol{x}$ vector of observed responses that incorporates the additive information about the decision maker's preferences coming from his/her responses to certain queries and/or from sample surveys. The updating process is continuous until a stop criterion is satisfied. The stop criterion is mainly based on the concept of *expected loss* that represents the regret associated with a decision based on partial, rather than on complete, preference information and incorporates the costs of additive information. Details about the expected loss and the value of information will be provided in Chapter 5. Here it is worthwhile to note that queries used in the adaptive utility elicitation significantly differ from queries used under the classical elicitation methods based on comparisons between lotteries, and this makes the partial information elicitation process much less cognitively demanding with respect to the complete information elicitation process.

Another approach based on partial preference information relies on cluster analysis [33]. First, standard techniques are used to fully elicit a high number of utility functions. Then, a cluster analysis is performed on the elicited utility functions and the most representative utility function (named *prototype function*) is identified for each group. Each prototype function is associated with an optimal action, according to the principle of maximum expected utility. Finally, each new individual is assigned to a cluster on the basis of a suitable classification criterion and the prototype function of that cluster is used as a proxy of his/her utility function.

In more detail, the procedure develops along the following subsequent steps:

1. One starts with a bank of n fully elicited utility functions.

2. n utility functions are clustered in G $(G < n)$ groups, according to suitable clustering algorithms and similarity criteria. As far as the choice of the similarity measure between two utility functions $u_h(\cdot)$ and $u_{h'}(\cdot)$ $(h, h' = 1, \ldots, n)$, there exist several proposals. For instance, Chajewska *et al.* [33] suggest minimizing a quantity based on the difference between the expected utilities of an action under $u_h(\cdot)$ and $u_{h'}(\cdot)$: the closer the expected utilities are, the more similar are the two utility functions. Differently, Ha and Haddawy [90] propose to adopt the probability that two individuals h and h' agree or disagree on the preference order of a pair of outcomes: in practice, this probability is obtained as the ratio between the number of pairs of outcomes with the same order divided by the total number of pairs of outcomes.

3. For each cluster, a prototype function is identified minimizing the total utility loss due to using $u_h(\cdot)$ instead of $u_{h'}(\cdot)$ to choose the optimal action, with $u_h(\cdot)$ and $u_{h'}(\cdot)$ belonging to the same cluster g and the utility loss given by the difference between the expected utility under $u_h(\cdot)$ and the expected utility under $u_{h'}(\cdot)$.

4. For each cluster g, the optimal action associated with the prototype utility function is identified according to the maximization of the expected utility principle: this is the optimal action for all utility functions (and, then, all individuals) belonging to the same cluster.

5. Each new individual, whose utility function is unknown, is asked to answer a questionnaire to find out the cluster to which his/her utility function likely belongs. Queries used during this phase are easier and less cognitively demanding than queries used with gamble methods.

6. After having classified the new individual in a cluster, the optimal action recommended corresponds to the one identified at step 4 above.

4.7 Combining multiple preferences

Decision makers can be represented by single individuals or collective bodies composed of a multitude of individuals (e.g., the management board of a company). In this second case, the preferences of the members of the collective decision maker have to be combined to obtain just one utility function (instead of a function per individual). We refer to two types of approaches, borrowed from the literature about the elicitation of probability functions [75]: the *mathematical approach* and the *behavioral approach*.

Under the mathematical approach, members of the team do not interact and a utility function is elicited for each of them in separated sessions. Then, these utility functions are mathematically combined in a unique function. The *opinion pools*, known also as the *axiomatic approach*, represent the most popular mathematical technique. In the *linear opinion pool*, the outcomes of the collective body are obtained from the weighted arithmetic mean of the individual outcomes

$$\bar{c}_i = \sum_{h=1}^{n} \omega_h c_{hj},$$

with $\bar{c}_i$ pooled outcome, ω_h weight assigned to individual h ($h = 1, \ldots, n$), and c_{hj} j-th outcome elicited by individual h. Differently, the *logarithmic opinion pool* is a normalized weighted geometric mean and is equivalent to applying a linear pool to the logarithms of individual values and, then, normalizing them

$$\bar{c}_i = C \prod_{h=1}^{n} c_{hi}^{\omega_h},$$

with the C normalization constant. As a main drawback of the logarithmic opinion pool, if an individual assigns value 0 to an outcome, then also the combined value will be null.[2]

An issue that deserves special attention is the assignment of weights to the pool members. Weights ω_h used in the computation of the combined values $\bar{c}_i$ should reflect the "true" weight that each member carries in the collective body in terms of charisma and leadership. For instance, in the management

---

[2]In the probability elicitation context, it has been shown that the linear opinion pool satisfies the marginalization property, whereas the logarithmic opinion pool satisfies the external Bayesian property. The marginalization property means that for multivariate states of nature the marginal probability obtained for a specific state of nature from the combined distribution is the same as what would be achieved if the individually elicited marginal distributions for that state of nature were combined. The external Bayesian property means that updating the individual prior distributions with new information and then combining them is the same as directly updating the combined prior distribution. These two properties have not yet been studied with specific reference to the utility elicitation context.

board of a firm the chief executive officer (CEO) can be associated with a double weight with respect to the other members.[3]

Under the behavioral approach, the collective body members interact during the elicitation process, discussing together and sharing their opinions to achieve a consensual view. One of the most commonly used behavioral technique is represented by the Delphi method [74, 148]. Each individual provides a utility assessment that is anonymously shared with the group. Then, individuals may change their assessments taking into account the new information and the new points of view that they have received from the team. The process iterates until the utility assessments of the individuals converge on a common utility function. Because often it is not possible to achieve a full consensus among the team members a mathematical approach (e.g., an opinion pool) is used after a few rounds to combine the different opinions.

4.8 Case study: Utility elicitation for banking foundations

To conclude this chapter, we illustrate a case of utility elicitation, where the decision maker is represented by a multitude of individuals whose preferences must converge on a common decision. Specifically, we focus on the banking foundations that were born in 1990 as part of a comprehensive reform of the Italian banking sector (see the work of Bacci and Chiandotto [11] for a detailed illustration of the case study). Banking foundations are non-profit organizations that derive from the spin-off of the banks and have to manage huge amounts of money resulting from the sale of the banking majority block of shares. According to the law, a banking foundation is a non-profit private legal entity that pursues social utility and economic development. A banking foundation has to decide, coherently with its preferences, which funding applications to accept and which projects to finance. For this aim, the utility function of a banking foundation represents a useful instrument to gather information about the attitude towards risk. Here we illustrate the procedure we followed to elicit the utility function of the "Ente Cassa di Risparmio di Firenze" (ECRF), the banking foundation born from the bank "Cassa di Risparmio di Firenze" (that was the main bank of Florence, Italy).

As the decisions of a banking foundation come out from the board of directors, which is a collective body composed of eight members, we decided to elicit the utility functions of all the members of the board separately from each other and, then, to combine them in accordance with a mathematical approach (Section 4.7). For the individuals' utility elicitation we adopted an *ad*

---

[3]In the probability elicitation context the assignment of weights takes into account different elements, such as the individual's previous experience and the individual's knowledge about the states of nature: individuals whose evaluations are believed to be more accurate receive a higher weight.

hoc questionnaire, formulated according to one of the elicitation approaches described above in this chapter. By virtue of its robustness to violations of the expected utility theory the TO method was used in its two variants, outward and inward (Section 4.4.2). Each board member was asked about chained queries concerning two types of decisional problems, which simulate finance-able projects.

The problem of type 1 is about the choice between two funding requests to build two care centers: center A to care for patients with multiple sclerosis and center B to care for patients with muscular dystrophy. The outcomes of the two financeable projects are measured in terms of number of cycles of therapy per year. In the first query, the board member has to choose the number c_1 of cycles of therapies that make him/her indifferent between center A and center B, where center A will be able to perform 3519 cycles of therapy per year with probability p and 5000 cycles of therapy otherwise ($c_{h0} = c_0 = 5000$ for all individuals)

$$\langle 3519 \; p \; 5000 (= c_0) \rangle,$$

whereas center B will be able to perform 847 cycles of therapy per year with probability p and c_{h1} otherwise

$$\langle 847 \; p \; c_{h1} (=?) \rangle.$$

The value of p is arbitrarily chosen (and set equal to 0.50) and the uncertainty of results of the two projects is anchored to the approval of a law about compliance with formal requirements (e.g., maximum number of beds per room).

As usual in the TO approach, the second comparison of problem of type 1 is chained to the first one through the substitution of $c_0 = 5000$ with the elicited value c_{h1}, such that the board member has to choose c_{h2} that makes him/her indifferent between

$$\langle 3519 \; p \; c_{h1} \rangle \sim \langle 847 \; p \; c_{h2} (=?) \rangle.$$

Then, the comparison is repeated two other times until a total of four values $c_{h1}, c_{h2}, c_{h3}, c_{h4}$ elicited. Note that, coherently with the outward TO approach, answers by a rational individual imply increasing values of c_{hi} ($i = 1, 2, \ldots$), that is, $c_0 = 5000 < c_{h1} < c_{h2} < c_{h3} < c_{h4}$.

The problem of type 2 is about the choice between two funding requests to present an exhibition of famous artists by two different museums. An amount of 26915 visitors at best and 20000 at worst is expected for exhibition A ($c_{h4} = c_4 = 20000$ for all individuals), whereas 33511 visitors at best and an unknown value c_{h3} (less than 20000) at worst are expected for exhibition B:

$$\langle 26915 \; p \; 20000 (= c_4) \rangle \sim \langle 33511 \; p \; c_{h3} (=?) \rangle.$$

After the elicitation of c_3, the following chained indifference relation is proposed:

$$\langle 26915 \; p \; c_{h3} \rangle \sim \langle 33511 \; p \; c_{h2} (=?) \rangle,$$

and so on, until c_{h1} and c_{h0} are elicited. As above, p is arbitrarily chosen and the uncertainty of the results depends on whether the restoration of a famous painting will be completed before the exhibition. Differently from the problem of type 1, now the board member has to elicit decreasing values that are coherent with the inward TO approach, that is, $c_4 = 20000 > c_{h3} > c_{h2} > c_{h1} > c_{h0}$.

In accordance with the suggestions from the literature about the elicitation of probability functions (see, among others, [75, 123]), the utility elicitation process was developed with the support of a statistician, following some subsequent phases:

- background phase: the statistician collects information about the usual activities of the ECRF banking foundation to be able to converse with the board members;

- motivating and training phase: the statistician illustrates the aims of the elicitation process and provides each board member with some basic knowledge in the fields of probability and utility theory; moreover, the initial queries of the questionnaire are preventively analyzed together. During this phase the statistician and the board member discuss the relation between the shape of a utility function and the attitude towards risk of the decision maker. They also analyze how some characteristics affect the riskiness level of a project, such as: the novelty level (the more innovative a project is, the more uncertain the results are); the duration; the sector of intervention (e.g., scientific and medical projects are usually more uncertain in their results than projects in the arts and human sciences); the geographical area of intervention; the typology of the proponent in terms of previous experiences, economic sources, dimensions, guarantees, and so on.

- elicitation phase: to make the elicitation process easier, queries are presented in a user-friendly format, such as

	approved law	notapproved law
center A	3519	5000
center B	847	$c_{h1} = ?$

Moreover, during the elicitation process the board member is warned of incoherent responses (e.g., non-monotonic elicited values) and is invited to correct them. Elicited values are normalized according to the following formula ($i = 0, 1, 2, 3, 4$)

$$c_{hi}^* = \frac{c_{hi} - c_{h0}}{c_{h4} - c_{h0}}$$

to make possible the comparison between the variants of the TO method. As a consequence, normalized outcomes c_{hi}^* range in $(0,1)$ with $c_{h0}^* = 0$ and $c_{h4}^* = 1$ for all individuals and, given $u(0) = 0$ and $u(1) = 1$ as usual, the interval of utilities is split into quartiles, that is, $u(c_{h1}^*) = 0.25$, $u(c_{h2}^*) = 0.50$, and $u(c_{h3}^*) = 0.75$. Therefore, normalized values will be observed such that:

$c_{hi}^* = i/4$ ($i = 1, 2, 3$) in the case of a risk neutral individual, corresponding to a linear utility function; $c_{hi}^* < i/4$ in the case of a risk adverse individual, corresponding to a concave utility function; and $c_{hi}^* > i/4$ in the case of a risk seeking individual, corresponding to a convex utility function. In the case of a mixed shaped utility function, some c_{hi}^* will be greater than $j/4$ and other c_{hi}^* will be smaller. In Table 4.1 the normalized elicited values of each board member are shown, separately for the outward and inward TO; the resulting shape of utility function is indicated, too.

TABLE 4.1
Elicited values c_{hi}^* for the board members of the ECRF banking foundation and utility function shape.

Member	Outward TO				Inward TO			
h	c_{h1}^*	c_{h2}^*	c_{h3}^*	shape	c_{h1}^*	c_{h2}^*	c_{h3}^*	shape
1 (CEO)	0.273	0.727	0.909	convex	0.133	0.466	0.800	concave-convex
2	0.100	0.300	0.600	concave	0.286	0.500	0.714	convex-concave
3	0.250	0.500	0.750	linear	0.167	0.375	0.667	concave
4	0.250	0.500	0.750	linear	0.250	0.500	0.750	linear
5	0.200	0.433	0.667	concave	0.028	0.055	0.278	concave
6	0.357	0.643	0.857	convex	0.100	0.300	0.600	concave
7	0.217	0.435	0.652	concave	0.133	0.333	0.666	concave
8	0.423	0.538	0.769	convex	0.062	0.323	0.635	concave

Coherently with the experiments of Fennema and van Assen [61], systematic and statistically significant[4] changes in the shape of the utility function and, then, in the attitude towards risk result from our experiment, according to the variant of TO: only three out of eight individuals maintain attitude towards risk constant, whereas in four out of the remaining five cases the shape of the utility function changes from convex, or at most linear, under the outward TO to concave under the inward TO.

- combining phase: the individuals' normalized outcomes are combined to elicit the utility function of the ECRF banking foundation. The combination of values is performed through the linear opinion pool approach, using three different systems of weights: (i) all board members have the same weights; (ii) the foundation's CEO (member $h = 1$ in Table 4.1) is weighted twice the other members; (iii) the foundation's CEO is weighted ten times the other members. Combined outcomes are shown in Table 4.2 and the resulting utility functions are displayed in Figure 4.6.

---

[4]Paired t-tests were performed for the differences between values c_i^* elicited under the outward TO and corresponding c_i^* elicited under the inward TO.

TABLE 4.2

Combined utility values $\bar{c}_i^*$ for the ECRF banking foundation and utility function shape.

Type of	Outward TO				Inward TO			
weights	$\bar{c}_1^*$	$\bar{c}_2^*$	$\bar{c}_3^*$	shape	$\bar{c}_1^*$	$\bar{c}_2^*$	$\bar{c}_3^*$	shape
(i)	0.259	0.510	0.744	convex-concave	0.145	0.357	0.639	concave
(ii)	0.260	0.534	0.763	convex	0.144	0.369	0.657	concave
(iii)	0.266	0.625	0.831	convex	0.139	0.414	0.724	concave

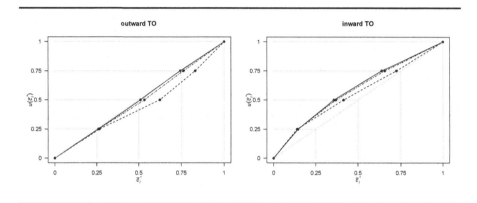

FIGURE 4.6

Utility functions elicited for ECRF: outward TO (left) and inward TO (right). Legend: weights of type (i): solid line; weights of type (ii): dashed-dotted line; weights of type (iii): dashed line; linear function: dotted grey line.

As above outlined about the individuals' utilities, the difference between outward and inward approaches clearly emerges also after the mathematical averaging of elicited values: under the outward TO a convex or at most convex-concave shape is observed, whereas under the inward TO a concave utility function results. Moreover, the leading role played by the CEO (or by any other influent member) is explicit in a combined utility function whose shape is closer to that of the CEO the more influential he/she is.

5

Classical and Bayesian statistical decision theory

CONTENTS

5.1 Introduction

In Chapters 3 and 4 we provided the basic elements of detecting the optimal solution of a decision problem when the decision maker operates in situations of certainty (value theory) or in situations of risk (utility theory). Under the utility theory the probability distribution of the states of nature is known, whereas under the subjective utility theory the decision maker explicates his/her beliefs about the states of nature in terms of prior probabilities. In both cases, the optimal decision is driven by the maximization of the expected utility, with the expected value computed with respect to the prior probabilities that are known or are subjectively assigned to the states of nature. The setting at issue, where the decision maker is given the prior information about the states of nature, in addition to the utilities of consequences, is also known as *Bayesian decision theory*.

In this chapter, we start with the illustration of the solutions proposed in the field of *classical decision theory* that deals with decision-making problems in situations of uncertainty, that is, when prior probabilities, as well as any other types of information, about states of nature are not available. We analyze again decision criteria illustrated in Section 1.4 taking into account losses instead of monetary outcomes.

Then, a typical statistical perspective is adopted throughout the illustration of the fundamentals of *classical statistical decision theory*, where the decision maker is given sample information to improve his/her knowledge about the states of nature, although prior information is not available. We introduce the concepts of decision function and risk (expected loss). We also illustrate that the sample information can be used to compute the risk associated with a certain decision function and state of nature.

Afterwards, the typical setting of *Bayesian statistical decision theory* is taken into account, which deals with decisional situations in which both sample information and prior information are available. In such a context, the concepts of expected risk and posterior probabilities are introduced and the equivalence between decisional analysis in normal form and in extensive form is illustrated. Basically, under the classical and the Bayesian decision theory settings, the traditional problems of statistical inference (i.e., estimation and hypothesis testing) are faced by a decisional perspective. Table 1.2, already illustrated in Section 1.5, provides a summary of the main frameworks characterizing the decision theory.

As concerns the acquisition of sample information, costs involved in a sample survey are also discussed and formalized in the decisional process: concepts of the expected value of perfect information, the expected value of sample information, and the net gain associated with a sample survey are introduced. Finally, the illustration of a case study will conclude the chapter.

5.2   Structure of the decision-making process

To provide a conceptual frame related to the topics that will be considered in the next sections, we here summarize the main elements that are involved in a decision-making process:

- a probabilistic model $f(x, \theta)$ that accounts for the joint distribution of the sample observations and of the states of nature;

- the space of the states of nature (or parametric space) Θ that in many situations has dimension $\mathbb{R}^k$ and can be either a discrete space or a continuous space;

- a discrete space of actions $A = (a_1, a_2, \ldots)$;

- a discrete space of experiments E;

- a sample space Ω consisting of the results of each experiment; for univariate random variables the sample results are expressed by n (sample size) real numbers;

- a discrete decision space $D = (d_1, d_2, \ldots)$.

In addition, we explicitly introduce the following functions:

- the utility function $U = u(a_i, \theta_j) = u_{ij}$ that associates a consequence, expressed in terms of utility, with each action and state of nature;

- the loss function $l_{ij} = l(a_i, \theta_j) = -u(a_i, \theta_j)$ that associates a consequence, expressed in terms of loss, with each action and state of nature;

- the decision function $d_h = \delta_h(\boldsymbol{x})$ that maps each point of the sample space $\boldsymbol{\Omega}$ into the space of actions A[1].

The relationships between the different elements defined above are illustrated in Figure 5.1, where the structure of a generic decision-making process is displayed. Different specifications of the decision-making process are characterized by the presence (or absence) of specific elements and related relationships, as will be clarified later in this chapter. For instance, under the classical decision theory (decisions in situations of uncertainty), only the space of the states of nature, the space of actions and the space of consequences (expressed in terms of utility or loss) are involved and the structure of the decision-making process simplifies as is shown in Figure 5.2.

---

[1] A decision function is said to be simple or non-random if each point $\boldsymbol{x} \in \boldsymbol{\Omega}$ is associated with only one point $a_i \in A$, whereas it is said to be mixed or random, if $\boldsymbol{x} \in \boldsymbol{\Omega}$ is associated with a range of elements of A with related probability distribution. It should be noted that the simple decision functions are obtained as a special case of the mixed ones, when the probability space is degenerate.

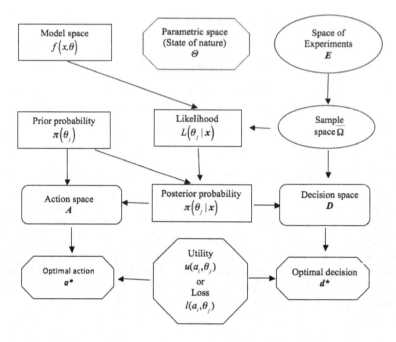

FIGURE 5.1
Conceptual frame of a decision-making process.

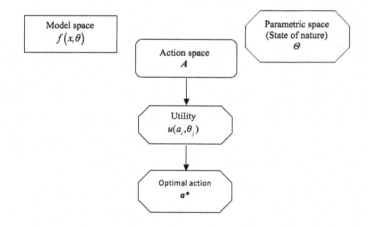

FIGURE 5.2
Conceptual frame of a decision-making process under uncertainty.

5.3 Decisions under uncertainty (classical decision theory)

From a statistical perspective the concept of utility function is often substituted with the complementary concept of *loss function* l_{ij},

$$l_{ij} = l(a_i, \theta_j) = -u(a_i, \theta_j),$$

to treat statistical decision theory according to the original setting given by Wald [217], which complies with typical statistical reasoning. The subject is extensively dealt with also in Ferguson [62], De Groot [53], Berger [14], Piccinato [171], and Robert [180]. For the sake of clarity, here we refer to the traditional instruments of statistical inference, that is, point estimation and hypothesis testing.

Under the point estimation setting, one of the properties required of an optimal estimator is the efficiency, which is traditionally measured in terms of mean simple error (Definition 2.20) or mean square error (Definition 2.21). We may interpret the mean simple error or the mean square error in terms of a loss function $l(a_i, \theta_j)$, with a_i denoting an estimator $\hat{\Theta}$ and θ_j denoting the true value of the unknown parameter. From this perspective, the decisional problem "choice of the optimal estimator $\hat{\Theta}^*$" solves in the search of the most efficient estimator, that is, the estimator that minimizes the mean simple error

$$\hat{\Theta}^* = \arg\left[\min_{\theta \in \Theta} l(\hat{\Theta}^*, \theta)\right] = \arg\left[\min_{\theta \in \Theta} mse(\hat{\Theta}^*)\right] = \arg\left[\min_{\theta \in \Theta} E(|\hat{\Theta}^* - \theta|)\right],$$

or the mean square error

$$\hat{\Theta}^* = \arg\left[\min_{\theta \in \Theta} l(\hat{\Theta}^*, \theta)\right] = \arg\left[\min_{\theta \in \Theta} MSE(\hat{\Theta}^*)\right] = \arg\left\{\min_{\theta \in \Theta} E[(\hat{\Theta}^* - \theta)^2]\right\}.$$

Similarly, under the hypothesis testing theory, we may interpret the probability of a II type error as the loss function to minimize, given the level of I type error (Section 2.9.4).

More in general, if we think in terms of losses, the decision table (Table 3.2) can be re-formulated expressing the consequences in terms of losses rather than in terms of utility (Table 5.1). Note also that no assumption about the probability distribution of the states of nature is here introduced.

The decision criteria introduced in Section 1.4 as a function of (monetary) outcomes can now be expressed in terms of losses:

- min-max criterion or Wald's criterion (extreme pessimistic perspective):

$$a^* = \arg\left[\min_i(\max_j l_{ij})\right] = \arg\left[\min_i(\max_j l(a_i, \theta_j))\right];$$

TABLE 5.1
Decision table under uncertainty
and with consequences expressed
in terms of losses, $l_{ij} = l(a_i, \theta_j)$
$(i = 1, \ldots, m; \, j = 1, \ldots, k)$.

Action	States of nature			
	θ_1	θ_2	$\ldots$	θ_k
a_1	l_{11}	l_{12}	$\ldots$	l_{1k}
a_2	l_{21}	l_{22}	$\ldots$	l_{2k}
$\vdots$	$\ldots$	$\ldots$	$\vdots$	$\ldots$
a_m	l_{m1}	l_{m2}	$\ldots$	l_{mk}

- min-min criterion (extreme optimistic perspective):

$$a^* = \arg\left[\min_i(\min_j \, l_{ij})\right] = \arg\left[\min_i(\min_j \, l(a_i, \theta_j))\right];$$

- Hurwicz's criterion:

$$a^* = \arg\left\{\min_i\left[(1-\alpha)\max_j \, l_{ij} + \alpha\min_j \, l_{ij}\right]\right\}$$
$$= \arg\left\{\min_i\left[(1-\alpha)\max_j \, l(a_i, \theta_j) + \alpha\min_j \, l(a_i, \theta_j)\right]\right\};$$

- Savage's criterion or min-max regret criterion:

$$a^* = \arg\left[\min_i(\max_j \, r_{ij})\right], \tag{5.1}$$

with $r_{ij} = l_{ij} - \min_i l_{ij} = l(a_i, \theta_j) - \min_i l(a_i, \theta_j)$.

Alternatively, these criteria can be expressed in terms of utilities: the difference consists of the maximization of utilities instead of the minimization of losses, but the results are the same. More precisely:

- max-min criterion or Wald's criterion (extreme pessimistic perspective):

$$a^* = \arg\left[\max_i(\min_j \, u_{ij})\right] = \arg\left[\max_i(\min_j \, u(a_i, \theta_j))\right];$$

- max-max criterion (extreme optimistic perspective):

$$a^* = \arg\left[\max_i(\max_j \, u_{ij})\right] = \arg\left[\max_i(\max_j \, u(a_i, \theta_j))\right];$$

- Hurwicz's criterion:

$$a^* = \arg\left\{\max_i\left[(1-\alpha)\min_j u_{ij} + \alpha\max_j u_{ij}\right]\right\}$$
$$= \arg\left\{\max_i\left[(1-\alpha)\min_j u(a_i,\theta_j) + \alpha\max_j u(a_i,\theta_j)\right]\right\};$$

- Savage's criterion or min-max regret criterion: the definition of the optimal action a^* is the same as in Eq. (5.1), with the regret now defined as

$$r_{ij} = \max_i u_{ij} - u_{ij} = \max_i u(a_i,\theta_j) - u(a_i,\theta_j).$$

As already stressed, these decision criteria have a certain level of acceptability: arguments that support one or the other criterion are of a different nature and depend on the decisional context in which one operates.

Example 5.1. *To illustrate the decision making process under uncertainty a decision problem is here described that has become classic in the literature, after the book of Grayson [86], because, despite its extreme simplicity, it allows us an in-depth discussion of all the aspects of interest. The decision problem is proposed in Table 5.2, where the payoffs are expressed in terms of utilities, and in Table 5.3, where the payoffs are expressed in terms of losses; all values are in thousands of US dollars. Note that, missing any other information, we are here assuming a linear utility function, as the utility of a consequence corresponds to the consequence itself (i.e., $u(c_{ij}) = c_{ij}$).*

TABLE 5.2
Decision table: drilling problem with utilities as payoffs (best results are in bold).

Action	Utilities		Decision criterion		
	θ_1: no oil	θ_2: oil	Wald	max-max	Savage
a_1: don't drill	0	-600	-600	0	**600**
a_2: drill	-300	600	**-300**	**600**	900

TABLE 5.3
Decision table: drilling problem with losses as payoffs (best results are in bold).

Action	Losses		Decision criterion		
	θ_1: no oil	θ_2: oil	Wald	min-min	Savage
a_1: don't drill	0	600	600	0	**600**
a_2: drill	300	-600	**300**	**-600**	900

The decision maker has to decide whether to not proceed (action a_1) or proceed (action a_2) to the drilling of an oil well, missing any information on

the presence of oil in the ground. Absence of oil denotes the state of nature
θ_1, *whereas presence of oil denotes the state of nature* θ_2. *Moreover, the cost*
of drilling is equal to $300,000 and the gross income, in the presence of oil,
amounts to $900,000, then the net income is equal to $600,000; obviously, the
gross income in case of absence of oil is $0.

If the decision maker knew the (prior) probabilities (Bayesian decision
theory), say $\pi(\theta_1) = 0.50 = \pi(\theta_2)$, *an expected utility of* $-\$300$ *would result*
for action a_1, *while the expected utility for action* a_2 *would be equal to $150,*
therefore the optimal action would be $a^* = a_2$. *Similarly, expected losses would*
be $300 and $-\$150$ *for action* a_1 *and* a_2, *respectively, resulting again in* a_2 *as*
the optimal action. The outcome is obvious if one thinks about the decision
problem in terms of a lottery: if the decision maker agrees on betting $300,000
in a lottery, he/she will have a 50% chance of receiving his/her bet tripled.
This conclusion is valid only assuming a linear utility function: an individual
strongly averse to risk could, in fact, decide not to play (not proceed with
drilling). It is worth noting that the optimal action detected under the Bayesian
decision setting is strongly associated with the prior probabilities of the states
of nature: changes in $\pi(\theta_1)$ *and* $\pi(\theta_2)$ *may lead to a conclusion in favor of*
a_1 *instead of* a_2, *as shown by results reported in Table 5.4. For values of*
$\pi(\theta_1)$ *strictly smaller than 0.8 the decision maker should prefer action* a_2,
whereas for values of $\pi(\theta_1)$ *strictly greater than 0.8 action* a_1 *is preferred;*
for $\pi(\theta_1) = 0.8$ *and* $\pi(\theta_2) = 0.2$ *indifference holds between* a_1 *and* a_2. *These*
results outline the relevance of additional information to support the decisional
process, as will be clear in the next sections, where sample information will be
introduced to update prior probabilities.

Aspects concerning the influence of individual attitudes towards decisional
choices also emerge in situations of uncertainty, when there is no information
on the probabilities of states of nature. For instance, in the above example both
the Wald's criterion (extreme pessimism) and the max-max criterion (extreme
optimism) identify a_2 *as the best action, both reasoning in terms of utilities*
(Table 5.2) and losses (Table 5.3). The substantial difference is that the ex-
treme pessimist expects a loss of $300, whereas the extreme optimist expects a
negative loss (i.e., utility) of $-\$600$.

Also the Hurwicz's criterion, which realizes a mixture between extreme
pessimism and extreme optimism criteria, leads to prefer action a_2 *over action*
a_1, *as shown in Table 5.5. More precisely, according to the value assumed by*
the coefficient of optimism α *we notice that (i) action* a_1 *is always associated*
with positive and increasing losses for α *decreasing from 1 to 0, (ii) action*
a_2 *is associated with negative losses for values of* $\alpha > 0.3\overline{3}$ *and with positive*
losses otherwise, (iii) losses of a_2 *are always smaller than losses of* a_1.

A different conclusion is reached under Savage's criterion (Tables 5.2 and
5.3, last column). In such a case the choice of the decision maker is driven
by a different reasoning, which is based on the concept of regret and leads to
prefer action a_1 *(don't drill) instead of action* a_2 *(drill).*

TABLE 5.4

Drilling problem with losses as payoffs and different values for
the prior probabilities $\pi(\theta_1)$ and $\pi(\theta_2)$ (best results are in bold).

$\pi(\theta_1)$	$\pi(\theta_2)$	Exp. Losses	
		a_1: don't drill	a_2: drill
0.10	0.90	540	**-510**
0.15	0.85	510	**-465**
0.20	0.80	480	**-420**
0.25	0.75	450	**-375**
0.30	0.70	420	**-330**
0.35	0.65	390	**-285**
0.40	0.60	360	**-240**
0.45	0.55	330	**-195**
0.50	0.50	300	**-150**
0.55	0.45	270	**-105**
0.60	0.40	240	**-60**
0.65	0.35	210	**-15**
0.70	0.30	180	**30**
0.75	0.25	150	**75**
0.80	0.20	120	120
0.85	0.15	**90**	165
0.90	0.10	**60**	210
0.95	0.05	**30**	255

TABLE 5.5

Decision table for the Hurwicz's criterion: drilling problem
with losses as payoffs and different values for the optimism
coefficient α.

Optimism coeff.	Losses	
α	a_1: don't drill	a_2: drill
1.00	0	-600
0.90	60	-510
0.80	120	-420
0.70	180	-330
0.60	240	-240
0.50	300	-150
0.40	360	-60
0.35	390	-15
0.34	396	-6
0.33	402	3
0.32	408	12
0.30	420	30
0.20	480	120
0.10	540	210
0.00	600	300

The example outlines the limits of criteria adopted under classical and Bayesian decision theory, as in the presence of poor (Bayesian decision setting) or null (classical decision setting) information about the true state of nature, the optimal decision may point to one direction as well as to its opposite. To avoid, or at least limit, this variability in the decisional process a statistical approach to decision theory has to be adopted, where sample information is explicitly taken into account, as will be illustrated in the next sections.

5.4 Decisions with sample information (classical statistical decision theory)

In the classical decision theory setting described in the previous section the decision maker ignores any kind of information about the states of nature. Instead, in the classical statistical decision theory setting, the decision maker can plan an experiment to gather information on the states of nature. Here, term experiment is intended in a broad sense that includes all cases of acquisition of information through sample surveys, case-control studies, collection of data already published, and so on.

Obviously, the acquisition of sample information can entail a cost that may not lead to a convenient reduction of uncertainty. Thus, a new decision problem arises: the question to be answered is whether or not to proceed with the acquisition of sample information whose value (and convenience) can only be measured retrospectively, that is, after having carried out the experiment and collected the information. The question concerning the value of sample information will be dealt in Section 5.6; here, the core of the classical statistical decision theory is illustrated, which deals with the optimal choice among different alternatives when only the sample information is available.

Let $x' = (x_1, x_2, \ldots, x_n)$ denote the outcome of an experiment resulting from the random variable $X \sim f(x; \theta)$, with unknown $\theta \in \Theta$. Moreover, let $d_h = \delta_h(x)$ be a *decision function*, which takes a dataset as input and gives a decision (i.e., an action) as output, that is, $d_h \implies a_i$ ($h = 1, \ldots, r$, $i = 1, \ldots, m$). The detailed specification of $d_h = \delta_h(x)$ depends on the problem at hand. Examples in classical statistical inference include:

- estimation problems: the decision consists of choosing the optimal estimation method;

- hypothesis testing problems: the decision consists of rejecting or not rejecting the null hypothesis;

- classification problems: the decision consists of classifying a new observation into a finite set of categories;

- model selection problems: the decision consists of selecting a set of significant covariates or the kind of model (e.g., linear model, logit, multilevel model, and so on).

If we consider r decision functions, the decision Table 5.1 becomes as in Table 5.6.

TABLE 5.6
Decision table with decisions
and losses.

Action	States of nature			
	θ_1	θ_2	...	θ_k
d_1	l_{11}	l_{12}	...	l_{1k}
d_2	l_{21}	l_{22}	...	l_{2k}
$\vdots$	...	...	$\vdots$	...
d_r	l_{r1}	l_{r2}	...	l_{rk}

Typically, an infinite number of decision functions are available; therefore, how can we determine which of these decision functions to use? A general criterion is based on the concept of loss function, which describes the loss (or cost) associated with all possible decisions. Different decision functions imply different types of mistakes. The loss function provides information about the type of mistakes one should be more concerned about. The best decision function is the function that yields the lowest expected loss.

In practice, if we refer to the solutions proposed in the classical statistical inference, a very limited number of decision functions is counted for a certain decisional problem, usually just one or two. For instance, let us assume that we are interested in estimating parameter μ (population mean) of a certain random variable X, relying on a random sample $\boldsymbol{X}' = (X_1, \ldots, X_n)$. If the loss function is the mean square error

$$MSE(T(\boldsymbol{X})) = E\left\{[T(\boldsymbol{X}) - \mu]^2\right\},$$

then the best decision $d_1 = \delta_1(X_1, \ldots, X_n)$ that minimizes the loss function $MSE(T(\boldsymbol{X}))$ is the sample mean, that is,

$$d_1 = \frac{1}{n}\sum_{i=1}^{n} X_i = \bar{X}.$$

If the loss function is the mean simple error

$$mse(T(\boldsymbol{X})) = E\left\{|T(\boldsymbol{X}) - \mu|\right\},$$

the best decision $d_2 = \delta_2(X_1, \ldots, X_n)$ that minimizes the loss $mse(T(\boldsymbol{X}))$ is the sample median. In the end, the final decision about the value of μ will be reasonably based on the sample mean or the sample median.

As outlined in Chapter 2, the sample mean $\bar{X}$ is the best estimator (according to the properties illustrated in Section 2.9.1), and consequently $\bar{x}$ is the best estimate, of μ, whatever the value of μ: we can say that $d_h = \delta_h(\boldsymbol{X}) = \bar{X}$ is a dominant strategy for the estimation of the population mean. However, the desirable properties of the sample mean do not usually hold when we consider unknown parameters different from μ. When this happens, the class of eligible decision functions is usually restricted in accordance with reasonable conditions, such as invariance and unbiasedness. For instance, in Section 2.11.2 we discussed how the best estimator of the vector of regression coefficients exists, and may be detected, in the restricted class of unbiased estimators (BUE estimator).

More generally, the *expected loss*, known also as *risk*, is calculated for each decision function to evaluate the effect of the sample information.

Definition 5.1. Expected loss or risk. *Let $\boldsymbol{x}' = (x_1, \ldots, x_n)$ be a random sample from a population $X \sim f(x; \theta)$ and $d_h = \delta_h(\boldsymbol{x})$ a decision function defined on $\boldsymbol{x}$. The risk $R(d_h, \theta)$ associated with the decision function d_i is the expected value of the loss function $l[\delta_h(\boldsymbol{x}), \theta_j]$ computed with respect to the sample observations:*

$$R(d_h, \theta_j) = R[\delta_h(\boldsymbol{x}), \theta_j] = E_{\boldsymbol{x}}\{l[\delta_h(\boldsymbol{x}), \theta_j]\}. \tag{5.2}$$

If X is a discrete random variable, the risk becomes

$$R(d_h, \theta_j) = E_{\boldsymbol{x}}\{l[\delta_h(\boldsymbol{x}), \theta_j]\} = \sum_{\boldsymbol{x}} l[\delta_h(\boldsymbol{x}), \theta_j] p(\boldsymbol{x}|\theta_j) =$$

$$= \sum_{x_1} \sum_{x_2} \cdots \sum_{x_n} l[\delta_h(x_1, x_2, \ldots, x_n), \theta] f(x_1, \theta_j) f(x_2, \theta_j) \ldots f(x_n, \theta_j),$$

where $f(x_i, \theta_j)$ $(i = 1, 2, \ldots, n)$ is the mass probability function of the random variable associated with the i-th sample.

Similarly, if X is a continuous random variable, the algebraic expression of the risk is

$$R(d_h, \theta_j) = E_{\boldsymbol{x}}\{l[\delta_h(\boldsymbol{x}), \theta_j]\} = \int_{\boldsymbol{x}} l[\delta_h(\boldsymbol{x}), \theta_j] f(\boldsymbol{x}|\theta_j) d\boldsymbol{x} =$$

$$= \int_{x_1} \int_{x_2} \cdots \int_{x_n} l[\delta_h(x_1, x_2, \ldots, x_n), \theta_j] f(x_1, \theta_j) f(x_2, \theta_j) \ldots f(x_n, \theta_j)$$
$$dx_1 dx_2 \ldots dx_n,$$

where $f(x_i, \theta_j)$ $(i = 1, 2, \ldots, n)$ is the density probability function of the random variable associated with the i-th sample.

The decision table, similar to Table 5.6 but now expressed in terms of risk functions, takes the form as in Table 5.7.

TABLE 5.7

Decision table with decisions and expected losses (risk functions).

Action	States of nature			
	θ_1	θ_2	$\dots$	θ_k
d_1	$R(d_1, \theta_1)$	$R(d_1, \theta_2)$	$\dots$	$R(d_1, \theta_k)$
d_2	$R(d_2, \theta_1)$	$R(d_2, \theta_2)$	$\dots$	$R(d_2, \theta_k)$
$\vdots$	$\dots$	$\dots$	$\vdots$	$\dots$
d_r	$R(d_r, \theta_1)$	$R(d_r, \theta_2)$	$\dots$	$R(d_r, \theta_k)$

The first step to analyze decision Table 5.7 requires the identification of the *dominant (or uniformly better) decision*, that is, the decision associated with the minimum risk, whatever the state of nature. If such a dominant decision exists the decision problem is solved. Otherwise, the decision maker proceeds with analyzing the decision table. The second step requires the elimination of decisions that are dominated by other decisions. The result is a list of *admissible (or eligible) decisions*.

Generally, as the dominant decision does not exists, the problem can be solved through the decision criteria under uncertainty, described in Section 5.3, applied on the risks instead of on the losses. Alternatively, the uniformly best decision may be searched in a constrained class of decision functions, in accordance with certain desirable requirements[2].

Example 5.2. *Let us consider again the decision problem illustrated in Example 5.1 and now let us assume that the decision maker can collect information on the presence or absence of oil in the ground using two independent seismographs. Let us also assume that the cost of information is null. The result of the use of a seismograph is denoted by $X_i = x$ (i = 1 for the first seismograph and i = 2 for the second seismograph) with $x = 0$ in the case of absence of oil and $x = 1$ in the case of presence of oil with probabilities depending on the state of nature as shown in Table 5.8. Using two independent seismographs, the sample space Ω may assume the following four values:*

- $X_1 = 0 \cap X_2 = 0$: *both seismographs indicate absence of oil;*

- $X_1 = 1 \cap X_2 = 1$: *both seismographs indicate presence of oil;*

- $X_1 = 0 \cap X_2 = 1$: *the first seismograph indicates absence of oil, while the second one indicates presence of oil;*

---

[2]One of these desirable requirements, which is usually introduced to restrict the class of eligible decision functions, is represented by the invariance. To justify this constraint one can say that if a decision problem is symmetrical or invariant with respect to certain operations, it seems reasonable to restrict the class of possible decisions to those that are symmetrical or invariant with respect to the same operations.

TABLE 5.8

Probabilities of the result of the use of a
seismograph, given the state of nature.

X_i	θ_1: no oil	θ_2: oil
	States of nature	
$x = 0$	$p(X_i = 0\|\theta_1) = 0.60$	$p(X_i = 0\|\theta_2) = 0.30$
$x = 1$	$p(X_i = 1\|\theta_1) = 0.40$	$p(X_i = 1\|\theta_2) = 0.70$

- $X_1 = 1 \cap X_2 = 0$: *the first seismograph indicates presence of oil, while the second one indicates absence of oil.*

Joint probabilities related to the use of the two seismographs follow:

$$p(0,0|\theta_1) = p(0|\theta_1)p(0|\theta_1) = 0.60 \cdot 0.60 = 0.36;$$
$$p(0,1|\theta_1) = p(0|\theta_1)p(1|\theta_1) = 0.60 \cdot 0.40 = 0.24;$$
$$p(1,0|\theta_1) = p(1|\theta_1)p(0|\theta_1) = 0.40 \cdot 0.60 = 0.24;$$
$$p(1,1|\theta_1) = p(1|\theta_1)p(1|\theta_1) = 0.40 \cdot 0.40 = 0.16;$$
$$p(0,0|\theta_2) = p(0|\theta_2)p(0|\theta_2) = 0.30 \cdot 0.30 = 0.09;$$
$$p(0,1|\theta_2) = p(0|\theta_2)p(1|\theta_2) = 0.30 \cdot 0.70 = 0.21;$$
$$p(1,0|\theta_2) = p(1|\theta_2)p(0|\theta_2) = 0.70 \cdot 0.30 = 0.21;$$
$$p(1,1|\theta_2) = p(1|\theta_2)p(1|\theta_2) = 0.70 \cdot 0.70 = 0.49. \tag{5.3}$$

The two extreme sampling points $(0,0)$ and $(1,1)$ indicate, respectively, the absence of oil and the presence of oil reported by both seismographs, while the two intermediate points $(0,1)$ and $(1,0)$ indicate that only one of the two seismographs indicates the presence of oil. Basically, there are three sample points for two different states of nature, so the number of decision functions is equal to $2^3 = 8$ (remember that the decision function is a function that matches each sample point with a specific action). The resulting eight decision functions are listed in Table 5.9, with a_1 corresponding to the decision of not drilling and a_2 to the decision of drilling. From the analysis of Table 5.9, the

TABLE 5.9

Decision functions and actions.

Sample points	**Decision functions**							
	d_1	d_2	d_3	d_4	d_5	d_6	d_7	d_8
$(0,0)$	a_1	a_2	a_1	a_1	a_2	a_1	a_2	a_2
$(1,0) \cup (0,1)$	a_1	a_2	a_1	a_2	a_1	a_2	a_1	a_2
$(1,1)$	a_1	a_2	a_2	a_1	a_1	a_2	a_2	a_1

poorness of some decision functions is evident: for instance, functions d_1 and d_2 identify action a_1 and a_2, respectively, whatever the sampling result. In addition, function d_5 suggests to proceed with drilling (action a_2) when both *seismographs register the absence of oil and to not proceed with drilling (action a_1) when one or both seismographs register the presence of oil.*

Let us ignore, for the moment, the reasonableness of some decision functions and focus on the computation of the expected losses (risks) associated with each decision function. For the sake of clarity, in Table 5.10 we illustrate the computational procedure for function d_6, which suggests to proceed with action a_1 if the sample result is $(0,0)$ and with action a_2 if the sample result is $(1,0) \cup (0,1) \cup (1,1)$. In detail, on the basis of the definition of risk in Eq. (5.2) and the joint probabilities in Eq. (5.3) the risk of d_6 under state θ_1 is given by

$$R(d_6, \theta_1) = 0 \cdot 0.36 + 300 \cdot 0.64 = 0 + 192 = 192$$

and the risk of d_6 under state θ_2 is

$$R(d_6, \theta_2) = 600 \cdot 0.09 + (-600) \cdot 0.91 = 54 - 546 = -492.$$

TABLE 5.10
Expected losses (risks) for decision function d_6.

Sample points	Action	**Losses**				Risk (AxB)	
		State θ_1		State θ_2			
		Loss (A)	Prob. (B)	Loss (A)	Prob. (B)	State θ_1	State θ_2
$(0,0)$	a_1	0	0.36	600	0.09	0	54
$(1,0) \cup (0,1) \cup (1,1)$	a_2	300	0.64	-600	0.91	192	-546
Risk						192	-492

The expected losses (risks) computed for all decision functions are listed in Table 5.11.

TABLE 5.11
Risks associated with the eight decision functions.

	Risks	
Decision d_h	θ_1: no oil	θ_2: oil
d_1	0	600
d_2	300	-600
d_3	48	-294
d_4	144	96
d_5	108	492
d_6	192	-492
d_7	156	-96
d_8	252	-12

First, from Table 5.11 we immediately notice that, for the decisional problem at issue, there does not exist any dominant decision function. In other

words, there is no decision function characterized by the minimum risk inde-
pendently of the state of nature. Second, we identify, and then eliminate, some
decision functions that are dominated by others: this is the case of decisions
d_4, d_5, d_7, and d_8, which are dominated by decision d_3. The list of remaining
decision functions that are eligible is provided in Table 5.12.

TABLE 5.12
Risks associated with the four
eligible decision functions.

Decision d_h	Risks	
	θ_1: no oil	θ_2: oil
d_1	0	600
d_2	300	-600
d_3	48	-294
d_6	192	-492

Note that from Table 5.11 decisions d_1 and d_2 are admissible; however, we
outlined above (Table 5.9) that they are anything but reasonable. In practice,
only decisions d_3 and d_6 should be taken seriously into account by the decision
maker.

More generally, we stress that, as outlined in the above example, not all
the eligible decision functions are substantially reasonable. Moreover, on the
basis of the Rao-Blackwell theorem (Theorem 2.11), it turns out that

Theorem 5.1. *Let $l[\delta_h(\boldsymbol{x}), \theta]$ be a convex[3] loss function, $T(\boldsymbol{x})$ a sufficient*
statistic for θ, and $\delta_h(\boldsymbol{x})$ a decision function depending on $T(\boldsymbol{x})$, that is,

$$\delta_h(\boldsymbol{x}) = \delta_h(t) = R[\delta_h(t), \theta] = E_{\boldsymbol{x}}\{l[\delta_h(\boldsymbol{x})|T(\boldsymbol{x}) = t, \theta]\},$$

then (Rao-Blackwell theorem)

$$R[\delta_h(\boldsymbol{x}), \theta] = R[\delta_h(t), \theta] \leq R[\delta_j(\boldsymbol{x}), \theta],$$

for any decision function $\delta_j(\boldsymbol{x})$ different from $\delta_h(\boldsymbol{x})$.

In practice, the above theorem states that, in the presence of convex loss
functions, only decision functions based on sufficient statistics can be eligible
(necessary but not sufficient condition).

---

[3] A function $f(\cdot) : \mathbb{R}^n \to \mathbb{R}$ is convex if

$$f[\alpha x_1 + (1 - \alpha)x_2] \leq \alpha f(x_1) + (1 - \alpha)f(x_2), \quad 0 \leq \alpha \leq 1,$$

and if its domain is a convex set. A set $\mathcal{C}$ is convex if all points $x = \alpha x_1 + (1 - \alpha)x_2$, with
$x_1, x_2 \in \mathcal{C}$ and $0 \leq \alpha \leq 1$, are elements of $\mathcal{C}$.

5.5 Decisions with sample and prior information (Bayesian statistical decisional theory)

Until now we assumed that the decision maker only had prior information about the states of nature (Bayesian decision theory) or he/she was in a situation of uncertainty without any type of information about the states of nature (classical decision theory; Section 5.3), or could use only sample information for the computation of the risk (classical statistical decision theory; Section 5.4). We also said that, usually, it is not possible to identify a decision function able to minimize the risk for each state of nature; in other words, there is no dominant decision.

A first possibility consists of applying one of the criteria for the decisions under uncertainty described in Chapter 1 and Section 5.3, taking into account the sub-set of eligible (and reasonable) decision functions. A second solution is based on formally introducing a probability distribution on the states of nature. In such a way, we may compute the expected value of risk and identify the decision that minimizes it (*analysis in normal form*).

An alternative procedure that leads to the same result consists of updating the prior information, represented by the probability distribution of the states of nature, with additional sample information through the application of the Bayes' formula (*analysis in extensive form*). In this regard, we emphasize that often the revision of prior probabilities through the Bayes' formula can be quite complex; and this is especially true if it is not possible to identify a statistic whose distribution is uniquely determined by the sample data. On the other hand, when there are a lot of states of nature and sample results, the definition of all the decision functions is very complicated, even impossible.

As outlined in Section 5.4, the availability of sample information makes it possible to compute the risk or expected loss (Definition 5.1) for each decision function and each state of nature. In addition, if the probability distribution of the states of nature is also known, it will be possible to synthesize the information about risk through the expected risk, which is the expected value of the expected loss.

Definition 5.2. Expected risk. *Given a random sample* $x' = (x_1, x_2, \ldots, x_n)$, *a decision function* $d_h = \delta(x_1, x_2, \ldots, x_n) = \delta(x)$, *a finite discrete set of states of nature* $\Theta = (\theta_1, \ldots, \theta_k)$ *with (objective or subjective) probabilities* $\pi(\theta_1), \ldots, \pi(\theta_k)$, *the expected risk is defined as*

$$E_\theta[R(d_h, \theta_j)] = \sum_{j=1}^{k} R(d_h, \theta_j)\pi(\theta_j). \tag{5.4}$$

Similarly, if the set Θ *of states of nature is continuous, the expected risk is defined as*

$$E_\theta[R(d_h, \theta)] = \int_\theta R(d_h, \theta)\pi(\theta)d\theta.$$

As the risk is defined as the expected value of losses, that is (Definition 5.1),

$$R(d_h, \theta_j) = E_x\{l[\delta_h(x), \theta_j]\},$$

then the expected risk is obtained computing a double expected value of the loss function, with respect to the the sample result and to the state of nature:

$$E_\theta[R(d_h, \theta_j)] = E_\theta E_x\{l[\delta_h(x), \theta_j]\}.$$

It is worth noting the analogy between the concept of expected risk and the concept of expected expected utility introduced in Section 4.6: the main differences are that, in the expected risk, losses instead of utilities are considered and sample results are explicitly taken into account.

The optimal decision is the one d^* that minimizes the expected risk

$$d^* = \arg\left\{\min_{d_h} E_\theta\left[R(d_h, \theta_j)\right]\right\}$$

$$= \arg\left\{\min_{\delta_h}\left[E_\theta E_x(l(\delta_h(x), \theta_j))\right]\right\}.$$

If the parametric space Θ and the sample space Ω are both discrete, the above expression becomes (*analysis in normal form*)

$$d^* = \arg\left\{\min_{d_h}\left[E_\theta\left(R(d_h, \theta_j)\right)\right]\right\}$$

$$= \arg\left\{\min_{d_h}\left[\sum_{j=1}^{k} R(d_h, \theta_j)\pi(\theta_j)\right]\right\}$$

$$= \arg\left\{\min_{\delta_h}\left[\sum_{j=1}^{k}\left(\sum_{h=1}^{r} l(\delta_h(x), \theta_j)f(x|\theta_j)\right)\pi(\theta_j)\right]\right\}. \qquad (5.5)$$

Similarly, if the parametric space Θ and the sample space Ω are both continuous, the optimal decision is provided by

$$d^* = \arg\left\{\min_{d_h}\left[E_\theta\left(R(d_h, \theta_j)\right)\right]\right\}$$

$$= \arg\left\{\min_{d_h}\left[\int_\theta R(d_h, \theta)\pi(\theta)d\theta\right]\right\}$$

$$= \arg\left\{\min_{\delta_h}\left[\int_\theta\left(\int_x l(\delta_h(x), \theta)f(x|\theta)dx\right)\pi(\theta)d\theta\right]\right\}. \qquad (5.6)$$

Products $f(x|\theta_j)\pi(\theta_j)$ and $f(x|\theta)\pi(\theta)$ appear in the above expressions (5.5) and (5.6), respectively. Recalling the Bayes' formula (2.22), these products represent how the sample information contributes to update the prior information providing the posterior probability distribution of the states of nature (*analysis in extensive form*). In more detail, if the order of the two sums (integrals) is exchanged and taking into account the relation $\delta_h(\boldsymbol{x}) = a_i$, from Eq. (5.5) it follows

$$d^* = \arg\left\{\min_{\delta_h}\left[\sum_{j=1}^{k}\left(\sum_{h=1}^{r}l(\delta_h(\boldsymbol{x}),\theta_j)f(\boldsymbol{x}|\theta_j)\right)\pi(\theta_j)\right]\right\}$$

$$= \arg\left\{\min_{\delta_h}\left[\sum_{h=1}^{r}\left(\sum_{j=1}^{k}l(\delta_h(\boldsymbol{x}),\theta_j)\pi(\theta_j)\right)f(\boldsymbol{x}|\theta_j)\right]\right\}$$

$$= \arg\left\{\min_{a_i}\left[\sum_{i=1}^{m}\left(\sum_{j=1}^{k}l(a_i(\boldsymbol{x}),\theta_j)\pi(\theta_j)\right)f(\boldsymbol{x}|\theta_j)\right]\right\}$$

$$= \arg\left[\min_{a_i}\left(\sum_{j=1}^{k}l(a_i,\theta_j)\pi(\theta_j|\boldsymbol{x})\right)\right] = a^{**},$$

and, similarly, from Eq. (5.6) it follows

$$d^* = \arg\left\{\min_{\delta_h}\left[\int_\theta\left(\int_x l(\delta_h(\boldsymbol{x}),\theta)f(\boldsymbol{x}|\theta)d\boldsymbol{x}\right)\pi(\theta)d\theta\right]\right\}$$

$$= \arg\left\{\min_{\delta_h}\left[\int_x\left(\int_\theta l(\delta_h(\boldsymbol{x}),\theta)\pi(\theta)d\theta\right)f(\boldsymbol{x}|\theta)d\boldsymbol{x}\right]\right\}$$

$$= \arg\left[\min_{a_i}\left(\int_\theta l(a_i(\boldsymbol{x}),\theta)\pi(\theta|\boldsymbol{x})d\theta\right)\right] = a^{**}.$$

From the above equations it results in the perfect equivalence between analysis in normal form and analysis in extensive form. Indeed, the two procedures are based on the same data (i.e., sample observations and prior probabilities) and algebra (i.e., computation of expected values), being the difference in the order of the analytical operations:

$$d^* = \arg\left\{\min_{d_h}\left[E_\theta\left(R(d_h,\theta_j)\right)\right]\right\}$$

$$= \arg\left\{\min_{\delta_h}\left[E_\theta E_x\left(l(\delta_h(\boldsymbol{x}),\theta_j)\right)\right]\right\}$$

$$= \arg\left\{\min_{a_i}\left[E_x E_\theta\left(l(a_i(\boldsymbol{x}),\theta_j)\right)\right]\right\} = a^{**}.$$

Given the equivalence of the two approaches, the decision maker can choose what is less demanding from an algebraic and operational point of view. However, it has to be noted that the procedure based on the introduction of the decision functions (analysis in normal form) is quite burdensome. Indeed, in the Example 5.2 where a simple decision problem with just 2 actions (drill or not drill) and 3 sampling results was involved, the number of decision functions was equal to 27 (even if there were only two "reasonable" eligible decision functions). From a practical perspective, the analysis in extensive form, based on the minimization of the expected loss computed on the basis of the posterior probabilities, is easier, even if computational difficulties cannot be excluded.

To avoid misunderstandings, it has to be stressed that the optimal action a^* identified using only prior information (Bayesian decision theory) is given by (assuming a discrete parametric space)

$$a^* = \arg \left[\min_{a_i} \left(\sum_{j=1}^{k} l(a_i, \theta_j) \pi(\theta_j) \right) \right], \tag{5.7}$$

and is substantially different from the optimal action a^{**} identified using both prior information and sample observations

$$a^{**} = \arg \left[\min_{a_i} \left(\sum_{j=1}^{k} l(a_i, \theta_j) \pi(\theta_j|x) \right) \right], \tag{5.8}$$

as happens under the Bayesian statistical decision theory setting. In such a setting the decision table (Table 5.1) in the presence of prior and sample information becomes as shown in Table 5.13, if losses are taken into account.

TABLE 5.13
Decision table in the presence of prior probabilities of the states of nature and sample information and with consequences expressed in terms of losses, $l_{ij} = l(a_i, \theta_j)$ $(i = 1, \ldots, m; j = 1, \ldots, k)$.

	Posterior probabilities			
Action	$\pi(\theta_1|x)$	$\pi(\theta_2|x)$	$\ldots$	$\pi(\theta_k|x)$
a_1	l_{11}	l_{12}	$\ldots$	l_{1k}
a_2	l_{21}	l_{22}	$\ldots$	l_{2k}
$\vdots$	$\ldots$	$\ldots$	$\vdots$	$\ldots$
a_m	l_{m1}	l_{m2}	$\ldots$	l_{mk}

In terms of utilities, the decision table is shown in Table 5.14 and the optimal action a^{**} corresponds to the one that satisfies the relationship

$$a^{**} = \arg \left[\max_{a_i} \left(\sum_{j=1}^{k} u(a_i, \theta_j) \pi(\theta_j|x) \right) \right], \tag{5.9}$$

which differs from Eq. (3.1) for the presence of the posterior probabilities $\pi(\theta_j|x)$ instead of the prior probabilities $\pi(\theta_j)$.

TABLE 5.14

Decision table in the presence of prior probabilities of the states of nature and sample information and with consequences expressed in terms of utilities, $u_{ij} = u(a_i, \theta_j)$ $(i = 1, \ldots, m; j = 1, \ldots, k)$.

Action	Posterior probabilities			
	$\pi(\theta_1|\boldsymbol{x})$	$\pi(\theta_2|\boldsymbol{x})$	$\ldots$	$\pi(\theta_k|\boldsymbol{x})$
a_1	u_{11}	u_{12}	$\ldots$	u_{1k}
a_2	u_{21}	u_{22}	$\ldots$	u_{2k}
$\vdots$	$\ldots$	$\ldots$	$\vdots$	$\ldots$
a_m	u_{m1}	u_{m2}	$\ldots$	u_{mk}

In what follows we discuss two examples that illustrate how the prior probabilities can be used to calculate the expected risk and can be updated in accordance with the sample results.

Example 5.3. *In Example 5.2, the decision to proceed or not to drill an oil well has been examined taking into account the sample information gathered through two seismographs.*

If we assume that the prior probabilities of the two states of nature are known, then the resulting expected risks associated with the eligible decision functions of Table 5.11 can now be calculated, according to Eq. (5.4). As an example, Table 5.15 shows the results for two different prior distributions:

- *case (i):* $\pi(\theta_1) = 0.20$ *and* $\pi(\theta_2) = 0.80$;

- *case (ii):* $\pi(\theta_1) = 0.55$ *and* $\pi(\theta_2) = 0.45$.

TABLE 5.15

Risk and expected risk associated with the four eligible decision functions (in bold the minimum expected risk).

Decision d_i	Risk		Expected risk	
	θ_1: no oil	θ_2: oil	case (i)	case (ii)
d_1	0	600	480	270
d_2	300	-600	**-420**	-105
d_3	48	-294	-226	-106
d_6	192	-492	-240	**-115**

In case (i) decision that minimizes the expected risk is d_2, which consists in drilling whatever the result of the sample survey: in such a specific case, the sample information is completely irrelevant. On the other hand, in case

(ii) *the optimal decision is d_6, according to which the decision maker should not proceed with drilling if both seismographs indicate the absence of oil and to proceed with drilling when one or both seismographs indicate the presence of oil (Table 5.9).*

Instead of proceeding as above, that is, through the introduction of the decision functions and the calculation of related expected risks (analysis in normal form), we can analyze the decision problem - and achieve the same final optimal solution - through the computation of the posterior probabilities (analysis in extensive form). In detail, the decision problem can be addressed according to the following steps:

- *fixing the (objective or subjective) prior probabilities on the states of nature;*

- *gathering sample data and recording the results;*

- *updating, through the Bayes' formula, the prior probabilities of the states of nature on the basis of the sample data;*

- *applying the Bayesian paradigm for the identification of the action corresponding to the minimum expected risk.*

If we apply this procedure, the decision Table 5.12 is no longer necessary, because the Bayesian decision criterion can be directly applied to the losses, using the posterior probabilities.

Indeed, recalling the Bayes' formula (2.22) and the values of joint probabilities computed in Eq. (5.3), the posterior probabilities obtained under case (ii) (the procedure is the same for case (i)) follow.

- *Posterior probabilities of θ_1 (no oil) and θ_2 (oil) given that both seismographs indicate the absence of oil:*

$$\pi[\theta_1|(0,0)] = \frac{p(0,0|\theta_1)\pi(\theta_1)}{p(0,0|\theta_1)\pi(\theta_1) + p(0,0|\theta_2)\pi(\theta_2)}$$
$$= \frac{0.36 \cdot 0.55}{0.36 \cdot 0.55 + 0.09 \cdot 0.45} = 0.83;$$
$$\pi[\theta_2|(0,0)] = 1 - \pi[\theta_1|(0,0)] = 1 - 0.83 = 0.17.$$

- *Posterior probabilities of θ_1 (no oil) and θ_2 (oil) given that one seismograph indicates the presence of oil and the other one indicates the absence of oil:*

$$\pi[\theta_1|(0,1) \cup (1,0)] = \frac{p[(0,1) \cup (1,0)|\theta_1]\pi(\theta_1)}{p[(0,1) \cup (1,0)|\theta_1]\pi(\theta_1) + p[(0,1) \cup (1,0)|\theta_2]\pi(\theta_2)}$$
$$= \frac{(0.24 + 0.24) \cdot 0.55}{(0.24 + 0.24) \cdot 0.55 + (0.21 + 0.21) \cdot 0.45} = 0.58;$$
$$\pi[\theta_2|(0,1) \cup (1,0)] = 1 - \pi[\theta_1|(0,1) \cup (1,0)] = 1 - 0.58 = 0.42;$$

- *Posterior probabilities of θ_1 (no oil) and θ_2 (oil) given that both seismographs indicate the presence of oil:*

$$\pi[\theta_1|(1,1)] = \frac{p(1,1|\theta_1)\pi(\theta_1)}{p(1,1|\theta_1)\pi(\theta_1) + p(1,1|\theta_2)\pi(\theta_2)}$$

$$= \frac{0.16 \cdot 0.55}{0.16 \cdot 0.55 + 0.49 \cdot 0.45} = 0.03;$$

$$\pi[\theta_2|(1,1)] = 1 - \pi[\theta_1|(1,1)] = 1 - 0.03 = 0.97.$$

Note that the same probabilities can also be obtained directly applying the mass probability distribution in Eq. (2.2) for a binomial random variable X that denotes the number of seismographs indicating the presence of oil out of $n = 2$ independent seismographs; possible values of X are 0, 1, and 2 and the probability of success (i.e., indication of presence of oil) of the binomial distribution is 0.45:

$$\pi[\theta_1|X = 0] = \pi[\theta_1|(0,0)] = 0.83;$$
$$\pi[\theta_2|X = 0] = \pi[\theta_2|(0,0)] = 0.17;$$
$$\pi[\theta_1|X = 1] = \pi[\theta_1|(0,1) \cup (1,0)] = 0.58;$$
$$\pi[\theta_2|X = 1] = \pi[\theta_2|(0,1) \cup (1,0)] = 0.42;$$
$$\pi[\theta_1|X = 2] = \pi[\theta_1|(1,1)] = 0.03;$$
$$\pi[\theta_2|X = 2] = \pi[\theta_2|(1,1)] = 0.97.$$

According to the observed sample result, the decision table is built, as shown in Tables 5.16-5.18. From Tables 5.16-5.18, the decision maker concludes in favor of not drilling when both seismographs indicate the absence of oil (Table 5.16) and to proceed with drilling when one (Table 5.17) or both (Table 5.18) seismographs indicate the presence of oil. Note that these choices exactly correspond to the optimal decision d_6 that minimizes the expected risk (Table 5.15, case (ii) in the last column).

TABLE 5.16

Decision table: drilling problem with posterior probabilities, when both seismographs indicate absence of oil (minimum risk in bold).

Action	Losses		Expected loss (risk)
	θ_1: no oil $\pi[\theta_1|X = 0] = 0.83$	θ_2: oil $\pi[\theta_2|X = 0] = 0.17$	
a_1: don't drill	0	600	**102**
a_2: drill	300	-600	147

TABLE 5.17
Decision table: drilling problem with posterior probabilities, when one
seismograph indicates presence of oil (minimum risk in bold).

Action	Losses		Expected loss (risk)
	θ_1: no oil $\pi[\theta_1\|X=1]=0.58$	θ_2: oil $\pi[\theta_2\|X=1]=0.42$	
a_1: don't drill	0	600	252
a_2: drill	300	-600	**-78**

TABLE 5.18
Decision table: drilling problem with posterior probabilities, when both
seismographs indicate presence of oil (minimum risk in bold).

Action	Losses		Expected loss (risk)
	θ_1: no oil $\pi[\theta_1\|X=2]=0.03$	θ_2: oil $\pi[\theta_2\|X=2]=0.97$	
a_1: don't drill	0	600	582
a_2: drill	300	-600	**-573**

**Example 5.4. Sampling from Bernoulli distribution and Bernoulli
prior probabilities for the states of nature.** *Let us assume that the
proportion θ_j of defective pieces in a batch can take four different values,
$\theta_1 = 0.02$, $\theta_2 = 0.03$, $\theta_3 = 0.04$, and $\theta_4 = 0.05$, and the related prior probabilities are $\pi(\theta_1) = 0.10$, $\pi(\theta_2) = 0.20$, $\pi(\theta_3) = 0.40$, and $\pi(\theta_4) = 0.30$. Let
us suppose we select a simple random sample of $n = 100$ pieces and observe
3 defective pieces. How should the prior information be revised to take into
account the sample results? Basically, what must be done to pass from $\pi(\theta_j)$
to $\pi(\theta_j|x)$, $j = 1,2,3,4$?*

*Let X be the random variable describing the number of defective pieces
in the batch out of the $n = 100$ pieces selected. X can assume values
$0, 1, 2, \ldots, x, \ldots, 100$, and the conditional probability - likelihood - of the sample result given a state of nature is provided by the binomial distribution (Eq.
(2.2))*

$$f(x|\theta_j) = \binom{100}{x}\theta_j^x(1-\theta_j)^{100-x}$$

and the related joint distribution is

$$p(X = x \cap \theta = \theta_j) = f(x|\theta_j)\pi(\theta_j) = \binom{100}{x}\theta_j^x(1-\theta_j)^{100-x}\pi(\theta_j).$$

Therefore, the unconditional probability $f(x)$ is

$$f(x) = p(X = x) = \sum_{j=1}^{4} p\left[(X = x) \cap (\theta = \theta_j)\right] = \sum_{j=1}^{4} f(x|\theta_j)\pi(\theta_j)$$

$$= \binom{100}{x} 0.02^x (1 - 0.02)^{100-x} 0.10 + \binom{100}{x} 0.03^x (1 - 0.03)^{100-x} 0.20 +$$

$$+ \binom{100}{x} 0.04^x (1 - 0.04)^{100-x} 0.40 + \binom{100}{x} 0.05^x (1 - 0.05)^{100-x} 0.30,$$

from which, by substituting $x = 3$, it results

$$f(3) = p(X = 3) = \sum_{j=1}^{4} p\left[(X = 3) \cap (\theta = \theta_j)\right] = \sum_{j=1}^{4} f(3|\theta_j)\pi(\theta_j)$$

$$= \binom{100}{3} 0.02^3 (1 - 0.02)^{100-3} 0.10 + \binom{100}{3} 0.03^3 (1 - 0.03)^{100-3} 0.20 +$$

$$+ \binom{100}{3} 0.04^3 (1 - 0.04)^{100-3} 0.40 + \binom{100}{3} 0.05^3 (1 - 0.05)^{100-3} 0.30$$

$$= 0.18 \cdot 0.10 + 0.23 \cdot 0.20 + 0.20 \cdot 0.40 + 0.14 \cdot 0.30 = 0.186.$$

The value 0.186 is the probability that the random variable X equals 3, that is, the probability of obtaining the observed sample result. On the basis of the sample result $X = 3$ and the prior probabilities, the posterior probabilities $\pi(\theta_j|X = 3)$ can now be computed as

$$\pi(\theta_j|X = 3) = \frac{f(3|\theta_j)\pi(\theta_j)}{\sum_{i=1}^{4} f(3|\theta_i)\pi(\theta_i)}.$$

Results are summarized in Table 5.19.

TABLE 5.19
Computation of posterior probabilities for $X = 3$.

State of nature θ_j	Prior prob. $\pi(\theta_j)$	Likelihood $f(3|\theta_j)$	Joint prob. $f(3|\theta_j)\pi(\theta_j)$	Posterior prob. $\pi(\theta_j|X = 3)$
0.02	0.10	0.18	0.018	0.097
0.03	0.20	0.23	0.046	0.248
0.04	0.40	0.20	0.080	0.430
0.05	0.30	0.14	0.042	0.225
	1.00		0.186	1.000

As can easily be inferred by data shown in Table 5.19, the sample results have produced changes in the probability distribution of the states of nature (from $\pi(\theta_j)$ to $\pi(\theta_j|X = 3)$). On the basis of these values, the decision maker can proceed with the calculation of the expected losses or with further sampling.

In this latter case, a further revision of probabilities will be possible, with posterior probabilities that now assume the role of prior probabilities. It can be easily verified that the procedure that updates the probabilities in two steps produces the same results as the procedure that adjusts the results of the two measurements in a single step.

In the examples discussed above, updating prior probabilities on the basis of the sample evidence is relatively simple; however in other cases the solution is not so immediate. Fortunately, there exist situations, where the process of updating is easily solved through the use of conjugated distributions (Section 2.10.1). We remind that a class of prior distributions $\pi(\theta)$ is conjugated to the class of likelihood functions $f(x|\theta)$, if the posterior probability $\pi(\theta|x)$ belongs to the same family as $\pi(\theta)$.

5.6   Perfect information and sample information

In Section 5.4, it was outlined how the decision of gathering sample information is a somewhat delicate matter. On the one hand, new information can reduce the state of uncertainty in which the decision maker operates. On the other hand, carrying out a survey to collect information has a cost and the extent of the reduction of the uncertainty is not known.

Moreover, as the decision to proceed with the acquisition of further information must be taken a priori, this circumstance represents an additional element of uncertainty related to the sampling results. In addition, it has to be taken into account that carrying out a sample survey requires time and may cause a delay in the decision-making process. For instance, in public investments a delay in the beginning of the work may be extremely expensive at least for two reasons, one related to political and social aspects and the other one related to technical issues. In both cases, it has to be taken into account that when information is acquired until when the decision becomes operative the actual context can be significantly changed.

To summarize, we should account for at least three elements when we consider whether to proceed with a sample survey:

- the survey involves a cost and it is not known with certainty to what extent this cost will be offset by the additional information;

- a limited period of time may be available to carry out the survey;

- the decision to acquire further information must be taken a priori.

It has to be outlined that the additional information can be used both to change the distribution of the probability on the states of nature and also to modify the set of consequences or the form of the utility function. Here we focus on the first aspect.

The measurement of goodness of additional information is given by comparing the expected payoff (utility/loss) calculated using the new information and the expected payoff available without sampling. In other words, carrying out a survey will be convenient whenever the cost of sampling is more than compensated for the reduction of the state of uncertainty. In this regard, the sample size plays a relevant role. On one hand, the higher the sample size is, the more information will be available; on the other hand, a higher reduction in the state of uncertainty will result in a higher cost. Then, the sample size should be increased until the cost of an additional unit of information balances the increase/decrease in the expected payoff (i.e., the marginal cost equals the marginal payoff).

Two relevant concepts involved in the cost-benefit analysis of a sample survey are now introduced:

- Expected Value of Perfect Information (EVPI);

- Expected Value of Sample Information (EVSI).

The EVPI is obtained by the difference between the expected payoff (utility/loss) corresponding to the optimal action (i.e., the maximum expected utility or the minimum expected loss) and the expected payoff corresponding to the perfect knowledge of the states of nature. If the decision-maker knew the true state of nature, say θ_j, he/she would have no difficulty in identifying the payoff corresponding to the best action a_p^*, that is,

$$\text{payoff}(a_p^*) = \max_i u(a_i, \theta_j) = \min_i l(a_i, \theta_j).$$

On the other hand, in situations of risk with a discrete space Θ the payoff related to the optimal action a^* is given by

$$\text{payoff}(a^*) = \max_i \left(\sum_{j=1}^{k} u(a_i, \theta_j)\pi(\theta_j) \right) = \min_i \left(\sum_{j=1}^{k} l(a_i, \theta_j)\pi(\theta_j) \right),$$

as it results from Eq. (3.1), when payoffs are expressed in terms of utilities, and from Eq. (5.7), when payoffs are expressed in terms of losses.

Then, the EVPI is defined as

$$EVPI = \text{payoff}(a_p^*) - \text{payoff}(a^*)$$

$$= \max_i u(a_i, \theta_j) - \max_i \left(\sum_{j=1}^{k} u(a_i, \theta_j)\pi(\theta_j) \right)$$

$$= \min_i l(a_i, \theta_j) - \min_i \left(\sum_{j=1}^{k} l(a_i, \theta_j)\pi(\theta_j) \right).$$

Moreover, when both prior information and sample information are available, the optimal action a^{**} is identified through the maximization of the expected utility or the minimization of the expected loss computed with respect to the posterior probabilities, as in Eqs. (5.9) and (5.8). Then, the EVSI is obtained by the difference between the expected payoff (utility/loss) based on the prior probabilities and the expected payoff (utility/loss) based on the posterior probabilities, that is,

$$EVSI = \text{payoff}(a^{**}) - \text{payoff}(a^*)$$

$$= \max_i \left(\sum_{j=1}^{k} u(a_i, \theta_j)\pi(\theta_j|\boldsymbol{x}) \right) - \max_i \left(\sum_{j=1}^{k} u(a_i, \theta_j)\pi(\theta_j) \right)$$

$$= \min_i \left(\sum_{j=1}^{k} l(a_i, \theta_j)\pi(\theta_j|\boldsymbol{x}) \right) - \min_i \left(\sum_{j=1}^{k} l(a_i, \theta_j)\pi(\theta_j) \right).$$

As already outlined, any sample survey involves a cost that can be defined by the relation

$$C(n) = C_f + C_v(n) = C_f + nC_v,$$

where the total cost of sampling C consists of a fixed quota C_f plus a variable quota $C_v(n)$ whose amount depends on the sample size n; under the assumption of linearity, $C_v(n)$ is equal to nC_v. The Expected Net Gain associated with the Sampling (ENGS) is given by the difference between the EVSI and the related costs, that is,

$$ENGS = EVSI - C(n).$$

Obviously, the optimal sample size n is the one that corresponds to the maximum value of $ENGS > 0$.

Example 5.5. *Let us consider again the decision-making problem of whether or not to proceed with drilling. The decision table of losses is shown in Table 5.20, where prior probabilities $\pi(\theta_1) = 0.55$ and $\pi(\theta_2) = 0.45$ of case (ii) considered in the Example 5.3 are here taken into account (see also Table 5.15 last column).*

On the basis of the prior information, the decision maker will choose the action corresponding to the minimum expected loss, that is, action a_2. However, the decision maker, if not completely persuaded of the values assigned to the priori probabilities, could decide to gather further information. In the Examples 5.2 and 5.3 the possibility of gathering information on the states of nature was examined using zero-cost seismographs. However, any sampling survey involves a cost, which is here explicitly taken into account: in the specific case, we assume a linear function without fixed costs.

TABLE 5.20

Drilling problem with losses as consequences and prior probabilities (in bold the minimum expected loss).

| | Losses | | Expected |
| Action | θ_1: no oil | θ_2: oil | loss |
	$\pi(\theta_1) = 0.55$	$\pi(\theta_2) = 0.45$	
a_1: don't drill	0	600	270
a_2: drill	300	-600	**-105**

The decision maker can proceed along two different paths: the first one, which is the most efficient but not easy to implement, consists in gathering sample information and, on the basis of the sample evidence, to decide whether to continue with sampling or make the final choice; the second path consists of carrying out a complete analysis (pre-posterior analysis), which allows us the identification of the optimal sample size.

Let us start with considering the use of $n = 1$ seismograph and the Bernoulli random variable X that assumes value $X = 0$ if the seismograph indicates absence of oil and value $X = 1$ if the seismograph indicates presence of oil.

On the basis of probabilities $p(X = x|\theta_j)$ ($j = 1, 2$) shown in Table 5.8, the unconditional probabilities for $X = 0$ (absence of oil) and $X = 1$ (presence of oil) are given by:

$$p(X = 0) = p(X = 0|\theta_1)\pi(\theta_1) + p(X = 0|\theta_2)\pi(\theta_2)$$
$$= 0.60 \cdot 0.55 + 0.30 \cdot 0.45 = 0.47;$$
$$p(X = 1) = p(X = 1|\theta_1)\pi(\theta_1) + p(X = 1|\theta_2)\pi(\theta_2)$$
$$= 0.40 \cdot 0.55 + 0.70 \cdot 0.45 = 0.54.$$

Then, through the Bayes' rule, the posterior probabilities of θ_1 and θ_2 follow:

$$\pi(\theta_1|X = 0) = \frac{p(X = 0|\theta_1)\pi(\theta_1)}{p(X = 0)} = \frac{0.60 \cdot 0.55}{0.47} = 0.71;$$
$$\pi(\theta_1|X = 1) = \frac{p(X = 1|\theta_1)\pi(\theta_1)}{p(X = 1)} = \frac{0.40 \cdot 0.55}{0.54} = 0.41;$$
$$\pi(\theta_2|X = 0) = 1 - \pi(\theta_1|X = 0) = 1 - 0.71 = 0.29;$$
$$\pi(\theta_2|X = 1) = 1 - \pi(\theta_1|X = 1) = 1 - 0.41 = 0.59.$$

Decision Table 5.20 is replaced by the new decision Table 5.21, where prior probabilities $\pi(\theta_j)$ are now substituted by posterior probabilities $\pi(\theta_j|\boldsymbol{x})$, $j = 1, 2$.

With $n = 1$ seismograph, the best action is again a_2 (drill) independent of the sample result: if $X = 0$ the expected loss associated with a_2 is \$39,000

TABLE 5.21

Drilling problem with posterior probabilities $\pi(\theta_j | X = x)$, sample results $X = x$ ($j = 1, 2$, $x = 0, 1$), and expected loss, for $n = 1$ (in bold the minimum expected loss).

	Losses			
	θ_1: no oil	θ_2: oil	**Sample**	**Expected**
Action	$\pi(\theta_1 \| X = 0) = 0.71$	$\pi(\theta_2 \| X = 0) = 0.29$	**result**	**loss**
	$\pi(\theta_1 \| X = 1) = 0.41$	$\pi(\theta_2 \| X = 1) = 0.59$		
a_1: don't drill	0	600	$X = 0$	174
a_1: don't drill	0	600	$X = 1$	354
a_2: drill	300	-600	$X = 0$	**39**
a_2: drill	300	-600	$X = 1$	**-231**

(against \$174,000 for a_1) and if $X = 1$ the expected loss associated with a_2 is −\$231,000 (against \$354,000 for a_1). Therefore, the expected unconditional loss is given by

$$\min_i \left[\sum_j l(a_i, \theta_j) \pi(\theta_j | x) \right] = 39,000 \cdot 0.47 + (-231,000) \cdot 0.54 = -105,450$$

and, under the assumption that a seismograph costs \$400, we obtain

$$EVSI = -105,000 - (-105,450) = 450$$

and

$$ENGS = EVSI - C(1) = 450 - 400 = 50.$$

In conclusion, the acquisition of further information through the use of just one seismograph ($n = 1$) is not totally disadvantageous, as $ENGS > 0$; however it does not appear particularly appealing, as the expected net gain amounts to just \$50.

However, if $n = 2$ seismographs are taken into account the decision maker achieves a different conclusion. Now, random variable X describing the sample results assumes three different values: $X = 0$ when both seismographs indicate absence of oil, $X = 1$ when only one seismograph indicates presence of oil, and $X = 2$ when both seismographs indicate presence of oil.

On the basis of probabilities provided in Eqs. (5.3), the unconditional probabilities of X are given by:

$$p(X = 0) = p(X = 0|\theta_1)\pi(\theta_1) + p(X = 0|\theta_2)\pi(\theta_2)$$
$$= 0.36 \cdot 0.55 + 0.09 \cdot 0.45 = 0.24;$$
$$p(X = 1) = p(X = 1|\theta_1)\pi(\theta_1) + p(X = 1|\theta_2)\pi(\theta_2)$$
$$= 0.48 \cdot 0.55 + 0.42 \cdot 0.45 = 0.45;$$
$$p(X = 2) = p(X = 2|\theta_1)\pi(\theta_1) + p(X = 2|\theta_2)\pi(\theta_2)$$
$$= 0.16 \cdot 0.55 + 0.49 \cdot 0.45 = 0.31.$$

Then, using the Bayes' rule,

$$\pi(\theta_1|X=0) = \frac{p(X=0|\theta_1)\pi(\theta_1)}{p(X=0)} = \frac{0.36 \cdot 0.55}{0.24} = 0.83;$$

$$\pi(\theta_1|X=1) = \frac{p(X=1|\theta_1)\pi(\theta_1)}{p(X=1)} = \frac{0.48 \cdot 0.55}{0.45} = 0.58;$$

$$\pi(\theta_1|X=2) = \frac{p(X=2|\theta_1)\pi(\theta_1)}{p(X=2)} = \frac{0.16 \cdot 0.55}{0.31} = 0.29;$$

$$\pi(\theta_2|X=0) = 1 - \pi(\theta_1|X=0) = 1 - 0.83 = 0.17;$$

$$\pi(\theta_2|X=1) = 1 - \pi(\theta_1|X=1) = 1 - 0.58 = 0.42;$$

$$\pi(\theta_2|X=2) = 1 - \pi(\theta_1|X=2) = 1 - 0.29 = 0.72.$$

Decision Table 5.21 is updated as shown in Table 5.22.

TABLE 5.22
Drilling problem with posterior probabilities $\pi(\theta_j|X=x)$, sample results $X = x$ ($j = 1, 2$, $x = 0, 1, 2$), and expected loss, for $n = 2$ (in bold the minimum expected loss).

	Losses			
	θ_1: no oil	θ_2: oil	**Sample**	**Expected**
Action	$\pi(\theta_1|X=0) = 0.83$	$\pi(\theta_2|X=0) = 0.17$	**result**	**loss**
	$\pi(\theta_1|X=1) = 0.58$	$\pi(\theta_2|X=1) = 0.42$		
	$\pi(\theta_1|X=2) = 0.29$	$\pi(\theta_2|X=2) = 0.72$		
a_1: don't drill	0	600	$X = 0$	**102**
a_1: don't drill	0	600	$X = 1$	250
a_1: don't drill	0	600	$X = 2$	429
a_2: drill	300	-600	$X = 0$	147
a_2: drill	300	-600	$X = 1$	**-76**
a_2: drill	300	-600	$X = 2$	**-344**

As shown in Table 5.22, the best action is a_1 (do not drill) if the sample result is $X = 0$ and a_2 (drill) if the sample result is $X = 1$ or $X = 2$. The corresponding conditional expected losses are 102,000 when $X = 0$, $-75,300$ when $X = 1$, and $-344,000$ when $X = 2$. Therefore, the expected unconditional loss is given by

$$\min_i \left[\sum_j l(a_i, \theta_j)\pi(\theta_j|x) \right] = 102,000 \cdot 0.24 + (-75,300) \cdot 0.45 \\ + (-344,000) \cdot 0.31 = -115,582$$

and, under the assumption that a seismograph costs $400, we obtain

$$EVSI = -105,000 - (-115,582) = 10,582$$

and

$$ENGS = EVSI - C(2) = 10,582 - 400 \cdot 2 = 9,782.$$

*The conclusion is that it is convenient to proceed with the use of two seis-
mographs, as the net gain associated with a sample size of $n = 2$ is definitely
positive. Note that the strategy suggested by the use of two seismographs taking
into account their cost is the same as the one identified in the Example 5.3
(use of two seismographs without accounting for the related cost), that is, a_1
if both seismographs indicate absence of oil and a_2 otherwise (i.e., decision
function d_6).*

*If the decision maker prosecutes the analysis with the use of $n = 3$ seis-
mographs, the sample results are $X = 0$ (no seismograph indicates presence of
oil), $X = 1$ (one seismograph indicates presence of oil), $X = 2$ (two seismographs indicate presence of oil), and $X = 3$ (all three seismographs indicate
presence of oil). The conditional probabilities of sample results follow from
Table 5.8, according to the same lines as in Eq. (5.3),*

$$p(X = 0|\theta_1) = p(0, 0, 0|\theta_1) = 0.60^3 = 0.22;$$

$$p(X = 0|\theta_2) = p(0, 0, 0|\theta_2) = 0.30^3 = 0.03;$$

$$p(X = 1|\theta_1) = p[(0, 0, 1) \cup (0, 1, 0) \cup (1, 0, 0)|\theta_1] = 3 \cdot 0.60^2 \cdot 0.40 = 0.43;$$

$$p(X = 1|\theta_2) = p[(0, 0, 1) \cup (0, 1, 0) \cup (1, 0, 0)|\theta_2] = 3 \cdot 0.30^2 \cdot 0.70 = 0.19;$$

$$p(X = 2|\theta_1) = p[(0, 1, 1) \cup (1, 1, 0) \cup (1, 0, 1)|\theta_1] = 3 \cdot 0.60 \cdot 0.40^2 = 0.29;$$

$$p(X = 2|\theta_2) = p[(0, 1, 1) \cup (1, 1, 0) \cup (1, 0, 1)|\theta_2] = 3 \cdot 0.30 \cdot 0.70^2 = 0.44;$$

$$p(X = 3|\theta_1) = p(1, 1, 1|\theta_1) = 0.40^3 = 0.06;$$

$$p(X = 3|\theta_2) = p(1, 1, 1|\theta_2) = 0.70^3 = 0.34.$$

Then, the unconditional probabilities $p(X = x)$ $(x = 0, 1, 2, 3)$ follow

$$p(X = 0) = p(X = 0|\theta_1)\pi(\theta_1) + p(X = 0|\theta_2)\pi(\theta_2)$$
$$= 0.22 \cdot 0.55 + 0.03 \cdot 0.45 = 0.13;$$

$$p(X = 1) = p(X = 1|\theta_1)\pi(\theta_1) + p(X = 1|\theta_2)\pi(\theta_2)$$
$$= 0.43 \cdot 0.55 + 0.19 \cdot 0.45 = 0.33;$$

$$p(X = 2) = p(X = 2|\theta_1)\pi(\theta_1) + p(X = 2|\theta_2)\pi(\theta_2)$$
$$= 0.29 \cdot 0.55 + 0.44 \cdot 0.45 = 0.36;$$

$$p(X = 3) = p(X = 3|\theta_1)\pi(\theta_1) + p(X = 3|\theta_2)\pi(\theta_2)$$
$$= 0.06 \cdot 0.55 + 0.34 \cdot 0.45 = 0.19,$$

and, using the Bayes' rule,

$$\pi(\theta_1|X=0) = \frac{p(X=0|\theta_1)\pi(\theta_1)}{p(X=0)} = \frac{0.22 \cdot 0.55}{0.13} = 0.91;$$

$$\pi(\theta_1|X=1) = \frac{p(X=1|\theta_1)\pi(\theta_1)}{p(X=1)} = \frac{0.43 \cdot 0.55}{0.33} = 0.74;$$

$$\pi(\theta_1|X=2) = \frac{p(X=2|\theta_1)\pi(\theta_1)}{p(X=2)} = \frac{0.29 \cdot 0.55}{0.36} = 0.44;$$

$$\pi(\theta_1|X=3) = \frac{p(X=3|\theta_1)\pi(\theta_1)}{p(X=3)} = \frac{0.06 \cdot 0.55}{0.19} = 0.19;$$

$$\pi(\theta_2|X=0) = 1 - \pi(\theta_1|X=0) = 1 - 0.91 = 0.09;$$

$$\pi(\theta_2|X=1) = 1 - \pi(\theta_1|X=1) = 1 - 0.74 = 0.26;$$

$$\pi(\theta_2|X=2) = 1 - \pi(\theta_1|X=2) = 1 - 0.44 = 0.56;$$

$$\pi(\theta_2|X=3) = 1 - \pi(\theta_1|X=3) = 1 - 0.19 = 0.81.$$

The conditional expected losses corresponding to each action and sample result are shown in Table 5.23, which updates Table 5.22 when $n = 3$.

TABLE 5.23
Drilling problem with posterior probabilities $\pi(\theta_j|X=x)$, sample results $X = x$ ($j = 1, 2$, $x = 0, 1, 2, 3$), and expected loss, for a sample size $n = 3$ (in bold the minimum expected loss).

	Losses			
Action	θ_1: no oil $\pi(\theta_1\|X=0) = 0.91$ $\pi(\theta_1\|X=1) = 0.74$ $\pi(\theta_1\|X=2) = 0.44$ $\pi(\theta_1\|X=3) = 0.19$	θ_2: oil $\pi(\theta_2\|X=0) = 0.09$ $\pi(\theta_2\|X=1) = 0.26$ $\pi(\theta_2\|X=2) = 0.56$ $\pi(\theta_2\|X=3) = 0.81$	**Sample result**	**Expected loss**
a_1: don't drill	0	600	$X=0$	**56**
a_1: don't drill	0	600	$X=1$	158
a_1: don't drill	0	600	$X=2$	334
a_1: don't drill	0	600	$X=3$	489
a_2: drill	300	-600	$X=0$	216
a_2: drill	300	-600	$X=1$	**63**
a_2: drill	300	-600	$X=2$	**-200**
a_2: drill	300	-600	$X=3$	**-433**

The optimal action, which depends on the sample results, corresponds to a_1 *(don't drill) when all the three seismographs indicate absence of oil, whereas action* a_2 *(drill) is preferred when at least one seismograph indicates presence of oil. In more detail, the conditional expected losses associated with the optimal actions are:*

- 56,000 *if the sample result is* $X = 0$;

- 63,000 *if the sample result is* $X = 1$;

- $-200,000$ *if the sample result is $X = 2$;*

- $-433,000$ *if the sample result is $X = 3$.*

Then, the expected unconditional loss is given by $56,000 \cdot 0.13 + 63,000 \cdot 0.33 + (-200,000) \cdot 0.36 + (-433,000) \cdot 0.19 = -\$126,200$ *and, under the assumption that a seismograph costs \$400, we obtain*

$$EVSI = -105,000 - (-126,200) = 21,200$$

and

$$ENGS = EVSI - C(3) = 21,200 - 400 \cdot 3 = 20,000.$$

Since the ENGS for $n = 3$ is higher than that obtained for $n = 2$ (20,000 versus 9,782), we again conclude that the decision maker should proceed with the use of at least three seismographs.

Proceeding along the same lines as above described, we can evaluate the convenience of gathering information through the use of $n = 4$ seismographs. In such a case, the ENGS obtained in correspondence with a sample of dimension $n = 4$ is lower than that obtained for $n = 3$, denoting a reduction of the marginal advantage given by one more sample unit, which leads to set the optimal sample size in $n = 3$.

The decisional problem of drilling is summarized in the decision tree shown in Figure 5.3. If one compares the best alternative without sample information, that is, action a_2 (drill) having an expected loss equal to $-105,000$, with the best alternative in the presence of sample information, that is, action a_2 (drill) having an expected loss equal to $-126,200$, we conclude that the best strategy consists of collecting sample information with an optimal sample size equal to $n = 3$. Then, the final decision (i.e., drill or not drill) depends on the observed sampling result: drill if at least one seismograph indicates the presence of oil, don't drill otherwise.

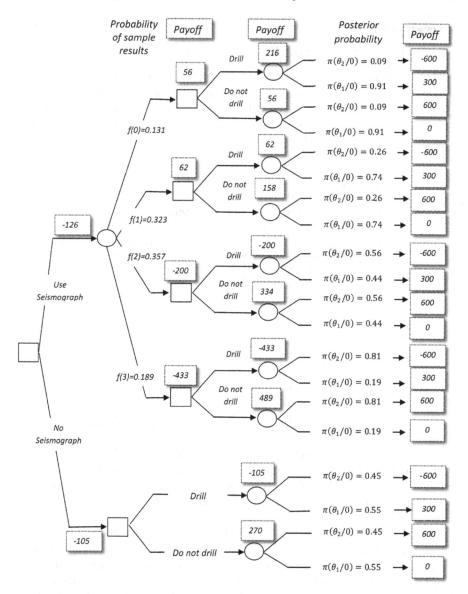

FIGURE 5.3
Drilling problem: decision tree chronologically displaying decisional nodes and chances nodes.

5.7   Case study: Seeding hurricanes

The emergence of hurricanes continues more and more to affect the United States as well as many other countries around the world. In Table 5.24 are reported the damages in billions of U.S. dollars of the most destructive hurricanes in the United States in the period 1900-2017. Notice that nine of them have occurred since 2004.

TABLE 5.24

Emergence of hurricanes in the United States: category and damage (in billions of U.S. dollars after accounting for inflation to 2017 dollars) for the ten most destructive hurricanes. (Source: National Hurricane Center, January, 26 2018)

Rank	Name	Year	Category	Damage
1	Katrina	2005	3	160.00
2	Harvey	2017	4	125.00
3	Maria	2017	4	90.00
4	Sandy	2012	1	70.20
5	Irma	2017	4	50.00
7	Andrew	1992	5	47.79
6	Ike	2008	2	34.80
8	Ivan	2004	3	27.06
9	Wilma	2005	3	24.32
10	Rita	2005	3	23.68

In such a context, any instrument suitable for reducing the destructive force of a hurricane is of relevant interest and can have important economic consequences. As for this point, in 1961 R. H. Simpson suggested to proceed with the seeding of hurricanes with silver iodide[4] to mitigate their destructive force [151].

The main perplexities about hurricane seeding were linked to the uncertainty about the evolution of the hurricane after seeding, which likely - but not surely - would reduce its destructive power, associated with natural changes in the wind speed with opposite effect. These two effects (reduction of wind speed due to seeding and increase of wind speed due to natural changes) cannot be separated. Therefore, if a seeded hurricane caused a very high amount of damage due to natural changes occurring after seeding, it could result in a strong disapproval reaction from public opinion, which could attribute the responsibility to the seeding itself and, therefore, to the U.S. government that authorized it. On the other hand, hurricane seeding - if actually effective -

---

[4]Hurricane seeding is a technique that aims to change the amount and type of precipitation through the dispersion of chemicals into the atmosphere to alter microphysical processes that allow the formation of clouds or ice cores.

would reduce the number of victims and material damages, saving millions of dollars.

Because of this uncertainty about the actual global effect of seeding hurricanes, in 1961 the U.S. government funded some experiments to weaken hurricanes by cloud seeding from aircraft and in 1962 started to fund the Stormfury project (whose first director was R. H. Simpson), which was aimed at investigating the effects of cloud seeding.

The first experimental evidence in favor of seeding was acquired in 1969, when a reduction of 31% and 15% of the maximum surface speed of wind was experimentally observed after seeding hurricane Debbie [76].

In 1970, an U.S. law was approved that allowed for hurricane seeding only in the presence of a probability of less than 10% that, within 18 hours from seeding, the center of the hurricane reaches less than 50 miles from an inhabited territorial area. In other words, the hurricane seeding could only be used for experimental purposes aimed at acquiring further information about the effectiveness of this technique.

Noting the existence of a trade off between accepting the political and social responsibilities deriving from seeding and the higher probability of material damages without seeding, Howard, Matheson and North [110] in 1972 studied the problem from a statistical perspective with the aim of identifying the optimal decision for a rational decision maker (i.e., the U.S. Gorvernment), which has to choose between modifying the law about seeding and maintaining the status quo[5].

---

[5]The immediate reaction to the article by Howard *et al.* are some letters published in Science (Vol. 179, No. 4075 and Vol. 181, No 4104).

In the first letter [174] (part of Power), the fact is outlined that some important detrimental effects of hurricanes, such as hurricane rainfall rate and losses of life, are completely ignored. Howard *et al.* recognize that certain important elements have not been considered in their analysis, but that results obtained retain all their validity.

In the second letter [174] (part of Kates), the danger of an improper use of the proposed methodology is recalled. Howard *et al.* reply that any instrument, from a hammer to medicine, can be misused or used for immoral purposes, but there is no evidence that the results of the analysis carried out by themselves are improperly used.

In the third letter [207], which is the most interesting one from the statistical point of view, the meteorologist Sundqvist, referring to their own research activity, attributes different prior probabilities to the assumption that hurricane seeding is more effective than not seeding and attains completely different conclusions than those of Howard *et al.*, outlining that the available data are too sparse to yield a statistical basis for conclusive statements. It is worth noting that the same conclusions were reached by the U.S. Government that in 1983 officially cancelled the Stormfury project. Howard *et al.* did not agree with these observations and they replied "Our analysis was based on the best information we could obtain from U.S. hurricane modification experts. As decision analysts we cannot comment on Sundqvist's differing opinion, except to say that our information sources were aware of his work and did not subscribe to his views".

The content of the original article and the subsequent discussion brings out the role played by prior information both in terms of probability distributions and in terms of the definition of a significant utility/loss function.

In this section, we propose a summary of the original work of Howard *et al.* taking into account the intervention of Sundqvist and adding some further statistical considerations.

In their work, Howard *et al.* analyze the decision of seeding a hypothetical hurricane within 12 hours from its impact with the coastal area. For the sake of clarity, the study ignores several geographic and meteorological details, which could influence the decision, and focuses only on the measure of the intensity of the hurricane (i.e., maximum surface wind speed), the evolution of which is random. In addition, consequences due to the intensity of the hurricane are represented by the economic damages (i.e., property damage) and the reactions of public opinion (i.e., government responsibility).

The decision consists of two alternative actions: seed or do not seed. The state of nature (the parametric space) is continuous and corresponds to changes in the intensity of the hurricane in the 12 hours after the choice whether to seed or not. Consequences are property damages and government responsibility. Both of them are uncertain and their exact quantification can only be carried out a posteriori.

Since there is no availability of data on the natural modifications of the maximum surface wind speed W over a time period of 12 hours, data on changes in time (12 hours) of central hurricane pressure are used, closely linked by a linear relationship at the maximum wind speed. On the basis of reasonable hypotheses, it is assumed that natural changes in the intensity of a hurricane over a 12-hour time period have a normal distribution, with mean equal to 100% and standard deviation to 15.6%

$$W \sim N(\mu = 100, \sigma = 15.6). \qquad (5.10)$$

In the case of seeding, changes in W would be due to a combined effect of nature and seeding itself. To simplify the illustration, Howard *et al.* assume that these two components are independent, so their effects are algebraically added (i.e., there are not interaction effects). Moreover, they discern three types of seeding effects, according to what state of nature holds:

- H_1 (*favorable hypothesis*): intensity of the seeded hurricane, measured by the maximum surface wind speed after seeding and denoted by W', is on average reduced with respect to W;

- H_2 (*null hypothesis*): no effect of seeding, that is, W' is on average equal to W;

- H_3 (*unfavorable hypothesis*): the intensity of the seeded hurricane W' on average increases with respect to W.

The favorable hypothesis is based on reasonable empirical support, whereas the null hypothesis appears plausible (for example, if the quantity of silver iodide is too low to have a positive effect on the intensity of the hurricane). Finally, the unfavorable hypothesis is included for completeness, even if it is not sustained by empirical evidence.

Under H_1, some studies have shown a reduction of the maximum wind speed W' between 8% and 22% (15% on average). Combining the uncertainty on the fluctuations with the presence of natural variation, the conditional distribution of W' given H_1 is

$$W'|H_1 \sim N(\mu = 85, \sigma = 18.6), \tag{5.11}$$

whereas the conditional distribution of W' in the case of null effect of seeding (H_2) does not change, that is,

$$W'|H_2 = W \sim N(\mu = 100, \sigma = 15.6). \tag{5.12}$$

Finally, with similar considerations, the three authors hypothesize

$$W'|H_3 \sim N(\mu = 110, \sigma = 18.6). \tag{5.13}$$

To properly evaluate the goodness of the seeding strategy, the unconditional distribution of W' is computed and, then, compared with the distribution of W. In detail, the unconditional distribution of W' is expressed by

$$f(W') = \sum_{i=1}^{3} f(W'|H_i)\pi(H_i), \tag{5.14}$$

where densities $f(W'|H_i)$ $(i = 1, 2, 3)$ are defined in eqs. (5.11)-(5.13) and probabilities $\pi(H_i)$ depend on the prior knowledge on states of nature H_1, H_2, and H_3. As far as this point, Howard *et al.* distinguish two situations that reflect different experts' opinions:

- scenario A: the three states of nature are equally plausible, then

$$\pi(H_1) = \pi(H_2) = \pi(H_3) = 1/3;$$

- scenario B: the seeding strategy is useless with a high probability, however, if seeding had any effect, a positive effect would be more likely than a negative one, that is,

$$\pi(H_1) = 0.15,$$
$$\pi(H_2) = 0.75,$$
$$\pi(H_3) = 0.10.$$

Differently from Howard *et al.*, the meteorologist Sundqvist [207] states that, if seeding had any effect, this effect would be definitely negative. In addition, he thinks that the average value of W' is underestimated under state H_1 in Eq. (5.11) and overestimated under state H_3 in Eq. (5.13). Therefore, a third scenario is introduced:

- scenario C:

$$W'|H_1 \sim N(\mu = 95, \sigma = 18.6), \tag{5.15}$$
$$W'|H_2 = W \sim N(\mu = 100, \sigma = 15.6), \tag{5.16}$$
$$W'|H_3 \sim N(\mu = 107, \sigma = 18.6) \tag{5.17}$$

with

$$\pi(H_1) = 0.0227,$$
$$\pi(H_2) = 0.7500,$$
$$\pi(H_3) = 0.2273.$$

The prior probabilities $\pi(H_i)$ $(i = 1, 2, 3)$ defined under the three scenarios (A, B, C) may be updated on the basis of the empirical evidence provided by seeding hurricane Debbie. Indeed, in 1969 two independent seeding experiments were conducted on hurricane Debbie, providing a reduction of 31% and 15% in the maximum surface wind speed. In other words, the values of W' observed on the seeded hurricane Debbie were equal to 69% and 85%. The probability of jointly observing these experimental results, that is, $\boldsymbol{w} = (w_1 = 69\%, w_2 = 85\%)$, strongly depends on the effectiveness of the seeding strategy and, therefore, on values assigned to $f(W' = w|H_i)$, as follows (independence assumption)

$$f(w_1 = 69\%, w_2 = 85\%|H_i) = f(w_1 = 69\%|H_i)f(w_2 = 85\%|H_i).$$

Under scenarios A, B, and C, values of $f(w_1 = 69\%, w_2 = 85\%|H_i)$ are

- scenarios A and B (obtained applying Eqs. (5.11)-(5.13)):

$$f(69\%, 85\%|H_1) = f(69\%|H_1)f(85\%|H_1) = 0.0148 \cdot 0.0214 = 0.0003,$$
$$f(69\%, 85\%|H_2) = f(69\%|H_2)f(85\%|H_2) = 0.0036 \cdot 0.0161 = 5.7 \cdot 10^{-5},$$
$$f(69\%, 85\%|H_3) = f(69\%|H_3)f(85\%|H_3) = 0.0019 \cdot 0.0087 = 1.6 \cdot 10^{-5};$$

- scenario C (obtained applying Eqs. (5.15)-(5.17)):

$$f(69\%, 85\%|H_1) = f(69\%|H_1)f(85\%|H_1) = 0.0081 \cdot 0.0186 = 0.0002,$$
$$f(69\%, 85\%|H_2) = f(69\%|H_2)f(85\%|H_2) = 0.0036 \cdot 0.0161 = 5.7 \cdot 10^{-5},$$
$$f(69\%, 85\%|H_3) = f(69\%|H_3)f(85\%|H_3) = 0.0027 \cdot 0.0107 = 2.8 \cdot 10^{-5}.$$

On the basis of the Bayes' rule, prior probabilities $\pi(H_i)$, which were formulated before Debbie experiments (pre-Debbie probabilities), are now updated as (posterior or post-Debbie probabilities)

$$\pi(H_i|\boldsymbol{w}) = \frac{f(w_1, w_2|H_i)\pi(H_i)}{\sum_{i=1}^{3} f(w_1, w_2|H_i)\pi(H_i)}$$

and, in detail for every scenario,

- scenario A:

$$\pi(H_1|\boldsymbol{w}) = 0.8119,$$
$$\pi(H_2|\boldsymbol{w}) = 0.1461,$$
$$\pi(H_3|\boldsymbol{w}) = 0.0420;$$

- scenario B[6]:

$$\pi(H_1|\boldsymbol{w}) = 0.5170,$$
$$\pi(H_2|\boldsymbol{w}) = 0.4652,$$
$$\pi(H_3|\boldsymbol{w}) = 0.0178;$$

- scenario C:

$$\pi(H_1|\boldsymbol{w}) = 0.0645,$$
$$\pi(H_2|\boldsymbol{w}) = 0.8133,$$
$$\pi(H_3|\boldsymbol{w}) = 0.1222.$$

As a result of the Debbie experiments, the unconditional distribution of the maximum surface wind speed under seeding, W', is now

$$f(W'|\boldsymbol{w}) = \sum_{i=1}^{3} f(W'|H_i)\pi(H_i|\boldsymbol{w}) \qquad (5.18)$$

which differs from Eq. (5.14) for the use of posterior probabilities $\pi(H_i|\boldsymbol{w})$ instead of the prior probabilities $\pi(H_i)$.

A first evaluation of the effectiveness of the hurricane seeding strategy is obtained by comparing the distribution of the maximum surface wind speed under the two alternatives (i.e., seeding and not seeding). For this aim, the complementary cumulative normal density distributions of W and W', without and with the sample information coming from experiments on hurricane Debbie, are plotted in Figure 5.4. For any value of $W = w$ (x-axis), the plotted curve (denoted by $1 - F(w)$) provides (y-axis) the probability that the wind speed is greater than w.

---

[6]It is worth noting that post-Debbie probabilities reported by Howard *et al.* [110], that is, $\pi(H_1|\boldsymbol{w}) = 0.49$, $\pi(H_2|\boldsymbol{w}) = 0.49$, and $p(H_3|\boldsymbol{w}) = 0.02$, are incompatible with pre-Debbie probabilities assumed under scenario B. Here we provide the right values of these posterior probabilities. We also outline that the logical reasoning followed by the three authors to link the post-Debbie probabilities to the pre-Debbie ones sounds a little anomalous from a Bayesian perspective: indeed, they first formulate the post-Debbie probabilities and, then, they deduce the corresponding prior ones.

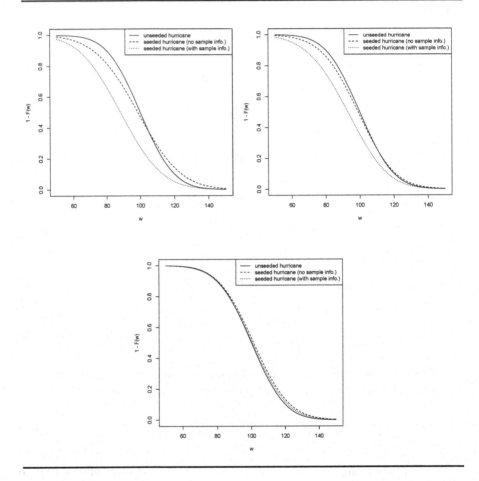

FIGURE 5.4
Probability distributions on 12-hour wind changes for the seeded and unseeded
hurricane, under the Bayesian approach: hypothesis A (top left), hypothesis
B (top right), and hypothesis C (bottom).

Plots in Figure 5.4 outline how the probability distribution of the states
of nature H_1, H_2, H_3 strongly affects the conclusion about the goodness of
hurricane seeding.

Under scenario A (H_1, H_2, H_3 are equally probable before the experiments
on Debbie; Figure 5.4, top left panel), a hypothetical seeded hurricane is likely
more intense than an unseeded hurricane for a high value of W, the opposite
being true for smaller values of W. However, after having incorporated the
information coming from empirical evidence, the seeding strategy is definitely
better than the unseeding strategy: the curve representing the complemen-

tary cumulative distribution of the wind speed for the seeding alternative is uniformly lower than the curve for the unseeding alternative.

The same type of conclusion is reached under scenario B (before the experiments on Debbie, H_2 is the most probable state of nature, followed by H_1). However, the distance among the three curves (Figure 5.4, top right panel) is now so much reduced that, in practice, no relevant difference is found between unseeding and seeding before the empirical evidence is incorporated in the analysis.

Finally, under scenario C (H_3 more probable than H_1) a definitely negative judgement about the hurricane seeding strategy occurs. As shown in Figure 5.4 (bottom panel), the decision of not seeding is associated with a wind speed that is almost everywhere smaller than the wind speed of a seeded hurricane.

Alternatively to the Bayesian analysis proposed by Howard *et al.* [110] and by Sundqvist [207], a *fully Bayesian analysis* can be performed, consisting of introducing a prior distribution on the unknown parameters of the density function of W instead of assuming constant (subjective) values for each state of nature H_1, H_2, H_3:

$$W'|\mu \sim N(\mu, \sigma = 18.6)$$
$$\mu \sim N(\mu_0, \sigma_0),$$

where the average speed μ is assumed to be normally distributed with mean μ_0 and standard deviation σ_0, and, for the sake of parsimony, σ is assumed to be known. Moreover, a finite set of plausible values for the hyperparameters μ_0 and σ_0 is taken into account, in order to evaluate their effects on the attractiveness of hurricane seeding: plausible values here adopted for μ_0 are 85, 100, and 110 and plausible values for the standard deviation σ_0 are 10, 15, 18.6, 25, and 30.

As illustrated in Section 2.10.1, the predictive prior distribution for W' is normal with mean equal to μ_0 and variance equal to $\sigma_0^2 + 18.6^2$, as in Eq. (2.24),

$$W' \sim N(\mu_0, \sigma_0^2 + 18.6^2)$$

and the predictive posterior distribution, which results after having accounted for the empirical evidence from hurricane Debbie (i.e., $w_1 = 69\%$ and $w_2 = 85\%$), is

$$W' \sim N(\mu_1, \sigma_1^2 + 18.6^2), \tag{5.19}$$

with μ_1 and σ_1^2 defined according to eqs. (2.25) and (2.26), respectively,

$$\mu_1 = \frac{(69 + 85)\sigma_0^2 + \mu_0 \cdot 18.6^2}{2\sigma_0^2 + 18.6^2},$$

$$\sigma_1^2 = \frac{18.6^2\sigma_0^2}{2\sigma_0^2 + 18.6^2}.$$

As shown in Figure 5.5, changes in the hyperparameters may substantially modify the probability of wind speed of a seeded hurricane.

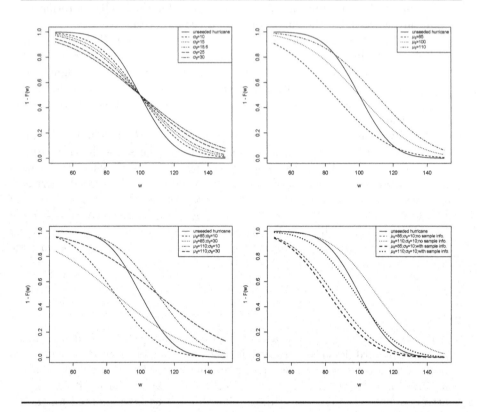

FIGURE 5.5
Probability distributions on 12-hour wind changes for the seeded and unseeded hurricane, under the fully Bayesian approach: $\mu_0 = 100$ and $\sigma_0 = 10, 15, 18.6, 25, 30$, no sample information (top left); $\mu_0 = 85, 100, 110$ and $\sigma_0 = 18.6$, no sample information (top right); $\mu_0 = 85, 110$ and $\sigma_0 = 10, 30$, no sample information (bottom left); $\mu_0 = 85, 110$ and $\sigma_0 = 10$ without and with sample information (bottom right).

Given μ_0, the increasing σ_0 (top left panel) gives rise to the distance between probability curves under seeding and not seeding strategies, with the not seeding alternative more attractive than seeding for values of wind speed greater than μ_0; moreover, high values of σ_0 reduce the slope of the complementary cumulative distribution of wind speed.

Similarly, high values of mean μ_0 (top right panel) favor the choice of not seeding. Mixed situations are observed when both the hyperparameters change (bottom left panel): the seeding alternative uniformly dominates the not seeding alternative only for small values of both the hyperparameters ($\mu_0 = 85$ and $\sigma_0 = 10$)).

Finally, the sample evidence from hurricane Debbie (bottom right panel) operates in favor of the hurricane seeding strategy: in fact, the curves of wind speed probability of a seeded hurricane move to the left side, that is, downward with respect to the curve of an unseeded hurricane.

The analysis of a hurricane wind speed is not interesting *per sé*, but as it is directly and indirectly related to the material damages to properties. Data available do not provide a detailed basis from which to induce a clear causal relationship between surface wind speed and property damages. Nevertheless, an estimate of property damages may be obtained introducing a specific mathematical function[7].

Now, it is possible to evaluate the two alternative strategies (i.e., seeding vs. not seeding) by referring to property damage rather than wind speed. For the sake of clarity, the continuous distribution of changes in wind speed is discretized, distinguishing five states of nature: increase of 25% or more (average value +32%), increase of 10% to 25% (average value +16%), little change between +10% to −10% (average value 0), reduction of 10% to 25% (average value -16%), reduction of 25% or more (average value -32%). The related probabilities of these states of nature are computed on the basis of Eq. (5.10) for an unseeded hurricane (Table 5.25, column 3), on the basis of post-Debbie probabilities in Eq. 5.18 for a seeded hurricane under scenarios A, B, or C (Bayesian approach; Table 5.25, columns 4-6), and on the basis of posterior predictive probabilities in Eq. (5.19) for a seeded hurricane under a fully Bayesian approach (Table 5.25, columns 7-8). In this last case only two extreme situations are considered: (*i*) $\mu_0 = 85$ (small value of mean) and $\sigma_0 = 10$; (*ii*) $\mu_0 = 110$ (high value of mean) and $\sigma_0 = 10$. Moreover, for each state of nature, the amount of property damage is calculated, resulting in respectively equal to (values in millions of U.S. dollars) 335.8, 191.1, 100.0, 46.7, 16.3 (Table 5.25, column 2).

Finally, all the necessary elements being available we may compute the expected loss for both decisions (seeding and not seeding). According to results displayed in Table 5.25, the alternative of not seeding is associated with an expected loss equal to 116.088 (million of U.S. dollars), whereas the expected loss associated with the seeding strategy depends on the adopted approach. Under the Bayesian approach, the decision of seeding should be preferred in two scenarios out of three, the expected loss under scenario C (120.900) being definitely greater than 116.088. Under the fully Bayesian approach, the seeding strategy is favored in case (*i*) (small value of μ_0), but is rejected in case (*ii*) (high value of μ_0), even if in this latter case the difference in the expected

---

[7]Howard *et al.* [110] base their study on the function $d = c_1 w^{c_2}$, where d represents the damages in millions of dollars, while c_1 and c_2 are two empirical constants determined using the available data related to 21 hurricanes supplied by the American Red Cross. However, since the same function is used to estimate the property damage in the case of seeding and not seeding, different specifications of the relation between d and w do not affect the conclusions about the effectiveness of the hurricane seeding strategy.

TABLE 5.25

Probabilities that a wind change occurs in the 12 hours before the hurricane landfall (discrete approximation) and related expected losses.

Avg. change in wind	Property damage	Unseeded hurricane	Bayesian A	Bayesian B	Bayesian C	Fully Bayesian (i)	Fully Bayesian (ii)
+32%	335.8	0.055	0.030	0.037	0.068	0.017	0.090
+16%	191.1	0.206	0.102	0.139	0.211	0.067	0.185
0	100.0	0.478	0.332	0.386	0.462	0.264	0.377
-16%	46.7	0.206	0.287	0.258	0.201	0.289	0.219
-32%	16.3	0.055	0.249	0.179	0.059	0.363	0.129
Expected loss							
prop.damage		116.088	81.386	93.988	120.900	65.501	116.920
prop.damage + govern.resp.		116.088	93.865	110.151	146.725	73.489	144.555

losses is small (116.920 for a seeded hurricane vs. 116.088 for an unseeded hurricane).

The above analysis does not take into account a relevant consequence related to the seeding strategy: the attribution by public opinion to the U.S. Government of the (legal and social) responsibility of the intensification of wind speed in a seeded hurricane. Indeed, there exists a not negligible probability that a hurricane - even if seeded - intensifies its destructive power due to natural causes. Therefore, the Government must decide between this assumption of responsibility or the higher probability of damage of an unseeded hurricane. The aim consists of evaluating the relative desirability of the reduction of material damage of a seeded hurricane compared to the assumption of responsibility for its intensification. After having incorporated this further element in the decisional analysis, the expected losses under Bayesian and fully Bayesian approaches are updated as shown in Table 5.25, last line. Obviously, the expected loss of an unseeded hurricane remains constant. Now, it can be concluded that the seeding strategy is still preferred under scenarios A and B (Bayesian approach) and under situation (i) of the fully Bayesian approach, even if the gap between the two alternatives is significantly reduced. On the other hand, if scenario C and situation (ii) of the fully Bayesian approach are considered more credible, then the seeding alternative should be definitely put aside in favor of the not seeding alternative.

The above analysis outlines the relevance for the decision maker of having information about the effectiveness of hurricane seeding. For this reason, Howard *et al.* complete their study with the computation of the value of information. The conclusion under the hypothesized scenario is in favor of collecting further information, even if the advantage in terms of expected loss is really small.

6

Statistics, causality, and decisions

CONTENTS

6.1 Introduction

As was illustrated in depth in the previous chapters, normative decision theory based on expected utility establishes principles of rational behavior. Empirical failures of normative theories favored the development of prescriptive [13] theories aimed at detecting rules of rational behavior that can be transformed into actual behaviors by the decision makers. From such a perspective, descriptive decision theories also assume a relevant role [25].

Alternatively to prescriptive and descriptive theories, in this chapter we consider a causal approach to decision theory. We provide an overview of the links between causality and decision theory, with a special focus on the structural equation modeling approach.

We start with a review of the philosophical debate about the relation between causality and correlation that began in the eighteenth century with Hume and Arbuthnot. Then, we review the concept of causality in the statistical setting, starting from the significance test of Fisher and the hypothesis testing of Neyman-Pearson, leading to the modern causal inference. We point out the relevance of the concept of "intervention" (or "manipulation") of a variable as opposed to the concept of mere conditioning as a distinctive element to interpret in causal sense the results of a statistical analysis. In such a context, approaches based on structural equation models and path analysis play a central role.

Next, decision theory is revised from a causal perspective. We introduce the idea that interventions on variables can modify the state of nature in which the decision maker operates, and this cannot be ignored. As a main consequence, the posterior probabilities of the states of nature change in accordance with the actions chosen by the decision maker and, then, the decision rule based on the maximization of expected utility has to be updated to account for this newly formulated posterior distribution of the states of nature.

The remainder of the chapter is devoted to illustrating two case studies related to the subscription fees to and satisfaction with the Italian Radiotelevision RAI, which is the State company that manages the TV programs in Italy. The first case study describes the application of recursive path models to identify the causes of the number of new subscriptions, whereas the second case study illustrates how structural equation models may be used to assess the TV users' satisfaction.

6.2   Causality and statistical inference

Since ancient times, the topic of causality has fascinated philosophers and scientists. Among the first main contributors, we remind Plato (428/427 BC - 348/347 BC), Aristotle (384/383 BC - 322 BC), Machiavelli (1469-1527), Copernicus (1473-1543), Bacon (1561-1626), Galileo (1564-1642), Kepler (1571-1630), Descartes (1596-1650), Huygens (1596-1687), Locke (1632-1704), Newton (1643-1727).

Broadly speaking, causality can be interpreted as the relationship between a first event, representing the cause, and a second event, representing the effect, where the second event is interpreted as a consequence of the first event. In common use, both the cause and the effect can have a different nature: they can be individual events or even objects, processes, properties, variables, facts, states of the world. Furthermore, both the cause and the effect can be of univariate or multivariate nature.

Common opinion among the statisticians is that causality cannot be investigated through statistical inference or, in other words, statistical inference is not *causal inference*. In this regard, Lindley [146] states that

"statisticians rarely refer to causality in their writings and, when they do, it is usually to warn of dangers in the concept."

Lindley (2002)

and he recalls the words of Speed [204]

"Considerations of causality should be treated as they have always been treated in statistics: preferably not at all but, if necessary, then with very great care."

Speed (1990)

Surely, decision makers regulate their behavior on sophisticated and often unconscious processes of causal reasoning that usually are not supported by adequate statistical data analyses. However, as it will be clarified later, not only modern statisticians suggest extreme caution in dealing with causality; philosopher Hume (1711-1776), who defined the fundamentals of inference in his critique of the principle of causation, excludes any possibility of an empirical validation of causal links between events. In more detail, in his "A Treatise on Human Nature" Hume [113] in 1739 states that

---

"When we look about us towards external objects, and consider the operation of causes, we are never able, in a single instance, to discover any power or necessary connection; any quality, which binds the effect to the cause, and renders the one an infallible consequence of the other.

[...] experience only teaches us, how one event constantly follows another; without instructing us in the secret connection, which binds them together, and renders them inseparable.

We then call the one object, cause; the other, effect. We suppose that there is some connection between them; some power in the one, by which it infallibly produces the other, and operates with the greatest certainty and strongest necessity.

I say then, that, even after we have experience of the operations of cause and effect, our conclusions from that experience are not founded on (a priori) reasoning, or any process of the understanding."

Hume (1739)

Moreover, in the subsequent volume "An Enquiry Concerning Human Understanding" of 1748 Hume [112] takes up the theme by saying that:

"All reasoning concerning cause and effect are founded on experience, and all the reasoning from experience are founded on the supposition that the course of nature will continue uniformly the same."

Hume (1748)

Summarizing, according to Hume causality can be adequately observed only in terms of empirical regularity, in the presence of

- contiguity or association (the cause and the effect occur together),

- succession (the cause precedes the effect),

- constant conjunction (the effect-cause relationship must be repeated over time),

but it cannot be verified.

Attempts to provide a causal response of observed phenomena in statistical terms can be found in the literature even before Hume. Arbuthnot (1667- 1735) in 1710 [8] states that

"Among innumerable Footsteps of Divine Providence to be found in the Works of Nature, there is a very remarkable one in the exact Balance that is maintained between the Numbers of Men and Women; for by this means it is provided, that the Species may never fail, nor perish, since every Male may have its Female, and of a proportional Age.

This Equality of Males and Females is not the Effect of chance but Divine Providence, working for a good End, which I thus demonstrate."

Arbuthnot (1710)

Mill (1806 - 1873) shares the conception of causality proposed by Hume; however he believes possible the empirical validation of causality. In this regard, he identifies a series of methods (canons) that allow us a valid inductive inference: method of agreement, method of difference, joint method of agreement and difference, method of residues, and method of concomitant variation (for details see [157]). Methods proposed by Mill have been the object of an innumerable series of criticisms. Among these, the most relevant one is about the assumption that the cause-effect relationship is unique and has a deterministic nature. In other words, according to Mill the presence of multiple causes and the interaction between causes are precluded and measurement errors with their interactions are ignored.

On the same topic, Pearson (1857- 1936) intervenes in 1892 saying [168]

"that a certain sequence has occurred and recurred in the past is a matter of experience to which we give expression in the concept of causation. [...] Science in no case can demonstrate any inherent necessity in a sequence, nor prove with absolute certainty that it must be repeated. Science for the past is a description, for the future a belief; it is not, and has never been, an explanation, if by this word is meant that science shows the necessity of any sequence of perceptions."

Pearson (1892)

Different conclusions were reached by Yule (1871- 1951) that in 1899 studied the main causes of the pauperism in England [222]. Through the use of statistical regression models, Yule concludes that pauperism is mainly due to changes in: politics and administration of the law; economic conditions (i.e., fluctuations in trade, wages, prices, and employment); social factors (e.g., density of population, overcrowding, or character of industry in a given district); moral factors (e.g., illustrated, for example, by the statistics of crime, illegitimacy, education, or death rates from certain causes); distribution of the population by age.

This was one of the first attempts to use statistical regression models to solve problems of causal statistical inference, passing from the measurement of the association between characters to the measurement of the relationship by introducing the directionality in the associative link. The methodology employed by Yule was the object of innumerable criticisms, above all, the attribution of a causal value to the measures of association obtained. However, Yule was fully aware of the weakness of his arguments and he clearly states that expression "due to" has strictly to be read as "associated with" (note 25 [222]).

After Yule, regression models (e.g., simple, multiple, and multivariate regression models, models of analysis of variance and covariance, simultaneous equation models, factor analysis models, structural equation models) have been a reference paradigm for many subsequent contributions, mainly in the observational setting. Despite the relevance of contributions present in the literature, it must be emphasized that the conclusions that many authors achieved (implicitly but often also explicitly) are weak due to interpreting in a causal sense, without any adequate verification, the results obtained through the use of statistical regression models. Among authors who have provided a relevant contribution regarding the improper use of statistical regression models in the analysis of causality we recall D. A. Freedman; see, for instance, his papers of 1991 [69] and 1999 [70].

6.3   Causal inference

Given that statistical inference in general and regression models in particular - *per sé* - are not suitable instruments to investigate the causal links between events, in this section we illustrate the basic outlines of causal inference. First, a general review of main contributions to different approaches to causality and causal inference is provided. Then, attention focuses on one of these approaches that is based on Structural Equation Models (SEMs; Section 2.12).

We premise that identification and estimation of causal links require the decision maker to follow a logical process:

- Make explicit the logic underlying the causal path he/she intends to investigate:

    - given the cause, do you want to identify the effect?

    - given the effect, do you want to go back to the cause?

- Clarify the nature of the cause:

    - necessary,

    - sufficient,

    - necessary and sufficient,

    - contributive or concurrent,

    - observable or latent (i.e., not directly observable).

- Define the causality path:

    - direct,

    - indirect (i.e., the causal relation between two variables is mediated by a third variable),

    - both direct and indirect.

- Define the size of the process:

    - the effect is determined by a single cause,

    - there is a plurality of causes.

Moreover,

- in the presence of a plurality of causes, evaluate if

  - the causes act independently,
  - the causes interact.

- Take into account the nature of the scientific setting:

  - experimental sciences,
  - observational sciences,
  - nearly experimental sciences.

- Take into account the disciplinary field, such as philosophy of science, epidemiology, biology, medicine, psychology, sociology, economics, management, jurisprudence, engineering, physics, history, politics, theology, and so on.

6.3.1 Statistical causality

As discussed in Section 6.2, Hume argues that inductive or inferential reasoning is irrational and reality can only be observed, whereas K. Pearson, even if sharing this point of view, proposes quantitative (statistical) tools to facilitate the comprehension of reality.

On the other hand, Fisher, Neyman and E. Pearson do not share the opinion of Hume and K. Pearson about the irrationality of inductive reasoning and introduce the framework of classical or frequentist statistical inference (Section 2.9) that, simplifying reality, not only makes its understanding easier but also allows us a rational resolution of induction problems, on the basis of "objective" information (i.e., sample results).

Approaches proposed by Fisher and Neyman-Pearson develop under the assumption of repeatability of an experiment and are based on sample data. In more detail,

- Fisher's approach emphasizes the relevance of the significance test (consisting of rejecting the null hypothesis), the likelihood function, and the principle of sufficiency;

- Neyman-Pearson's approach (Section 2.9.4) introduces an alternative hypothesis, in addition to the null hypothesis, and it drives the decision maker to decide in favor of the null or the alternative hypothesis, according to the sample evidence.

The introduction of Fisher's significance test represents a revolution in the theory and practice of statistics, similar to the impact of the principle of falsification treated by the philosopher Popper (1902-1994) in the context of the philosophy of science. Both these approaches provide a solution to the problem of rational learning from reality, being based on the same principle: a theoretical hypothesis can be rationally accepted if it "survives" (according to Popper's terminology) rigorous tests.

Different from classical inference, the perspective originated by Bayes and Laplace supports the idea that rational learning from reality proceeds through the assignment of (prior) probabilities to the causes, and the subsequent updating of these probabilities (posterior probabilities) in light of the experience.

In the work of Laplace the meaning to be attributed to the prior probabilities is not entirely clear and their determination is based on what is usually called the principle of insufficient reason. Keynes (1883-1946) [132] and Carnap (1891-1970) [28] try to overcome the criticism of Laplace's approach by interpreting the prior probability as a measure of the logical relationship between events, but fail in their intent as the logical probability defined by them is not univocal.

An alternative definition of logical probability is proposed by Jeffreys who adopts an approach to inference similar to that of Fisher. However, differently from Fisher, Jeffreys argues that to achieve a reasonable conclusion about an hypothesis on the basis of observed data a different conception of probability is needed. He introduces (Section 2.10.2) the idea that probability is a measurement of the lack of knowledge; this is essentially an updated revision of Laplace's non-disclosure prior distribution.

Ramsey [179] and de Finetti [52], to whom the authorship of the Bayesian-subjective approach (Section 2.10) to statistical inference is attributed, criticize independently of each other both the classical as well as the logical definition of probability, proposing the subjective definition of probability. Their definition solves the problem of the non-uniqueness of the logical probability assuming that different individuals with the same information can express different probability measures for the same event.

The dissemination of the contributions of Ramsey and, above all, de Finetti in the scientific field is due to the work of Savage [191] who also proposes and develops the subjective utility theory (Section 3.4.2).

6.3.2   Modern causal inference

In the statistical literature, different methods and models have been proposed for the analysis of causality in the context of observational studies, from the classic approach of the simultaneous equations of the Cowles Commission (Haavelmo, 1943 [91]; see also Haavelmo, 1944 [92]) to the methods of the potential outcomes proposed by Neyman [159] and developed by Rubin [188], Rosenbaum and Rubin [186], and Holland [108, 109]. In relation to these approaches, it is also worth remembering methods for the estimation of causal effects proposed in the econometrics context, with special reference to Hirano, Imbens and Ridder [106] and Heckman and Vytlacil [102].

Another research line has been developed in parallel in the context of artificial intelligence, especially by Pearl [163, 164, 165] and by Spirtes, Glymour,

Scheines [205]. In such a context, the contributions of Dawid [46, 47, 48, 49], Dawid, Musio and Fienberg [51], Lauritzen [136, 137], Lauritzen and Nilsson [139] and, above all, Woodward [220] are also very relevant.

In his volume of 2003, Woodword collects the result of a thirty-year research activity and illustrates a theory of causality that contrasts with the counterfactual theory developed by Lewis [141]. Woodword's contribution originates from the studies of Spirtes, Glymour and Scheines [205], Pearl [164] and Heckman [100] but, while these authors focus on the theoretical-methodological aspects, Woodword focuses on the philosophical foundations of causal reasoning. In particular, he introduces a simple but clear definition of causality: event C causes event E if and only if the value of E changes *due to some intervention* on C.

Woodword presents the tools for causal analysis (graphs and equations) and illustrates his theory of intervention. The idea of an intervention (manipulation) on the variables represents the focal point from which originate the theoretical developments of Haavelmo [91, 92, 93], Heckman and Pinto (the *fix* operator) [101], Pearl (the *do* operator) [166], but also of Holland ("No causation without manipulation") [108].

In particular, one of the most relevant contributions due to Haavelmo consists in the formalization of the difference between correlation and causation. As far as this point is concerned, the distinction of the concept of *fixing* assumes a significant role from the concept of *conditioning*. Indeed, the causal effects of inputs (i.e., variables, events, processes) on outputs are determined by the impact of interventions (fixing) on the inputs, the simple conditioning to draw conclusions in causal sense being not sufficient. Formally, we have

$$p(Y|X = fix(x)) \neq p(Y|X = x).$$

Moreover, the $fix(\cdot)$ operator of Haavelmo is conceptually equivalent to the $do(\cdot)$ operator of Pearl, in the sense that both of them denote the intervention on a variable X, that is,

$$p(Y|X = fix(x)) = p(Y|X = do(x)).$$

To stress the difference between conditioning and intervention on variables, Pearl uses the operator $see(\cdot)$, that is,

$$p(Y|X = x) = p(Y|X = see(x)). \tag{6.1}$$

It is also worth recalling that many other authors from different disciplines have provided fundamental contributions to causal inference. Among the others,

- the statisticians Dempster [54], Shafer [194], Lauritzen and Richardson [140];

- the epidemiologist Robins [182, 183] and VanderWeele [211];

- the econometrician Granger [85];

- the philosophers of science Gibbard and Harper [77], Lewis [141], Skyrms [196, 197], Cartwright [29, 30, 31], and Sobel [200, 201, 202].

In the observational setting, the different approaches to causal inference can be summarized in the following main different types:

- path analysis;

- graphical models;

- methods of potential outcomes;

- counterfactual approach;

- SEMs;

- Dempster-Shafer method;

- Granger causality.

As concerns the wide variety of approaches, the words of Dawid [50] are emblematic:

"after decades of neglect, recent years have seen a flowering of the field of statistical Causality, with an impressive array of developments both theoretical and applied. However there is as yet no one fully accepted foundational basis for this enterprise. Rather, there is a variety of formal and informal frameworks for framing and understanding causal questions, and hot discussion about their relationships, merits and demerits: we might mention, among others, structural equation modeling and path analysis (Wright 1921), potential response models (Rubin 1978), functional models (Pearl 2009), and various forms of graphical representation. This plethora of *foundations* leaves statistical causality in much the same state of confusion as probability theory before Kolmogorov."

Dawid (2015)

The causality approaches above mentioned are characterized by their own peculiarities that, according to the authors themselves, do not make them compatible with each other. In addition, many authors tend to consider their own approach superior to the others. However, diversity and superiority are not here shared, both because the different approaches are characterized by common fundamental traits and because a universal statistical-methodological tool able to provide the most satisfactory answer in all research situations does not exist.

On the other hand, the combined use of different approaches often allows to achieve the most interesting and significant results, as clearly highlighted by White and Chalak [219] that unify the approaches based on structural equations by the Cowles Commission, treatment effects by Rubin [188] and Rosenbaum and Rubin [186], and Directed Acyclic Graphs (DAGs) by Pearl [163] to define, identify, and estimate causal effects.

With regard to the combined use of different approaches, Lauritzen [138] in the discussion in the article "Direct and Indirect Causal Effects via Potential Outcomes" by Rubin [189] states

---

"Professor Rubin's paper advocates the use of potential responses in contrast to graphical models, illustrated with a discussion of direct and indirect effects in connection with the use of surrogate endpoints in clinical trials. Although discussions of this nature can be used to sharpen the minds and pinpoint important issues, I find them generally futile. Personally I see the different formalisms as different 'languages'. The French language may be best for making love whereas the Italian may be more suitable for singing, but both are indeed possible, and I have no difficulty accepting that potential responses, structural equations, and graphical models coexist as languages expressing causal concepts each with their virtues and vices.

It is hardly possible to imagine a language that completely prevents users from expressing stupid things. Personally I am much more comfortable with the language of graphical models than that of potential responses, which I, as also Dawid (2000), find just as seductive and potentially dangerous as Professor Rubin finds graphical models. I certainly disagree with Professor Rubin's statement that graphical models tend to bury essential scientific and design issues. It is always true that concepts that strongly support intuition in many cases, can seduce one to incorrect conclusions in other cases. Each of us speaks and understands our own language most comfortably but, to communicate in a fruitful way, it is indeed an advantage to learn other languages as well".

Lauritzen (2004)

The reader who is interested in learning more about Rubin's approach can consult, among others, the volume "Causal Inference for Statistics, Social, and Biomedical Sciences" by Imbens and Rubin [115]. For a more in-depth look at Pearl's approach, the reference volumes are "Causality: Models, Reasoning, and Inference" by Pearl [164] and "The Book of Why: The New Science of Cause and Effect" by Pearl and Mckenzie [167]. Moreover, as concerns the contribution of Robins the reader can consult the recent volume "Causal Inference" by Hernàn and Robins [103].

Here we will consider only the structural equation approach. As illustrated in Section 2.12, the combination of simultaneous equation models from the econometric field and factor analysis from the psychometric field is commonly described as SEM. Since their first formulation, SEMs have been often interpreted in a causal sense [82], similarly to path analysis [221, 55] and to the graphs of influence [111].

6.3.3   Structural equation approach to causal inference

As clarified in Section 2.12, the class of SEMs comprises a series of statistical models that allow complex relationships between one or more independent variables and one or more dependent variables. Variables that are not affected by any other variable in the model are called exogenous variables, whereas variables that are affected by other variables in the model are called endogenous variables. Endogenous variables may play the role of dependent variables in some model equations and the role of independent variables in other equations. Moreover, endogenous variables that are affected by some (exogenous or endogenous) variables and affecting other variables are also called mediators. Another distinction is between directly observable or manifest variables and unobservable or latent variables.

Although similar in their appearance, SEMs are different from regression models (even if regression models represent a special case of SEM). In a regression model, there exists a clear distinction between independent and dependent variables, while in a SEM a dependent variable in one model equation can become an independent variable in another equation. Note that only endogenous variables may act both as dependent and independent variables, whereas exogenous variables act always as independent variables. This type of reciprocal role that endogenous variables can play enables SEM to infer causal relationships between variables.

It is worth noting that the introduction of causal relationships between endogenous and exogenous variables and between endogenous variables cannot be dissociated from theoretical or empirical prior knowledge. In other words, causal analysis through SEMs requires not only suitable data and algorithms, but also knowledge from experts [184]. This methodology *per sé* does not allow the demonstration of the existence of causal relationships, but it simply represents an instrument to verify the compliance of hypothesized causal links with observed data. In turn, causal links are not suggested by data, but come from prior knowledge on the phenomenon by experts: at most, results of the estimation of a SEM may suggest if and how to review and update the initially formulated links among variables.

An ancestor of SEM that was from the beginning focused on the causal interpretation of relationships between variables is *path analysis*. Path analysis is based on multiple regression equations, as in a SEM, where only manifest variables are involved: in other words, path analysis corresponds to the structural part of a SEM, without any specification of the measurement part[1].

Path analysis is usually used to describe the direct dependencies among a set of observable variables, including as special cases models for multiple and multivariate regression, canonical correlation analysis, discriminant analysis, as well as more general models for variance and covariance multivariate analysis. In such models, each equation expresses a relationship among variables that is interpreted in the sense that, intervening on a certain variable X, it is possible to affect the determination of another variable Y. In other words, it is possible to control Y through the intervention of X. Similar considerations hold also for SEMs, where the presence of latent variables, in addition to observed variables, and of (causal) relationships between them has to be taken into account. Once the model of path analysis or SEM has been formulated, it will be necessary to verify the plausibility of the causal hypotheses, as above outlined.

From a practical perspective, a relevant distinction in the class of SEMs is between recursive and non-recursive (or interdependent) models (Section 2.12). In a recursive model all causal effects are uni-directional and all disturbances are uncorrelated, whereas in a non-recursive model one or more reciprocal effects (feedback loops) are present. Recursive models are always identified and are simple to estimate; they can be represented by Directed Acyclic Graphs (DAGs), also known as Bayesian networks [44, 136, 205, 164]. On the other hand, non-recursive models are more flexible, but their identification is more problematic. Moreover, even if a non-recursive model is identifiable the parameter estimation process is not always simple.

As concerns the causal relationships, even if methodology based on DAGs is limited to recursive models, the relationships defined in a non-recursive SEM may be fully represented through path diagrams (introduced in Section 2.12). Some examples of path diagrams for recursive and non-recursive models are provided in Figures 6.1 - 6.3, whose symbols are defined in accordance with Figure 2.8.

Figure 6.1 shows a recursive SEM with three observed exogenous variables and two observed endogenous variables, whose relations are formalized in accordance with notation of Eqs. 2.30 and 2.31, as follows (error terms are omitted in the graph diagram for the sake of parsimony):

$$y_{1i} = \beta_0 + \beta_1 x_{1i} + \beta_2 x_{2i} + \beta_3 x_{3i} + e_{1i}, \quad \beta_1 < 0, \beta_2 > 0, \beta_3 > 0,$$
$$y_{2i} = \gamma_0 + \beta_4 x_{3i} + \gamma_1 y_{1i} + e_{2i}, \quad \beta_4 > 0, \gamma_1 > 0.$$

---

[1] A useful examination of logical and methodological relations between correlation and causality and among causal inference, path analysis, and SEMs is provided, among others, by Hauser and Goldberger [98] and by Shipley [195].

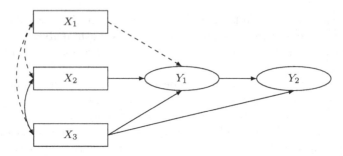

FIGURE 6.1
Example of path diagram for a recursive SEM (observed exogenous variables:
X_1, X_2, X_3; observed endogenous variables: Y_1, Y_2).

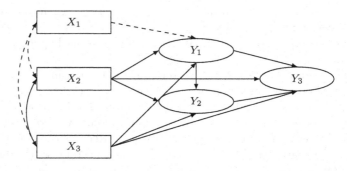

FIGURE 6.2
Example of path diagram for a recursive SEM (observed exogenous variables:
X_1, X_2, X_3; observed endogenous variables: Y_1, Y_2, Y_3).

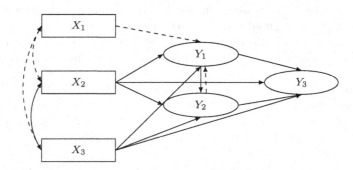

FIGURE 6.3
Example of path diagram for a non-recursive SEM (observed exogenous variables: X_1, X_2, X_3; observed endogenous variables: Y_1, Y_2, Y_3).

Similarly, SEM in Figure 6.2 is a slightly more complex recursive model, involving one more endogenous variable:

$$y_{1i} = \beta_0 + \beta_1 x_{1i} + \beta_2 x_{2i} + \beta_3 x_{3i} + e_{1i}, \quad \beta_1 < 0, \beta_2 > 0, \beta_3 > 0,$$

$$y_{2i} = \gamma_0 + \beta_4 x_{2i} + \beta_5 x_{3i} + \gamma_1 y_{1i} + e_{2i}, \quad \beta_4 > 0, \beta_5 > 0, \gamma_1 > 0,$$

$$y_{3i} = \gamma_0 + \beta_6 x_{2i} + \beta_7 x_{3i} + \gamma_2 y_{1i} + \gamma_3 y_{2i} + e_{2i}, \quad \beta_6 > 0, \beta_7 > 0, \gamma_2 > 0, \gamma_3 > 0.$$

Both recursive SEMs in Figures 6.1 and 6.2 can be estimated through the ordinary least squares approach.

Finally, the model of Figure 6.3, corresponding to the system

$$y_{1i} = \beta_0 + \beta_1 x_{1i} + \beta_2 x_{2i} + \beta_3 x_{3i} + \gamma_1^* y_{2i} + e_{1i}, \; \beta_1 < 0, \beta_2 > 0, \beta_3 > 0, \gamma_1^* < 0,$$

$$y_{2i} = \gamma_0 + \beta_4 x_{2i} + \beta_5 x_{3i} + \gamma_1 y_{1i} + e_{2i}, \quad \beta_4 > 0, \beta_5 > 0, \gamma_1 > 0,$$

$$y_{3i} = \gamma_0 + \beta_6 x_{2i} + \beta_7 x_{3i} + \gamma_2 y_{1i} + \gamma_3 y_{2i} + e_{2i}, \quad \beta_6 > 0, \beta_7 > 0, \gamma_2 > 0, \gamma_3 > 0.$$

is interdependent, due to the presence of the reciprocal effect of y_1 on y_2, measured by positive coefficient γ_1, and of y_2 on y_1, measured by negative coefficient γ_1^*, and it cannot be consistently estimated by the ordinary least squares. However, if the equation system is identifiable coefficients of the structural sub-model can be obtained from the coefficients of the reduced form of the model. A simple example of a non-recursive model is illustrated in Example 6.1.

Example 6.1. Demand and supply relationship (part I). *Let us denote by Q_d the quantity of a commodity (or service) demanded by potential buyers and Q_s the quantity of the same commodity (or service) supplied by sellers.*

Many elements affect the amounts of demanded and supplied quantities: surely, one of the most important is the price, P. The relation between price and quantity demanded (which usually holds) is of type[2] (demand function)

$$Q_d = \alpha_d + \gamma_d P + e_d \quad \gamma_d < 0,$$

that is, when price P goes up, the quantity demanded Q_d decreases and, when price P goes down, the quantity demanded Q_d increases. In other words, usually there exists an inverse relationship between price and quantity demanded.

On the other hand, a direct relationship between price and quantity supplied is usually observed (supply function)

$$Q_s = \alpha_s + \gamma_s P + e_s \quad \gamma_s > 0.$$

---

[2] It is worth observing that usually a non-linear relationship between price and demanded or supplied quantities is adopted in the economic theory. Here, we are not specifically interested in the exact form of the relationship and, therefore, we assume linearity for the sake of parsimony and clarity.

In such a case, when price P goes up, the quantity supplied Q_s increases and, when price P goes down, the quantity supplied Q_s decreases, too.

The actual price of a commodity is determined by the interaction between supply and demand in the market. The price that results from the intersection between the demand curve and the supply curve is named equilibrium price *and represents a sort of agreement between buyers and sellers. At the equilibrium point the allocation of the commodity in the market is the most efficient, because quantity supplied by sellers equals quantity demanded by consumers (Figure 6.4). Thus, both sellers and buyers are satisfied with the current economic state: at the equilibrium price, sellers are selling all the amount that they have produced and consumers are buying all the amount that they have demanded. In practice, due to the fluctuations in demand and supply, the equilibrium point cannot be really reached and the price of a commodity (or service) tends to continuously change.*

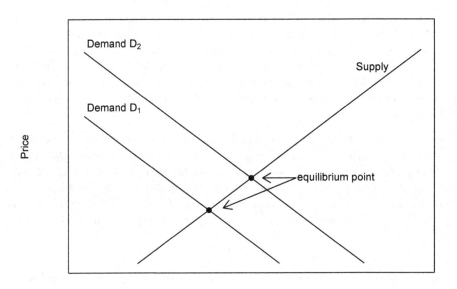

FIGURE 6.4
Demand and supply relationship.

Obviously, demanded and supplied quantities depend not only on the price, but also on other elements. For instance, quantity demanded is often affected by the buyers' income, the prices of related commodities (i.e., substitutes or complementary commodities), the consumers' preferences and tastes, and so on. Thus, accounting for the buyers' income (or any other variable) the demand function modifies as follows

$$Q_d = \alpha_d + \gamma_d P + \beta I + e_d \quad \gamma_d < 0; \; \beta > 0$$

and, assuming n observations on Q_d, Q_s, and P, the SEM is

$$Q_{di} = \alpha_d + \gamma_d P_i + \beta I_i + e_{di}$$
$$Q_{si} = \alpha_s + \gamma_s P_i + e_{si}, \quad i = 1, \dots, n. \tag{6.2}$$

SEM in Eq. 6.2 cannot be estimated using the ordinary least squares approach. First, the endogenous variable P_i in the supply equation is negatively correlated with the error term e_{si}; then the least squares estimator of γ_s tends to be underestimated. Second, if variable income I_i changes, the demand curve shifts (e.g., from D_1 to D_2 as shown in Figure 6.4), while the supply curve remains fixed. In such a situation, there are infinite demand curves passing through the equilibrium point and, therefore, infinite values for the slope γ_d. Thus, parameters of the demand model γ_d and β cannot be consistently estimated.

SEM in Eq. 6.2 is identified and can be solved if the equation system is suitably expressed in the reduced form, as in Eq. 2.31. Let variable Q_e denote the demanded and supplied quantity at the equilibrium point, that is,

$$Q_d = Q_s \simeq Q_e.$$

If Q_e is substituted in Eq. 6.2, the following equality is obtained:

$$\alpha_d + \gamma_d P_i + \beta I_i + e_{di} = \alpha_s + \gamma_s P_i + e_{si}$$

or, ignoring the constant terms,

$$\gamma_d P_i + \beta I_i + e_{di} = \gamma_s P_i + e_{si}.$$

After some algebra, the system in Eq. 6.2 can be solved in terms of exogenous variable I_i and error terms e_{di} and e_{si} and the following reduced form is obtained

$$P_i = \delta_1 I_i + u_{1i}$$
$$Q_{ei} = \delta_2 I_i + u_{2i}, \quad i = 1, \dots, n, \tag{6.3}$$

where δ_1 and δ_2 are the reduced form coefficients depending on the structural equation coefficients (β, γ_s, and γ_d) as follows:

$$\delta_1 = \frac{\beta}{\gamma_s - \gamma_d},$$

$$\delta_2 = \frac{\gamma_s \beta}{\gamma_s - \gamma_d}.$$

Similarly, u_{1i} and u_{2i} denote the reduced form errors that depend on the structural equation errors (e_{di}, e_{si}) through

$$u_{1i} = \frac{e_{di} - e_{si}}{\gamma_s - \gamma_d},$$

$$u_{2i} = \frac{\gamma_s e_{di} - \gamma_d e_{si}}{\gamma_s - \gamma_d}.$$

The model in Eq. 6.3 is now identified, as the endogenous variables $(P_i$ and $Q_{ei})$ are separated from the exogenous one (I_i), and its coefficients can be estimated through the ordinary least squares approach:

$$\hat{\delta}_1 = \frac{\sum_{i=1}^{n} P_i I_i}{\sum_{i=1}^{n} I_i^2},$$

$$\hat{\delta}_2 = \frac{\sum_{i=1}^{n} Q_i I_i}{\sum_{i=1}^{n} I_i^2}.$$

In terms of structural equation coefficients, we obtain from above

$$\gamma_s = \frac{\delta_2}{\delta_1}$$

and, for $\delta_1 \neq 0$ and $\delta_2 \neq 0$, the supply equation is identified and γ_s can be consistently estimated by the indirect least squares

$$\hat{\gamma}_s = \frac{\hat{\delta}_2}{\hat{\delta}_1}.$$

The demand equation remains unidentified.

Since in many research situations decision makers can intervene on observable (usually endogenous) variables (see, for instance, the case study in Section 6.5), a further diversification of the observable variables is useful to make possible the interpretation in the causal sense of the links among variables. Thus, we introduce the distinction between *background variables* and *policy variables*. The background variables can be observed, but no external intervention is possible; on the other hand, policy variables can be manipulated through interventions. To integrate this distinction in the path diagrams, symbols of Figure 2.8 are extended as shown in Figure 6.5; in addition, the symbol used to identify the outcome or target of the causal analysis is specifically denoted, too.

As an example of intervention we can consider the model shown in Figure 6.3 and hypothesize the possibility of intervention on the exogenous variable X_3 and on the endogenous variable Y_2. The effect of this intervention results in the following system

$$y_{1i} = \beta_0 + \beta_1 x_{1i} + \beta_2 x_{2i} + \beta_3 x_{3i} + \gamma_1^* y_{2i} + e_{1i}, \quad \beta_1 < 0, \beta_2 > 0, \beta_3 > 0, \gamma_1^* < 0,$$

$$y_{3i} = \gamma_0 + \beta_6 x_{2i} + \beta_7 x_{3i} + \gamma_2 y_{1i} + \gamma_3 y_{2i} + e_{2i}, \quad \beta_6 > 0, \beta_7 > 0, \gamma_2 > 0, \gamma_3 > 0,$$

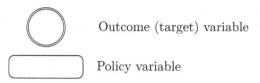

Outcome (target) variable

Policy variable

FIGURE 6.5
(Continuous of Figure 2.8) Symbols used in the path diagrams to distinguish observed variables in accordance with their role in the causal analysis: outcome or target variable and policy variable.

and is graphically shown in Figure 6.6. Note that policy variable Y_2 loses its nature of endogenous variable and is transformed into an exogenous variable; consequently, all links that point to it are eliminated from the path diagram (compare Figure 6.3 with Figure 6.6).

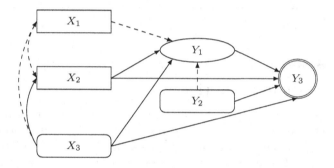

FIGURE 6.6
Example of path diagram for a non-recursive SEM with causal links (observed exogenous variables: X_1, X_2, X_3; observed endogenous variables: Y_1, Y_2, Y_3).

Use of graphical tools, as in the path analysis and in its generalizations (SEMs), makes it easier to disentangle spurious effects from causal effects (total effect)[3]. In turn, the total causal effects may be decomposed in direct and indirect effects [5, 66, 68, 22][4], whose distinction is of special interest in using SEMs to predict consequences of actions, interventions, and policy decisions.

As illustrated in Section 2.12, a SEM may be represented through Eqs. 2.33 for the structural sub-model, and Eqs. 2.34 for the measurement sub-

---

[3]A correlation between two variables, say X and Y, is spurious when it is created by a third variable, say Z, that causes both X and Y. In such a situation, correlation between X and Y disappears when we condition on Z and, then, the path from X to Y (or vice-versa) cannot be interpreted in a causal sense.

[4]A causal effect is indirect when the causal path between X and Y is fully or partly mediated by a third variable Z. For instance, X affects Z and, in turn, Z affects Y: hence, X has an indirect effect on Y.

model. On the basis of these equations, direct, indirect, and total causal effects of variables ξ and η on variables η, y, and x are disentangled as shown in Table 6.1 [22].

TABLE 6.1
Decomposition of total causal effects in a SEM.

Variable	Type of effect	Effect on η	Effect on y	Effect on x
ξ				
	direct	Γ	0	Λ_x
	indirect	$(I-B)^{-1}\Gamma-\Gamma$	$\Lambda_y(I-B)^{-1}\Gamma$	0
	total	$(I-B)^{-1}\Gamma$	$\Lambda_y(I-B)^{-1}\Gamma$	Λ_x
η				
	direct	B	Λ_y	0
	indirect	$(I-B)^{-1}-I-B$	$\Lambda_y[(I-B)^{-1}-I]$	0
	total	$(I-B)^{-1}-I$	$\Lambda_y[(I-B)^{-1}]$	0

6.4 Causal decision theory

As outlined in the previous sections, the causal approach to statistical inference, is characterized by two elements: (i) the prior definition of variables of interest and the specification of causal links among themselves; (ii) the identification of observable - exogenous or endogenous - variables on which to intervene through manipulation, that is, fixing (or doing) their values.

The methodological setting proposed by Pearl has all the prerequisites to provide a theoretical justification to interpret in a causal sense the coefficients of a SEM; however its insertion into the statistical classical or Bayesian inferential context is not completely satisfactory. Indeed, interventions on variables can modify the state of nature in which one operates and this cannot be ignored. Therefore, the appropriate context to deal with causal analysis is the decision-making one. This opinion is shared by, among others, the statisticians Heckerman and Shachter [99] and Dawid [50] and by the psychologists Hagmayer and Sloman [94]. Heckerman and Shachter substantially reiterate the definition of cause and effect of Pearl from a decisional perspective. Dawid presents the main features of an innovative decisional theoretical approach. Hagmayer and Sloman propose a decisional theoretical approach to causal analysis taking into account the generalizations of utility theory, mainly the prospect theory.

All these contributions present interesting elements; however we here prefer to address the issue of causality from a decision-making perspective taking into particular consideration the contribution of some philosophers [9, 10, 77, 107, 141]. In particular, we will refer to the *causal decision theory* of Lewis [141].

We remark that also the expected utility theory and the subjective expected utility theory discussed in Chapter 3 are not entirely adequate to support the decision-making processes. Indeed, despite their numerous generalizations, these theories deal with situations in which consequences of the actions have no effect on the states of nature. In many decision-making contexts this assumption is unrealistic: often, actions chosen by decision makers cause changes in the probability distribution of the states of nature [121, 127, 128, 129].

Similar to what was discussed in Chapter 3, the causal decision theory is based on a set of axioms of rational preference and on the existence of a utility function representing the preferences of the decision maker. In addition, posterior probabilities of the states of nature modify with the actions, other than being dependent on the prior probabilities as well as on the sample evidence. For this reason, we represent the posterior probabilities of the states of nature with the expression $p(\theta_j|fix(\boldsymbol{x}), a_i)$, which is the same as $p(\theta_j|do(\boldsymbol{x}), a_i)$. Therefore, the decision table Table 5.14, which refers to the Bayesian statistical decisional theory setting, is now extended as shown in Table 6.2 (and, similarly, if losses instead of utilities are taken into account).

TABLE 6.2

Decision table under the causal decision theory: posterior probabilities conditioned to prior probabilities, sample information, action, and intervention; consequences expressed in terms of utilities, $u_{ij} = u(a_i, \theta_j)$ $(i = 1, \ldots, m; j = 1, \ldots, k)$.

	Posterior probabilities			
Action	$\pi(\theta_1|fix(\boldsymbol{x}), a_i)$	$\pi(\theta_2|fix(\boldsymbol{x}), a_i)$	$\ldots$	$\pi(\theta_k|fix(\boldsymbol{x}), a_i)$
a_1	u_{11}	u_{12}	$\cdots$	u_{1k}
a_2	u_{21}	u_{22}	$\cdots$	u_{2k}
$\vdots$	$\cdots$	$\cdots$	$\vdots$	$\cdots$
a_m	u_{m1}	u_{m2}	$\cdots$	u_{mk}

Hence, under the causal decision setting the optimal action a^{***} satisfies

$$a^{***} = \arg\left[\max_{a_i}\left(\sum_{j=1}^{k} u(a_i, \theta_j)\pi(\theta_j|fix(\boldsymbol{x}), a_i)\right)\right], \qquad (6.4)$$

if we consider utilities, or

$$a^{***} = \arg\left[\min_{a_i}\left(\sum_{j=1}^{k} l(a_i, \theta_j)\pi(\theta_j|fix(\boldsymbol{x}), a_i)\right)\right],$$

if we consider losses.

It is worth stressing again that optimal action a^{***} distinguishes from optimal action a^{**} in Eq. 5.9 for the definition of the posterior probability of the state of nature θ_j. In the Bayesian statistical decision theory, the posterior probability of θ_j, denoted by $\pi(\theta_j|x)$, depends only on the example evidence. Instead, in the context of causal decision theory the posterior probability of θ_j, denoted by $\pi(\theta_j|fix(x), a_i)$, also depends on the intervention on the sample information and on the action itself. If we look at probability $\pi(\theta_j|fix(x), a_i)$ as a sort of generalization of $\pi(\theta_j|x)$, the causal decision theory may be thought of in terms of the causal approach to the Bayesian statistical a decision theory.

The case study illustrated in Section 5.7 provides an example of a decision problem in the causal decision theory setting. The decision problem consists of deciding whether to proceed with seeding hurricanes with silver iodide to mitigate their destructive force. Two alternative actions are possible: seeding hurricanes through the insemination or not seeding hurricanes. Consequences of the actions are expressed in terms of material damage (loss) and depend on the state of nature that is represented by the intensity of the hurricane, measured through the maximum surface speed of the wind (which has been assumed to be normally distributed). Note that the state of nature, in turn, depends on the action, other than on prior information coming from previous studies on hurricanes and sample evidence related to two independent experiments of seeding. Indeed, in the case of seeding the probability distribution of the maximum surface speed changes in accordance with three plausible scenarios (beneficial, detrimental, neutral effect).

The decisional situation illustrated in Eq. 6.4 assumes the availability of both prior information and sample information but, as illustrated in Chapter 5, there may also be situations in which only prior information is available and situations in which only the sample information is available. In the first case, we introduce the conditional prior probability $\pi(\theta_j|a_i)$ and the optimal action results

$$a^{***} = \arg\left[\max_{a_i}\left(\sum_{j=1}^{k} u(a_i, \theta_j)\pi(\theta_j|a_i)\right)\right].$$

Note that this solution is different from the one proposed in the Bayesian decision theory context (Section 5.5), where the probability of the states of nature does not depend on the action (compare with Eq. 5.7).

In the second case, when only sample information is available, the procedure described under the classical statistical decision theory setting (Section 5.4) is still valid, but the decision maker has to remember that policy variables are fixed and, therefore, he/she has to take care of computing expected utilities (or expected losses) only with respect to the background variables.

A situation that is intermediate with respect to those above described is when both sample information and prior information are available, as assumed in Eq. 6.4, but the possibility of intervention is ignored. This is the frame proper of the *evidential decision theory* of Jeffrey [121]. In such a frame, the optimal action is detected by maximizing the expected utility, as follows

$$a^{***} = \arg\left[\max_{a_i}\left(\sum_{j=1}^{k} u(a_i, \theta_j)\pi(\theta_j|see(\boldsymbol{x}), a_i)\right)\right],$$

where the $see(\cdot)$ operator denotes the conditioning on sample data, as in Eq. 6.1.

Summarizing, any branch of decision theory prescribes taking the action that maximizes the expected utility (or minimizes the expect loss). The difference among the approaches depends on how the probabilities of the states of nature are computed. Table 6.3 displays a synthesis.

TABLE 6.3
Decision theory settings, type of information available on the states of nature, and probabilities of the states of nature.

Decision theory setting	Type of information	Probability of θ_j
Bayesian	prior	$\pi(\theta_j)$
Bayesian statistical	prior and sample	$\pi(\theta_j|\boldsymbol{x})$
causal	prior	$\pi(\theta_j|a_i)$
causal Bayesian statistical	prior and sample	$\pi(\theta_j|fix(\boldsymbol{x}), a_i)$
evidential	prior and sample	$\pi(\theta_j|see(\boldsymbol{x}), a_i)$

In conclusion, the causal perspective has the relevant advantage of leading to substantial simplifications in the search for satisfactory solutions in many decision-making issues, as illustrated in the following example.

Example 6.2. Demand and supply relationship (part II). *In Example 6.1, a non-recursive SEM (Eq. 6.2) concerning the interaction between supply and demand of a commodity (or service) has been analysed; we have shown that the system is identified only for the supply equation. The graphical representation of the model is shown in Figure 6.7.*

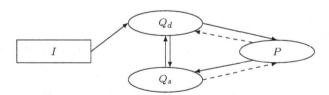

FIGURE 6.7
Path diagram for the supply-demand system of equations.

Let us introduce a simple utility function for the seller of a certain commodity

$$u(\theta_j) = Q_e \cdot P, \quad j = 1, 2, \ldots,$$

with θ_j representing the market condition (e.g., θ_1 for favorable market conditions and θ_2 for adverse market conditions) and Q_e and P the supplied and demanded quantity and the related price at the equilibrium point.

On the basis of the trend of sales observed in the past (say during year $t = 1$), at a certain time occasion (say year $t = 2$) the seller can decide to intervene by fixing quantity Q_s to be supplied to the market in order to evaluate the effect on the price P.

Therefore, in time $t = 2$ two possible alternative actions are: no intervention (a_1, evidential causal theory) and intervention by fixing $Q_s = Q_s^*$ (a_2, causal decision theory). The path diagram in Figure 6.7 is now updated to account for the intervention as shown in Figure 6.8.

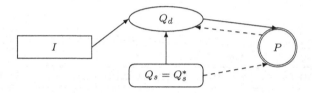

FIGURE 6.8
Path diagram for the supply-demand system of equations, after intervention on Q_s.

In an alternative but equivalent way, the producer can decide to fix the price. In such a case, the two alternative actions are no intervention (a_1, evidential causal theory) and intervention by fixing $P = P^*$ (a_2, causal decision theory) and the path diagram modifies as shown in Figure 6.9.

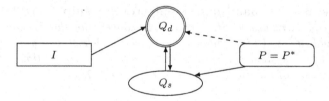

FIGURE 6.9
Path diagram for the supply-demand system of equations, after intervention on P.

Moreover, in both cases the resulting utility function is

$$u(a_i, \theta_j) = Q_s \cdot P, \quad j = 1, 2, \dots,$$

where a prior probability $\pi(\theta_j)$ is associated with the state of nature θ_j under the (unrealistic) assumption that the market conditions are constant, whereas a prior probability $\pi(\theta_j|a_i)$ is associated under the realistic assumption that the market conditions are affected by the supplied quantity Q_s (or the price P) fixed by the seller.

If we consider the simple case in which the market conditions may be favorable (θ_1) or adverse (θ_2), the decision table at time $t = 2$ assuming intervention on Q_s is shown in Table 6.4.

TABLE 6.4

Decision table: demand and supply relationship with prior probabilities of the states of nature depending on the intervention on the supplied quantity.

Action	Utilities		Expected utility
	θ_1: favorable market	θ_2: adverse market	
	$\pi(\theta_1\|a_1)$	$\pi(\theta_2\|a_1)$	
	$\pi(\theta_1\|a_2)$	$\pi(\theta_2\|a_2)$	
a_1: don't fix Q_s	$u(a_1, \theta_1)$	$u(a_1, \theta_2)$	$u(a_1, \theta_1)\pi(\theta_1\|a_1) +$ $+u(a_1, \theta_2)\pi(\theta_2\|a_1)$
a_2: fix $Q_s = Q_s^*$	$u(a_2, \theta_1)$	$u(a_2, \theta_2)$	$u(a_2, \theta_1)\pi(\theta_1\|a_2) +$ $+u(a_2, \theta_2)\pi(\theta_2\|a_2)$

At the next year $t = 3$, after observing the values assumed by Q_d and P, say $\hat{Q}_d$ and $\hat{P}$, a new decision table is built that is similar to Table 6.4, where prior probabilities $\pi(\theta_j|a_i)$ are replaced with posterior probabilities $\pi(\theta_j|a_i, \hat{Q}_s, \hat{Q}_d, \hat{P})$. Note that $\hat{Q}_s$ is not necessarily the same as Q_s^*, as the quantity supplied on the market in $t = 2$ is generally different from the quantity $\hat{Q}_d$ bought by the market at the same time.

6.5 Case study: Subscription fees of the RAI - Radiotelevisione Italiana

Current legislation in Italy requires that anyone who has one or more devices suitable to receive television programs must pay a subscription fee (law R.D.L. 21/02/1938 n. 246) to the RAI - Radiotelevisione Italiana; the fee for private use is due from households owning a television device. However, there exist two main types of individuals who evade the payment of the subscription fee: (*i*) the *illegals* (*abusiveness*) who have a TV device but are not registered in the RAI archive and (*ii*) the *defaulters* that are registered in RAI archive but, nevertheless, omit the payment.

Until 31 December 2015[5], the RAI department for subscription fees was responsible for management of registered members and promotion of new subscriptions. The attention of the RAI subscription fees department was focused on registered members, paying members, defaulters (obtained as difference between registered and paying members), new subscriptions and cancellations

---

[5]From 1 January 2016 the subscription fee has been paid through the electricity bill.

from the archive. The principal aims of the department consist of maintaining the number of registered members (i.e., reducing the unjustified cancellations from the RAI archive), reducing the defaulters, and reducing the abusiveness.

As outlined in Section 6.3, when we want to carry out a causal analysis we must make explicit the logic underlying the causal path we intend to investigate, that is,

- given the cause, do you want to identify the effect?,

- given the effect, do you want to go back to the cause?

In this case study[6], we focus on the second path: given the number of new subscriptions (NA) observed in a year, what are the causes?

Factors that may reasonably act as causes of new subscriptions can be distinguished in external and internal factors of the RAI. External factors are related with the institutional, political, economic, demographic, and psychosocial context and are out of RAI control. On the other hand, internal factors consist in actions promoted by the RAI subscription fees department and with respect to which the department may define specific interventions. External and internal factors used in the analysis are listed with their acronyms in Table 6.5.

In addition, acquisition of new subscriptions can be classified as direct and indirect. Direct acquisition is due to the activity carried out by an office within the RAI department empowered to identify unregistered owners of a TV device. Indirect acquisition is due to communications from retailers, as concerns the names of TV buyers, and from inspectors, as concerns the names of TV owners that are ascertained through home visits. Another relevant element that affects the number of new subscriptions is the quality of TV programs: it will be extensively dealt in the next case study (Section 6.6).

The model employed by the RAI subscription fees department since 1986 to explain the new subscriptions is a recursive SEM, or, better, a recursive

---

[6]In what follows the results of a causal analysis to explain the yearly number of new subscriptions are illustrated. This work was presented for the first time by one of the author, B. Chiandotto, at a workshop organized by RAI - Radiotelevisione Italiana in Bologna (Italy) in 1986. The approach here proposed based on SEMs has been used again and again by RAI in almost all the subsequent years. Results here summarized concern year 1992 and can be found in an internal report of RAI [38, 118]. For the sake of clarity, path diagrams follow notation introduced in Figures 2.8 and 6.5 instead of that adopted in the original reports.

TABLE 6.5
Description of variables used in the RAI's subscription fees analysis.

Variable	Description
External factors	
RED	Per capita income
TAT	Working rate
	(working population out of resident population)
TDI	Unemployed rate
Internal factors	
AIU	Direct subscriptions
CUR	Complaints coming from TV users
GEO	Ordinary and coactive subscription management
PAE	Inspectors: reports about families owning a TV device
RAE	Dealers: reports about families that buy a TV device
RP	Average number of families enrolled through inspectors
RR	Average number of families enrolled through retailers
TAB	Rate of abusiveness
	(number of registered families out of number of families
	that, according to current legislation, must pay the fee)
UAD	Payment notices sent by RAI offices
FNC	Other factors not considered (error term)

path model (Model 1 in the following), defined by the following nine equations in reduced form:

$$RED = f_1(TAT, TDI),$$
$$RR = f_2(TAT, TDI, TAB),$$
$$RR = f_3(TAT, TDI, TAB, RED),$$
$$RP = f_4(TAT, TDI, TAB),$$
$$RP = f_5(TAT, TDI, TAB, RED),$$
$$NA = f_6(TAT, TDI, TAB),$$
$$NA = f_7(TAT, TDI, TAB, RED),$$
$$NA = f_8(TAT, TDI, TAB, RED, RR),$$
$$NA = f_9(TAT, TDI, TAB, RED, RR, RP),$$

with $f_1(\cdot), \ldots, f_9(\cdot)$ formulated in accordance with Eq. 2.31. The corresponding path diagram is shown in Figure 6.10.

Results here described concern data collected in 1992 and refer to the 19 Italian regions (Valle d'Aosta, Piemonte, Lombardia, Veneto, Friuli Venezia Giulia, Liguria, Emilia-Romagna, Toscana, Umbria, Marche, Lazio, Abruzzo, Molise, Campania, Puglia, Basilicata, Calabria, Sicilia, Sardegna) and the 2 autonomous provinces of Bolzano and Trento. Table 6.6 shows the correlation coefficients between pairs of variables.

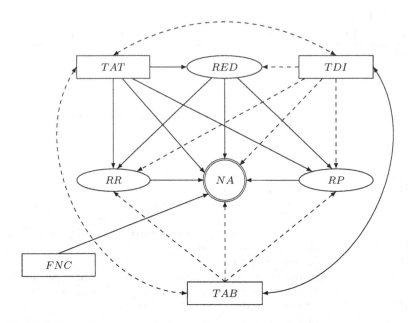

FIGURE 6.10
Model 1: causal path diagram for the acquisition of new TV subscriptions.

At first glance, some correlations between pairs of variables shown in Table 6.6 are not coherent with the hypotheses of causality initially assumed by the managers of RAI subscription fees department.

TABLE 6.6
Correlation between variables.

	TAT	RED	TDI	TAB	CUR	RAE	PAE	UAD	GEO	RR	RP	NA	AIU
TAT	1												
RED	0.845	1											
TDI	-0.904	-0.921	1										
TAB	-0.638	-0.623	0.802	1									
CUR	0.027	0.027	-0.031	0.092	1								
RAE	-0.219	-0.125	0.267	0.373	-0.451	1							
PAE	-0.257	-0.058	0.295	0.449	0.384	-0.040	1						
UAD	-0.384	-0.241	0.444	0.460	0.410	0.358	0.179	1					
GEO	-0.070	0.120	-0.011	-0.045	0.084	0.013	0.495	0.070	1				
RR	0.288	0.263	-0.375	-0.502	0.135	-0.157	-0.209	-0.052	-0.259	1			
RP	0.063	0.302	-0.120	0.044	0.072	-0.267	0.626	0.013	0.128	0.018	1		
NA	0.574	0.659	-0.752	-0.866	-0.031	-0.465	-0.275	-0.303	0.119	0.485	0.280	1	
AIU	0.258	0.101	-0.230	-0.350	0.193	-0.064	-0.273	-0.016	-0.327	0.453	-0.284	0.054	1

Among the variables used in Model 1, the rate of abusiveness (TAB) covers an important role. Initially, as the rate of abusiveness was interpreted as an indicator of higher probability of acquisition of new subscriptions, a positive correlation between TAB and NA was expected. The idea was that in a relatively "rich" market (i.e., a market with a high level of abusiveness), the probability of capturing abusive families should be higher with respect to markets with a small level of abusiveness. However, the correlation coefficient between TAB and NA is equal to -0.866, denoting a high and negative relation between the two variables. In light of this result (and of results related with the path coefficients illustrated below), TAB variable is a measure of the propensity to abusiveness and interpreted as a proxy of contextual (economic, psychosocial, and behavioral) elements that negatively affect the TV users.

Another apparently counterintuitive relationship that nevertheless supports this interpretation is observed between the average number of new subscriptions acquired through communications from retailers (RR) and TAB. Again, the sign of the correlation coefficient (-0.502) is negative, making reasonable the interpretation that TAB is a synthesis of those elements, not easily identifiable and measurable, which negatively affect the TV users, favoring the propensity to abusiveness. An alternative interpretation, which was not shared by RAI department's management, consists in assuming that the cause-effect relationship between RR and TAB is the opposite of that assumed in Model 1, that is, not from TAB to RR, but from RR to TAB.

On the other hand, the positive sign of the correlation between variable TAB and the average number of new subscriptions from inspectors (RP) may be interpreted in support of the very first hypothesis: higher rates of abusiveness make the activity of inspectors who operate in a relatively "rich" market easier; however, this correlation is very weak (0.040).

The signs of the correlation coefficients between the other variables confirm the expectations and do not require further remarks.

As concerns the estimation of path coefficients of Model 1, as the system of simultaneous equations is recursive the estimation is performed separately for each single equation through the ordinary least squares approach. Estimated path coefficients of Model 1 are shown in Table 6.7, while the decomposition of total effects, performed according to the approach of Alwin and Hauser [5], is reported in Table 6.8.

TABLE 6.7
Estimated path coefficients of Model 1 ($R^2 = 0.860$).

	RED	RR	RR	RP	RP	NA	NA	NA	NA
TAT	0.070	-0.018	-0.033	-0.493	-0.379	-0.257	-0.218	-0.215	-0.106
TDI	-0.858	0.056	-0.146	-0.977	0.591	-0.445	0.094	0.107	-0.070
TAB		-0.559	-0.505	0.513	0.091	-0.673	-0.818	-0.776	-0.817
RED			-0.157		1.224		0.421	0.434	0.072
RR								0.083	0.055
RP									0.292

The variable that mainly seems to affect the acquisition of new subscriptions is variable TAB. In front of a negative correlation coefficient equal to -0.866, the highest, and negative, total (-0.673) and direct (-0.817) effects result. Variable TDI (unemployment rate) and RED (per capita income) follow

TABLE 6.8

Decomposition of effects of Model 1.

Policy	Effect						Corr.
variables	direct	indirect			total	spurious	
		RED	RR	RP			
TAT	-0.106	-0.039	-0.003	-0.109	-0.257	0.891	0.574
TDI	-0.070	-0.539	-0.013	0.177	-0.445	-0.307	-0.752
TAB	-0.817	0.145	-0.042	0.041	-0.673	-0.193	-0.866
RED	0.072	–	-0.013	0.362	0.421	0.238	0.659
RR	0.055	–	–	–	0.055	0.430	0.485
RP	0.292	–	–	–	0.292	-0.012	0.280

with total effects of some relevance (-0.445 and 0.421, respectively), but with negligible direct effects (-0.070 and 0.072, respectively).

As far as the other variables are concerned, positive correlation coefficients between NA and variables TAT (working rate; 0.574) and RR (0.485) substantially resolve in spurious links (i.e., correlations without causal implications). Finally, the effect of variable RP is positive but weak.

Results of the causal analysis applied to data of year 1992 only partially confirm results that were obtained in the previous years with the same path model (Model 1), when variables RR and RP were identified as the most relevant explanatory factors and, therefore, the ones on which to focus interventions by RAI management for the acquisition of new subscriptions. For this reason, a second path model (Model 2) was formulated to address the presence of new plausible sources of new subscriptions, that is, variables GEO (ordinary and coactive subscription management), AIU (direct subscriptions), CUR (complaints coming from TV users), UAD (payment notices sent by RAI offices), RAE (reports by dealers about families that buy a TV device), and PAE (reports by inspectors about families that own a TV device). Variables UAD and CUR are exogenous, whereas RAE, PAE, AIU, and GEO are endogenous. In particular, AIU and GEO represent direct and strong activities that the management may carry out to challenge abusiveness.

Model 2 is entirely characterized by the presence of variables that are directly related to the activity of the RAI subscription fees department and with respect to which the management can intervene; therefore, variables RED,

TAT, and TDI, which in the last years lost their significance, are eliminated. A path model consisting of twenty recursive equations results:

$$RR = f_1(TAB)$$
$$RP = f_2(TAB)$$
$$RAE = f_3(RR)$$
$$PAE = f_4(RP)$$
$$GEO = f_5(TAB)$$
$$GEO = f_6(TAB, RR)$$
$$GEO = f_7(TAB, RR, RP)$$
$$AIU = f_8(TAB, CUR, UAD)$$
$$AIU = f_9(TAB, CUR, UAD, RR)$$
$$AIU = f_{10}(TAB, CUR, UAD, RR, RP)$$
$$AIU = f_{11}(TAB, CUR, UAD, RR, RP, GEO)$$
$$AIU = f_{12}(TAB, CUR, UAD, RR, RP, GEO, RAE)$$
$$AIU = f_{13}(TAB, CUR, UAD, RR, RP, GEO, RAE, PAE)$$
$$NA = f_{14}(TAB, CUR, UAD)$$
$$NA = f_{15}(TAB, CUR, UAD, RR)$$
$$NA = f_{16}(TAB, CUR, UAD, RR, RP)$$
$$NA = f_{17}(TAB, CUR, UAD, RR, RP, GEO)$$
$$NA = f_{18}(TAB, CUR, UAD, RR, RP, GEO, RAE)$$
$$NA = f_{19}(TAB, CUR, UAD, RR, RP, GEO, RAE, PAE)$$
$$NA = f_{20}(TAB, CUR, UAD, RR, RP, GEO, RAE, PAE, AIU).$$

The related causal path diagram is shown in Figure 6.11; a variant of this model where GEO (ordinary and coercive subscription management) explicitly is represented as a policy variable is illustrated in Figure 6.12.

Estimated path coefficients of Model 2 are displayed in Table 6.9 and the decomposition of total effects is reported in Table 6.10. Variable TAB is confirmed as the variable with the highest influence on the new subscriptions, with a negative total effect equal to -0.926 and a negative direct effect equal to -0.706. Another relevant variable is RP, with a positive effect (total effect equal to 0.318 and direct effect equal to 0.502). On the other hand, the behavior of variable PAE does not correspond to expectations, as both the total effect (-0.641) and the direct effect (-0.488) are opposite to what is expected. In such a case, a cognitive analysis is necessary to ascertain the causes of this anomalous behavior.

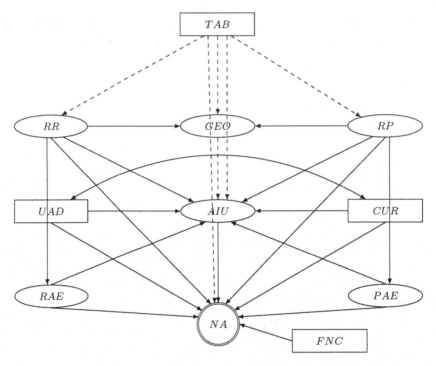

FIGURE 6.11
Model 2: causal path diagram for the acquisition of new TV subscriptions.

TABLE 6.9
Estimated path coefficients of Model 2 ($R^2 = 0.950$).

	AIU	AIU	AIU	AIU	AIU	AIU	NA	NA	NA	NA	NA	NA	NA
TAB	-0.455	-0.443	-0.247	-0.369	-0.447	-0.799	-0.926	-0.908	-0.930	-0.915	-0.851	-0.535	-0.706
CUR	0.228	0.247	0.190	0.240	0.107	0.107	0.094	0.044	0.025	0.019	-0.053	0.144	0.166
UAD	0.184	0.181	0.110	0.177	0.247	0.247	0.121	0.115	0.121	0.113	0.136	0.047	0.100
RR		-0.284	-0.294	-0.252	-0.604	-0.604		0.030	0.016	0.033	0.061	0.136	0.147
RP			0.314	0.171	0.034	0.034			0.318	0.313	0.274	0.631	0.502
GEO				-0.299	0.078	-0.653				0.037	0.058	0.351	0.212
RAE					0.717	0.078					-0.139	-0.058	-0.041
PAE						0.717						-0.641	-0.488
AIU													-0.213

6.6   Case study: Customer satisfaction for the RAI - Radiotelevisione Italiana

From the path analysis illustrated in the previous section, the rate of abusiveness (i.e., number of registered families out of number of families that, according to current legislation, must pay the fee) becomes the main determinant

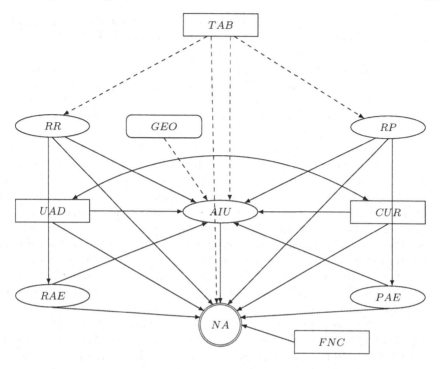

FIGURE 6.12
Model 2: causal path diagram for the acquisition of new TV subscriptions
with a policy variable (GEO).

TABLE 6.10
Decomposition of effects in Model 2.

Policy variables	direct	Effect indirect						total	spurious	Corr.
		RR	RP	GEO	RAE	PAE	AIU			
TAB	-0.706	-0.018	0.022	-0.015	-0.064	-0.316	0.171	-0.926	0.060	-0.866
CUR	0.166	0.050	0.019	0.006	0.072	-0.197	-0.022	0.094	-0.125	-0.031
UAD	0.100	0.006	-0.006	0.008	-0.023	0.089	-0.053	0.121	-0.424	-0.303
RR	0.147	–	0.014	-0.017	-0.028	-0.075	-0.011	0.030	0.455	0.485
RP	0.502	–	–	0.005	0.039	-0.357	0.129	0.318	-0.038	0.280
GEO	0.212	–	–	–	-0.021	-0.293	0.139	0.037	0.082	0.119
RAE	-0.041	–	–	–	–	-0.081	-0.017	-0.139	-0.326	-0.465
PAE	-0.488	–	–	–	–	–	-0.153	-0.641	0.366	-0.275
AIU	-0.213	–	–	–	–	–	–	-0.213	0.267	0.054

for the acquisition of new subscriptions. As already outlined, this variable is
interpreted as a proxy of latent socio-economic contextual elements that negatively affect the acquisition of new subscriptions as well as other user accounts
(such as retention rates and delinquency), both directly and indirectly through
the mitigation of effects of the RAI management's policies. In this section we

propose a case study aimed at identifying the latent variables related to attitudes and behavior of TV users towards the payment of the Rai-TV fee[7]. For
this aim we will use *customer satisfaction* models.

Customer satisfaction is commonly considered as a latent variable that
assesses the overall performance (e.g., productivity levels, competitive advan-
tages) of a company. This concept is strongly related to the quality of the com-
modity/service as perceived by consumers. Therefore, suitable interventions
to improve the quality of the commodity/service affect customer satisfaction
and, then, contribute to improve the overall performance of the company.

The most commonly used approach to measure the level of customer satisfaction is based on some suitable indices, generically known as Customer
Satisfaction Indices (CSI), and barometers that may be transaction-specific
or cumulative. In the first case, the index is based on a single consumption
episode, whereas in the second case attention is paid to all those psychological
aspects that lead to defining satisfaction as a complex experience that the con-
sumer/user matures with the commodity/service or with the producer/seller
over time. Evaluations of this type derive from several transactions and are
not limited to a single consumption episode, given that the consumer/user
proceeds to a continuous updating of his/her consumption/use experience.

The first index proposed in 1989 was the Swedish barometer (SCSB -
Swedish Customer Satisfaction Barometer), followed by the American index
(ACSI - American Customer Satisfaction Index) in 1994, by the Norwegian
barometer (NCSB - Norwegian Customer Satisfaction Barometer) in 1996, and
by the European Community index (ECSI - European Customer Satisfaction
Index) in 1999. All these indices are built starting from specific theories, developed and validated over the years, concerning the behavior of consumers, their
satisfaction and the perceived quality of the commodities/services purchased.
They are based on statistical models characterized by causal links between
a number of latent variables, which have a role in determining the level of
customer satisfaction and are measured through a specific set of directly observable indicators (manifest variables). Differences between the different proposals depend both on the number of latent variables involved in the analysis
and on the number of causal links.

The SEM setting represents the most appropriate methodological framework to estimate the effects of CSI models, by virtue of the presence and the
relationships among observable and latent variables. In more detail, both ACSI
and ECSI models take into account similar configurations of latent variables
involved, such as quality, perceived value, and loyalty. On the other hand, the
ECSI model is an evolution of the ACSI model. Differently from ACSI model,

[7]The case study here illustrated resumes the main results of a project developed by one
of the authors of this book, B. Chiandotto, in collaboration with B. Bertaccini (University
of Florence, IT), for the RAI - Radiotelevisione Italiana. The original study [39] had the aim
of identifying possible causes of evasion of subscription fees and was based on the collection
of opinions from a large sample of TV users. For other similar case studies see the works of
Chiandotto, Bini, and Bertaccini [41] and Chiandotto and Masserini [42].

the ECSI model does not account for the complaint behavior as a consequence of (dis)satisfaction and it includes among the latent variables, the corporate image assuming that it has a direct effect on consumer expectations, satisfaction and loyalty. In more detail, in the ECSI SEM the following exogenous and endogenous latent variables are considered:

- overall satisfaction (SATI; endogenous latent variable) is the most relevant factor with respect to which causes and effects are studied;

- perceived quality (exogenous latent variable) refers to the evaluation of recent experiences of consumption/use related to a certain commodity/service (perceived quality of hardware - QUAHW) and to the support provided during and after the consumption/use experience (perceived quality of humanware - QUAUW); both factors are assumed to have a direct and positive effect on overall satisfaction.

- Value (VALU; endogenous latent variable) refers to the ratio between perceived quality and price; it is positively affected by the perceived quality and, in turn, positively affects the overall satisfaction.

- Image (IMAG; exogenous latent variable) concerns the sensations derived from the association between product, brand, and company; in the classical ECSI model it is assumed to have a positive effect on value, overall satisfaction, and loyalty, even if there exist variants that assume a direct effect also on the perceived quality.

- Expectation (EXPE; exogenous latent variable) reflects what the user expects to receive from the consumption/use experience and generally depends on previous similar experiences. It is assumed to have a positive effect on both value and overall satisfaction.

- Complaints (COMP; endogenous latent variable) refer to the type and intensity of complaints coming from users and to the way in which they are dealt with. This factor is treated as a consequence of satisfaction: low levels of satisfaction are expected to result in an increase in complaints.

- Loyalty (LOYA; endogenous latent variable) is considered a proxy of the organization's profitability, in terms of repurchase intentions, tolerance to price changes, and intention to recommend the commodity or service to others. Loyalty is assumed to be affected by image and overall satisfaction.

The path diagram reproducing these latent variables and expected structural relationships in terms of causal links is shown in Figure 6.13 (the measurement part of the model is not shown).

A simplified version of the above ECSI SEM is here adopted that ignores variable COMP and aggregates the two dimensions of perceived quality in a single dimension (QUA); the path diagram of the structural part of this model is plotted in Figure 6.14.

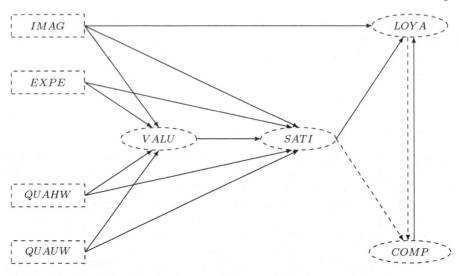

FIGURE 6.13
ACSI/ECSI model: structural part and expected causal links.

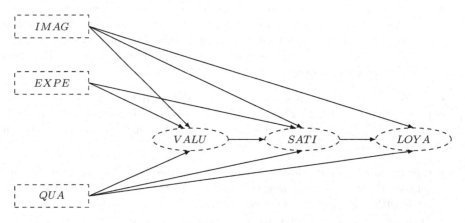

FIGURE 6.14
ECSI model (structural part) for the evaluation of the quality of RAI service.

Data used in the analysis comes from a survey that was not specifically tailored to study customer satisfaction, involving a random sample of 3,079 families living in eight municipalities (Milan, Naples, Rome, Palermo, Latina, Salerno, Trapani, and Varese) and representing the Italian population; the sample includes both families registered in the RAI archive and families that are not registered. The survey was conducted using CATI (Computer Aided Telephone Interviewing) with telephone interviews lasting up to twenty minutes. After eliminating observations with incomplete responses, the sample size was reduced to 1,240 families. Observed variables used in the measurement part of the ECSI model are listed in Table 6.11.

TABLE 6.11

Description of the measurement part of the ECSI model used in the RAI customer satisfaction analysis.

Latent variables	Observed variables
IMAG	
	1. In your opinion, is the RAI service useful?
	2. In your opinion, does RAI provide a public service?
	3. Are you in favor of privatization of RAI?
	4. Why should RAI not broadcast advertising?
EXPE	
	5. What do you expect from privatization of RAI?
	5.1 Subscription fees would no longer be paid
	5.2 RAI would no longer be conditioned on political power
	5.3 Efficiency of RAI service would increase
	5.4 There would be an increase in competitiveness
QUA	
	6. Are you aware that RAI broadcasts advertising?
	7. What is the quality of viewing of RAI with respect to the other broadcasters?
	8. How do you judge the overall quality of RAI programs?
VALU	
	9. How do you consider the cost of the subscription fee (high, medium, low)?
	10. What do you think about reducing the subscription fee to reduce the rate of abusiveness?
	11. What do you think about improving the quality of TV programs to reduce the rate of abusiveness?
SATI	
	12. Do you think it is fair to pay the RAI subscription fee?
LOYA	
	13. Has your family got a subscription to private pay TVs?
	14. Does your family use a satellite device?

An exploratory factor analysis has been performed on the model specified according to Figure 6.14 (structural part) and Table 6.11 (measurement part), in order to identify the optimal number of latent variables as well as the subset of manifest variables suitable for their measurement. Exploratory factor analysis confirmed the six latent factors identified in the simplified ECSI model of Figure 6.14, whereas some changes were suggested for the measurement part, as results from Table 6.12 (numbering of manifest variables is the same as in Table 6.11 to make the comparison easier; note that a further variable, variable 15, has been added to the model).

TABLE 6.12
Description of the measurement part of the ECSI model used in the RAI customer satisfaction analysis, as suggested by the exploratory factor analysis.

Latent var.	Observed var.
IMAG	
	1. In your opinion, is the RAI service useful?
	3. Are you in favor of privatization of RAI?
EXPE	
	5. What do you expect from privatization of RAI?
	5.1 Subscription fees would no longer be paid
	5.2 RAI would no longer be conditioned on political power
	5.3 Efficiency of RAI service would increase
	5.4 There would be an increase in competitiveness
	15. Why are you against the privatization of RAI?
	15.1 Because RAI provides a public service
	15.2 Because it would result in a reduction of the quality of TV programs
QUA	
	2. In your opinion, does RAI provide a public service?
	6. Are you aware that RAI broadcasts advertising?
	8. How do you judge the overall quality of RAI programs?
VALU	
	9. How do you consider the cost of subscription fee (high, medium, low)?
	10. What do you think about reducing the subscription fee to reduce the rate of abusiveness?
	11. What do you think about improving the quality of TV programs to reduce the rate of abusiveness?
SATI	
	12. Do you think it is fair to pay the RAI subscription fee?
LOYA	
	13. Has your family got a subscription to private pay TVs?
	14. Does your family use a satellite device?

Starting from results of the exploratory factor analysis, a complete ECSI SEM that explicitly takes into account the temporal sequencing of causal links has been specified. In the complete ECSI model each latent variable on the left of the path diagram is potentially able to directly (and indirectly) affect all the latent factors on the right of the diagram. In other words, variables IMAG, EXPE, and QUA are assumed to be theoretically able to explain both the components of satisfaction (i.e., VALU and SATI) as well as variable loyalty (LOYA). In addition, it is also reasonable to assume that the consumption/use experience activates a cognitive process that assigns a value to the commodity/service that directly affects loyalty, other than overall satisfaction: therefore, a direct effect of VALU on LOYA is introduced. Therefore,

Figure 6.14 is modified as shown in Figure 6.15, where two new links are added (in boldface).

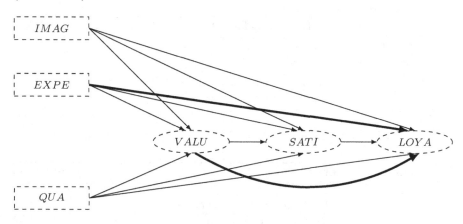

FIGURE 6.15
Complete ECSI model (structural part) for the evaluation of the quality of RAI service: in boldface the new hypothesised links.

The complete SEM specified according to the path diagram in Figure 6.15 has been estimated using the WLSMV (Weighted Least Square Mean and Variance) estimator. Results of the estimation process suggest the elimination of some causal links, leading to the final model shown in Figure 6.16. See also Table 6.13 for the estimates of coefficients related to the links between latent variables and Table 6.14 for the estimates of coefficients related to the links between latent variables and manifest variables.

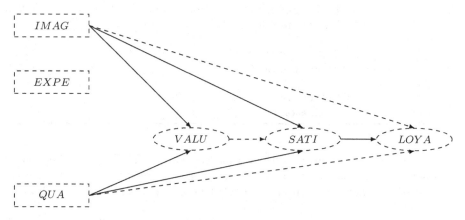

FIGURE 6.16
Final ECSI model (structural part) for the evaluation of the quality of RAI service.

TABLE 6.13

Final ECSI model (structural part): estimates of coefficients between latent variables.

Latent var.	IMAG	EXPE	QUA	VALU	SATI	LOYA
VALU	0.153	–	0.855	–	–	–
SATI	0.158	–	1.217	-0.025	–	–
LOYA	-0.169	–	-0.193	–	0.101	–

TABLE 6.14

Final ECSI model (measurement part): estimates of coefficients between latent variables and observed variables.

Latent var. Observed var.	Coefficients
IMAG	
1. Utility of a public TV service	1.000
3. Favourable to privatisation	1.288
EXPE	
5. Privatisation: elimination of subscriptions fees	1.000
5. Privatisation: increase of efficiency	0.893
5. Privatisation: increase of competitiveness	0.700
15. No privatisation: RAI as a public service	−1.159
15. No privatisation: reduction in the quality	−0.617
QUA	
6. Amount of advertising	1.000
2. RAI is a public service	1.829
8. Quality of RAI programs	−1.370
VALU	
9. Adequacy of the subscription fee	1.000
10. Reduction of the subscription fee	0.053
11. Improving the quality of TV programs	−0.022
SATI	
12. Fairness of the RAI subscription fee?	1.000
Actual fee payment	0.430
LOYA	
13. Subscription of private pay TVs	1.000
14. Possession of a satellite device	0.946

From the examination of the estimated final ECSI SEM some interesting conclusions can be drawn about the relationships between latent factors and, most of all, about the causal effects that latent factors have on the observed variables.

Coherently with the initial assumptions, latent factor value (VALU) is directly and positively affected by the RAI image (IMAG) and by the perceived quality (QUA). In turn, overall satisfaction (SATI) is directly and positively affected by perceived quality and image and, to a negligible negative extent, by variable value.

As concerns loyalty (LOYA), which is usually interpreted as a proxy of the organization profitability (customers recommend the product, bear price changes, etc.) and in the specific case as a measure of "affection" for RAI, not very high and counterintuitive coefficients result. Indeed, we would expect all negative coefficients (according to how the variables are defined in the questionnaire); instead the coefficient of SATI is positive, even if really small.

As far as the indirect effects, image affects satisfaction through value and loyalty through satisfaction and value; a similar causal pattern holds for quality. In turn, value has an indirect influence on loyalty through satisfaction.

Finally, a statistically significant impact of the latent factors on the manifest variables is observed (Table 6.14) in the following cases:

- IMAG on the usefulness of a public radio and television service and on the attitude towards a possible privatization of RAI;

- EXPE on the possible consequences of the privatization of RAI in terms of elimination of the subscription fee, greater competition and higher efficiency, and, on the other hand, loss of RAI as a public service and decline in the TV programs' quality;

- QUA on the amount of the advertising, on the quality of the TV programs and on whether RAI is considered as a public service;

- VALU on the adequacy of the amount of the subscription fee and on the efficacy of actions to reduce the rate of abusiveness (i.e., reduction of the fee and improving of TV programs' quality);

- SATI on the fairness of the subscription fee and its actual payment;

- LOYA on having a pay-TV subscription and a satellite device.

Summarizing, the application of an ECSI SEM to the analysis of TV users' satisfaction with the RAI service allowed us to identify the latent factors affecting those individual behaviors that contribute to the abusiveness.

Overall satisfaction proved to be the factor that mainly affects the payment of the subscription fee. In turn, satisfaction is related to the image of RAI and value and perceived quality determine certain attitudes towards RAI service as a public service and towards the fairness of paying a subscription fee and the adequacy of its amount. All these results provide clear suggestions to the RAI management: interventions aimed at reducing the rate of abusiveness must be directed in the sense of improving the levels of users' overall satisfaction which, in turn, depends on intervention aimed at improving the image, the value, and the perceived quality.

Following these suggestions, in the past RAI management carried out interventions on some aspects. Among others, the subscription fee was reduced, the quality of TV programs was improved, some advertising was offered to clarify the role of RAI as a public service. However, the worsening of the Italian economic situation and the impossibility to directly intervene with the abusive users by RAI management brought about an increase in the abusiveness. As a consequence of this situation, the Italian Government intervened changing the legislation and incorporating the RAI subscription fee in the bill for the domestic supply of electricity. This intervention led to a reduction of the abusiveness rate from 33% to 10%.

Our Knowledge depends on the Tools we use to interpret the World.

References

[1] M. Abdellaoui. Parameter-free elicitation of utility and probability weighting functions. *Management Science*, 46:1497–1512, 2000.

[2] M. Abdellaoui, H. Bleichrodt, and O. L'Haridon. A tractable method to measure utility and loss aversion under prospect theory. *Journal of Risk and Uncertainty*, 36:245–266, 2008.

[3] M. Abdellaoui, H. Bleichrodt, and C. Paraschiv. Loss aversion under prospect theory: a parameter-free measurement. *Management Science*, 53:1659–1674, 2007.

[4] M. Allais. Le comportement de l'homme rationnel devant le risque: critique des postulats et axiomes de l'école Américaine. *Econometrica*, 21:503–546, 1953.

[5] D.F. Alwin and R.M. Hauser. The decomposition of effects in path analysis. *American Sociological Review*, 40:37–47, 1975.

[6] R. D. Anderson and H. Rubin. Statistical inference in factor analysis. *Proceedings of the Third Berkeley Symposium of Mathematical Statistics and Probability*, 5:111–150, 1956.

[7] F. J. Anscombe and R. J. Aumann. A definition of subjective probability. *Annals of Mathematical Statistics*, 34:199–205, 1963.

[8] J. Arbuthnot. An argument for divine providence, taken from the constant regularity observed in the births of both sexes. *Philosophical Transactions of the Royal Society of London*, 27:186–190, 1710. Reprinted in M.G. Kendall and R.L. Plackett (eds.), Studies in the History of Statistics and Probability Volume II, pp. 30-34, 1977.

[9] B. Armendt. A foundation for causal decision theory. *Topoi*, 5:3–19, 1986.

[10] B. Armendt. Conditional preference and causal expected utility. In W.L. Harper and B. Skyrms, editors, *Causation in Decision, Belief Change, and Statistics, Vol. II*, pages 3–24. Kluwer Academic Publishers, Dordrecht, NL, 1988.

[11] S. Bacci and B. Chiandotto. Struttura di preferenze e decisioni razionali nelle fondazioni bancarie. *Studi e Note di Economia*, 2:19–56, 2002.

[12] V. Barnett. *Comparative Statistical Inference*. John Wiley, New York, 1999.

[13] D. Bell, H. Raiffa, and A. Tversky. *Decision Making: Descriptive, Normative, and Prescriptive Interactions*. Cambridge University Press, Cambridge, UK, 1988.

[14] J. O. Berger. *Statistical Decision Theory and Bayesian Analysis*. Springer Science+Business Media. Inc., New York, 1985.

[15] J. O. Berger. The case for objective Bayesian analysis. *Bayesian Analysis*, 1:385–402, 2006.

[16] J. O. Berger, J. M. Bernardo, and D. Sun. The formal definition of reference priors. *The Annals of Statistics*, 37:905–938, 2009.

[17] J. M. Bernardo. Reference posterior distributions for Bayesian inference (with discussion). *Journal of the Royal Statistical Society, Series B*, 41:113–147, 1979.

[18] D. Bernoulli. Exposition of a new theory on the measurement of risk. *Econometrica*, 22:23–36, 1954.

[19] P. Blavatskyy. Error propagation in the elicitation of utility and probability weighting functions. *Theory and Decision*, 60:315–334, 2006.

[20] H. Bleichrodt and J. L. Pinto. A parameter-free elicitation of the probability weighting function in medical decision analysis. *Management Science*, 46:1485–1496, 2000.

[21] H. Bleichrodt, J. L. Pinto, and P. P. Wakker. Making descriptive use of prospect theory to improve the prescriptive use of expected utility. *Management Science*, 47:1498–1514, 2001.

[22] K. A. Bollen. Total, direct, and indirect effects in structural equation models. *Sociological Methodology*, 17:37–69, 1987.

[23] K. A. Bollen. *Structural Equations with Latent Variables*. John Wiley and Sons, New York, 1989.

[24] W. M. Bolstad. *Introduction to Bayesian Statistics*. John Wiley & Sons, New York, 2007.

[25] L. Buchack. *Risk and Rationality*. Oxford University Press, Oxford, UK, 2013.

[26] C. F. Camerer. Recent tests of generalizations of expected utility theory. In W. Edwards, editor, *Utility Theories: Measurement and Applications*, pages 207–251. Kluwer Academic Publishers, Boston, 1992.

[27] B. P. Carlin and T. A. Louis. *Bayesian Methods for Data Analysis.* Chapman and Hall/CRC Press, Boca Raton, FL, 2009.

[28] R. Carnap. *Logical Foundations of Probability.* University of Chicago Press, Chicago, IL, 1950.

[29] N. Cartwright. Causal laws and effective strategies. *Noûs*, 13:419–437, 1979.

[30] N. Cartwright. Causation: One word, many things. *Philosophy of Science*, 71:805–819, 2004.

[31] N. Cartwright. Where is the theory in our "theories" of causality? *The Journal of Philosophy*, 103:55–66, 2006.

[32] G. Casella and R. L. Berger. *Statistical Inference.* Thomson Learning, Pacific Grove, CA, 2006.

[33] U. Chajewska, J. Getoor, J. Normal, and Y. Shahar. Utility elicitation as a classification problem. In *Proceedings of the Fourteenth Conference on Uncertainty in Artificial Intelligence (UAI 1998), July 24-26, 1998*, pages 79–88. Madison, WI, 1998.

[34] U. Chajewska, D. Koller, and R. Parr. Making rational decisions using adaptive utility elicitation. In *Proceedings of the Seventeenth National Conference on Artificial Intelligence (AAAI-00), August, 2000*, pages 363–369. Austin, TX, 2000.

[35] A. Chateauneuf and M. Cohen. Risk seeking with diminishing marginal utility in a non-expected utility model. *Journal of Risk and Uncertainty*, 9:77–91, 1994.

[36] A. Chateauneuf, M. Cohen, and J.-Y. Jaffray. Decision under uncertainty: the classical models. *Documents de travail du Centre d'Economie de la Sorbonne*, 2008.86, 2008.

[37] B. Chiandotto. L'approccio bayesiano empirico alla problematica dell'inferenza statistica. In *Proceedings of the Conference "I fondamenti dell'inferenza statistica", Firenze 28-30 aprile 1977*, pages 257–268. Università degli Studi di Firenze, Firenze, IT, 1978.

[38] B. Chiandotto. L'atteggiamento dei cittadini nei confronti del canone TV. Technical Report, Italian Radiotelevision RAI, University of Florence (IT), 2000.

[39] B. Chiandotto. Utenza TV: un modello per la misura della customer satisfaction. Technical Report, Italian Radiotelevision RAI, University of Florence (IT), 2004.

[40] B. Chiandotto. Bayesian and non-bayesian approaches to statistical inference: a personal view. In Crescenzi, F. and Mignani, S. (eds), Statistical Methods and Applications from a Historical Perspective: Selected Issues, pages 3–13. Springer, Heidelberg, 2014.

[41] B. Chiandotto, M. Bini, and B. Bertaccini. Evaluating the quality of the university educational process: an application of the ECSI model. In F. Fabbris, editor, *Effectiveness of University Education in Italy: Employability, Competences, Human Capital*, pages 43–54. Physica-Verlag, Heidelberg, 2007.

[42] B. Chiandotto and L. Masserini. Two-level structural equation model for evaluating the external effectiveness of phd. *Journal of Applied Quantitative Methods*, 6:72–87, 2011.

[43] R. A. Cohn, W. G. Lewellen, R. C. Lease, and G. C. Schlarbaum. Individual investor risk aversion and investment portfolio composition. *The Journal of Finance*, 30:605–619, 1975.

[44] R. G. Cowell, A. P. Dawid, S. L. Lauritzen, and D. J. Spiegelhalter. *Probabilistic Networks and Expert Systems: Exact Computational Methods for Bayesian Networks*. Springer Verlag, New York, 1999.

[45] D. R. Cox. Some problems connected with statistical inference. *Annals of Mathematical Statistics*, 29:357–372, 1958.

[46] A. P. Dawid. Conditional independence in statistical theory. *Journal of the Royal Statistical Society, Series B*, 41:1–31, 1979.

[47] A. P. Dawid. Causal inference without counterfactuals. *Journal of the American Statistical Association*, 95:407–424, 2000.

[48] A. P. Dawid. Influence diagrams for causal modelling and inference. *International Statistical Review*, 70:161–189, 2002.

[49] A. P. Dawid. Counterfactuals, hypotheticals and potential responses: a philosophical examination of statistical causality. In F. Russo and J. Williamson, editors, *Causality and Probability in the Sciences*, pages 503–532. College Publications, London, UK, 2007.

[50] A. P. Dawid. Statistical causality from a decision-theoretic perspective. *Annual Review of Statistics and Its Application*, 2:273–303, 2015.

[51] A.P. Dawid, M. Musio, and S.E. Fienberg. From statistical evidence to evidence of causality. *Bayesian Analysis*, 11:725–752, 2016.

[52] B. de Finetti. La previon: ses lois logiques ses sources subjectives. *Annales de l'Institut Henry Poincaré*, 7:1–68, 1937. English translation in H. E. Kyburg and H. E. Smokler, eds., Studies in Subjective Probability, Foresight, its logical laws, its subjective sources, pp. 97- 156, 1964.

[53] M. H. DeGroot. *Optimal Statistical Decisions.* MacGraw-Hill, New York, 1970.

[54] A. P. Dempster. A generalization of Bayesian inference. *Journal of the Royal Statistical Society, Series B*, 30:205–247, 1968.

[55] O. D. Duncan. Path analysis: Sociological examples. *American Journal of Sociology*, 72:1–16, 1966.

[56] O. D. Duncan. *Introduction to structural equation models.* Academic Press, New York, 1975.

[57] J. S. Dyer. MAUT - Multiattribute utility theory. In S. Greco, M. Ehrgott, and J. R. Figueira, editors, *Multiple Criteria Decision Analysis: State of the Art Surveys*, pages 265–292. Springer, New York, 2016.

[58] D. Ellsberg. Risk, ambiguity, and the Savage axioms. *Quarterly Journal of Economics*, 75:643–669, 1961.

[59] T. Eppel, D. Matheson, J. Miyamoto, G. Wu, and S. Eriksen. Old and new roles for expected and generalized utility theories. In W. Edwards, editor, *Utility Theories: Measurement and Applications*, pages 271–294. Kluwer Academic Publishers, Boston, 1992.

[60] P. H. Farquhar. Utility assessment methods. *Management Science*, 30:1283–1296, 1984.

[61] H. Fennema and M. Van Assen. Measuring the utility of losses by means of the tradeoff method. *Journal of Risk and Uncertainty*, 17:277–295, 1999.

[62] T. S. Ferguson. *Mathematical Statistics: A Decision Theoretic Approach.* Academic Press, New York, 1967.

[63] P. Fishburn and G. A. Kochenberger. Two-piece Von Neumann-Morgenstern utility functions. *Decision Sciences*, 10:503–518, 1979.

[64] P. C. Fishburn. Normative theories of decision making under risk and under uncertainty. In J. Kacprzyk and M. Roubens, editors, *Non-Conventional Preference Relations in Decision Making*, pages 1–21. Springer, Berlin, Heidelberg, 1988.

[65] P. C. Fishburn. Decision making under risk and uncertainty. In J. Geweke, editor, *Nontransitive Preferences and Normative Decision Theory*, pages 3–10. Kluwer Academic Publishers, Dordrecht, 1992.

[66] J. Fox. Effect analysis in structural-equation models: Extensions and simplified methods of computation. *Sociological Methods & Research*, 9:3–28, 1980.

[67] J. Fox. *Linear Statistical Models and Related Methods: With Applications to Social Research*. John Wiley and Sons, New York, 1984.

[68] J. Fox. Effect analysis in structural-equation models. II: Calculation of specific indirect effects. *Sociological Methods & Research*, 14:81–95, 1985.

[69] D. A. Freedman. Statistical models and shoe leather. *Sociological Methodology*, 21:291–313, 1991.

[70] D. A. Freedman. From association to causation: Some remarks on the history of statistics. *Statistical Science*, 14:243–258, 1999.

[71] S. French. *Decision Theory: An Introduction to the Mathematics and Rationality*. Ellis Horwood Limited, Chichester, UK, 1986.

[72] M. Friedman and L. J. Savage. The utility analysis of choices involving risk. *The Journal of Political Economy*, 56:279–304, 1948.

[73] I. Friend and M. E. Blume. The demand for risky assets. *American Economic Review*, 65:900–922, 1975.

[74] D. G. Garson. *The Delphi Method in Quantitative Research*. Statistical Associates Publishers, Asheboro, NC, 2014.

[75] P. H. Garthwaite, J. B. Kadane, and A. O'Hagan. Statistical methods for eliciting probability distributions. *Journal of the American Statistical Association*, 100:680–701, 2005.

[76] R. C. Gentry. Debbie modification experiments. *Science*, 168:473–475, 1969.

[77] A. Gibbard and W. L. Harper. Counterfactuals and two kinds of expected utility. In W. L. Harper, R. Stalnaker, and G. Pearce, editors, *IFS*, pages 153–190. Springer, Dordrecht, 1978.

[78] I. Gilboa. *Theory of Decision under Uncertainty*. Cambridge University Press, Cambridge, UK, 2009.

[79] C. Gini. I pericoli della statistica. In *Proceedings of the First Conference of the Italian Statistical Society, Pisa 9 ottobre 1939*. Pisa, IT, 1939.

[80] C. Gini. I test di significatività. In *Proceedings of the Seventh Conference of the Italian Statistical Society, Roma 27-30 giugno 1943*. Roma, IT, 1943.

[81] G. Gini. Considerazioni sulle probabilità a posteriori e applicazioni al rapporto dei sessi nelle nascite umane. *reprinted in Metron*, 15:133–172, 1949 (first edition 1911).

[82] A. S. Goldberger. Structural equation methods in the social sciences. *Econometrica*, 40:979–1001, 1972.

[83] M. Goldstein. Subjective Bayesian analysis: Principles and practice. *Bayesian Analysis*, 1:403–420, 2006.

[84] R. Gonzales and G. Wu. On the shape of the probability weighting function. *Cognitive Psychology*, 38:129–166, 1999.

[85] C. W. J. Granger. Investigating causal relations by econometric models and cross-spectral methods. *Econometrica*, 37:424–438, 1969.

[86] C. Grayson. *Decisions Under Uncertainty: Drilling Decisions by Oil and Gas Operators.* Harvard Business School, Division of Research, Cambridge, MA, 1960.

[87] W. H. Greene. *Econometric Analysis.* Prentice Hall, Englewood Cliffs, N. J., 1999.

[88] M. Grether and C. R. Plott. Economic theory of choice and the preference reversal phenomenon. *The American Economic Review*, 69:623–638, 1979.

[89] D. N. Gujarati and D. Porter. *Basic Econometrics.* McGraw-Hill Education, New York, 2017.

[90] V. Ha and P. P. Haddawy. Towards case-based preference elicitation: Similarity measures on preference structures. In *Proceedings of the Fourteenth Conference on Uncertainty in Artificial Intelligence (UAI 1998), July 24-26, 1998*, pages 193–201. Madison, WI, 1998.

[91] T. Haavelmo. The statistical implications of a system of simultaneous equations. *Econometrica*, 11:1–12, 1943.

[92] T. Haavelmo. The probability approach in econometrics. *Econometrica Supplement*, 12:iii–vi, 1–115, 1944.

[93] T. Haavelmo. Structural models and econometrics. *Econometric Theory*, 31:85–92, 2005.

[94] Y. Hagmayer and S. A. Sloman. Decision makers conceive of their choices as interventions. *Journal of Experimental Psychology: General*, 138:22–38, 2009.

[95] P. R. Halmos and L. J. Savage. Applications of the Radon-Nikodyn Theorem to the theory of sufficient statistics. *Annals of Mathematical Statistics*, 20:225–241, 1949.

[96] P. Hammond. Consequentialist foundations for expected utility. *Theory and Decision*, 25:25–78, 1988.

[97] D. W. Harless and C. F. Camerer. The predictive utility of generalized expected utility theories. *Econometrica*, 62:1251–1289, 1994.

[98] R.M. Hauser and A.S. Goldberger. The treatment of unobservable variables in path analysis. *Sociological Methodology*, 3:81–117, 1971.

[99] D. Heckerman and R. Shachter. Decision-theoretic foundations for causal reasoning. *Journal of Artificial Intelligence Research*, 3:405–430, 1995.

[100] J. J. Heckman. The scientific model of causality. *Sociological Methodology*, 35:1–97, 2005.

[101] J. J. Heckman and R. Pinto. Causal analysis after Haavelmo. *Econometric Theory*, 31:115–151, 2015.

[102] J. J. Heckman and E. J. Vytlacil. Structural equations, treatment effects and econometric policy evaluation. *Econometrica*, 73:669–738, 2005.

[103] M. A. Hernán and J. M. Robins. *Causal Inference.* Chapman & Hall/CRC, Boca Raton, FL, 2019 (forthcoming).

[104] J. C. Hershey, H. C. Kunreuther, and P. J. H. Schoemaker. Sources of bias in assessment procedures for utility functions. *Management Science*, 28:936–953, 1982.

[105] J. C. Hershey and P. J. H. Schoemaker. Probability versus certainty equivalence methods in utility measurement: are they equivalent? *Management Science*, 31:1213–1231, 1985.

[106] K. Hirano, G. W. Imbens, and G. Ridder. Efficient estimation of average treatment effects using the estimated propensity score. *Econometrica*, 71:1161–1189, 2003.

[107] C. Hitchcock. What is the "cause" in causal decision theory? *Erkenntnis*, 78:129–146, 2013.

[108] P. W. Holland. Statistics and causal inference. *Journal of the American Statistical Association*, 81:945–960, 1986.

[109] P. W. Holland. Causal inference, path analysis, and recursive structural equations models. *Sociological Methodology*, 18:449–484, 1988.

[110] R. A. Howard, J. E. Matheson, and D. W. North. The decision to seed hurricanes. *Science*, 176:1191–1202, 1972.

[111] R.A. Howard and J.E. Matheson. Influence diagrams. In R.A. Howard and J.E. Matheson, editors, *Readings on the Principles and Applications of Decision Analysis*. Strategic Decisions Group, Menlo Park, CA, 1984.

[112] D. Hume. *An Enquiry Concerning Human Understanding.* Oxford University Press, New York, 1999 (first edition 1748).

[113] D. Hume. *A Treatise of Human Nature.* Oxford University Press, New York, 2000 (first edition 1739).

[114] C. Huygens. De ratiociniis in ludo aleae. In *Exercitationum Mathematicarum.* Ex Officina Johannis Elsevirii Academiae Typographi, 1657.

[115] G. W. Imbens and D. B. Rubin. *Causal Inference for Statistics, Social, and Biomedical Sciences. An Introduction.* Cambridge University Press, Cambridge, UK, 2015.

[116] T. Z. Irony and N. D. Singpurwalla. Non-informative priors do not exist. a dialogue with José M. Bernardo. *Journal of Statistical Planning and Inference,* 65:159–189, 1997.

[117] A. Ishizaka and P. Nemery. *Multi Criteria Decision Analysis: Methods and Software.* John Wiley & Sons, New York, 2013.

[118] Italian Radiotelevision RAI. Sviluppo dell'utenza e proiezioni di tendenza 1994-1998. Technical Report Internal Report, 1994.

[119] E. D. Jaynes. Prior probabilities. *IEEE Transactions on Systems Science and Cybernetics,* 4:227–241, 1968.

[120] E. T. Jaynes. *Probability Theory: The Logic of Science.* Cambridge University Press, Cambridge, UK, 2003.

[121] R. C. Jeffrey. *The Logic of Decision.* University of Chicago Press, Chicago, IL, 1983, 2nd ed.

[122] H. Jeffreys. *Theory of Probability.* Oxford University Press, Oxford, UK, 1961.

[123] D. Jenkinson. The elicitation of probabilities: A review of the statistical literature. Technical Report BEEP Working Paper 2005, University of Sheffield, 2005.

[124] K. G. Jöreskog. A general method for estimating a linear structural equation. In A. S. Golderberger and O. D. Duncan, editors, *Structural Equation Models in the Social Sciences.* Seminar Press, New York, 1973.

[125] J. M. Joyce. *The Foundations of Causal Decision Theory.* Cambridge University Press, Cambridge, UK, 1999.

[126] D. Kahneman and A. Tversky. Prospect theory: an analysis of decision under risk. *Econometrica,* 47:263–293, 1979.

[127] E. Karni. A definition of subjective probabilities with state-dependent preferences. *Econometrica,* 61:187–198, 1993.

[128] E. Karni. States of nature and the nature of states. *Economics and Philosophy*, 33:73–90, 2017.

[129] E. Karni, D. Schmeidler, and K. Vind. On state dependent preferences and subjective probabilities. *Econometrica*, 51:1021–1031, 1983.

[130] R. E. Kass and L. Wasserman. The selection of prior distributions by formal rules. *Journal of the American Statistical Association*, 91:1343–1370, 1996.

[131] R. L. Keeney and H. Raiffa. *Decisions with Multiple Objectives. Preferences and Value Tradeoffs*. Cambridge University Press, Cambridge, UK, 1993.

[132] J. M. Keynes. *A Treatise on Probability*. Macmillan, London, UK, 1921.

[133] A. N. Kolmogorov. *Foundations of the Theory of Probability*. Chelsea Publishing Company, New York, 1956.

[134] T. Koopmans. Statistical estimation of simultaneous economic relations. *Journal of the American Statistical Association*, 40:448–466, 1945.

[135] D. Kreps. *Notes on the Theory of Choice. Underground Classics in Economics*. Westview Press, Boulder, CO, 1988.

[136] S. L. Lauritzen. Causal inference from graphical models. In O.E. Barndorff-Nielsen, D. R. Cox, and C. Klüppelberg, editors, *Complex Stochastic Systems*, pages 63–107. Chapman and Hall/CRC, Boca Raton, FL, 2000.

[137] S. L. Lauritzen. Chain graph models and their causal interpretations. *Journal of the Royal Statistical Society, Series B*, 64:321–361, 2002.

[138] S. L. Lauritzen. Comment on "Direct and indirect causal effects via potential outcomes", by D. B. Rubin. *Scandinavian Journal of Statistics*, 31:189–192, 2004.

[139] S. L. Lauritzen and D. Nilsson. Representing and solving decision problems with limited information. *Management Science*, 47:1235–1251, 2001.

[140] S. L. Lauritzen and T. S. Richardson. Chain graph models and their causal interpretation (with discussion). *Journal of the Royal Statistical Society, Series B*, 64:321–361, 2002.

[141] D. Lewis. Causal decision theory. *Australasian Journal of Philosophy*, 59:5–30, 1981.

[142] S. Lichtenstein and P. Slovic. Reversals of preference between bids and choices in gambling decisions. *Journal of Experimental Psychology*, 89:46–55, 1971.

[143] S. Lichtenstein and P. Slovic. Response-induced reversals of preference in gambling: an extended replication in Las Vegas. *Journal of Experimental Psychology*, 101:16–20, 1973.

[144] D. V. Lindley. *Making Decisions*. John Wiley, New York, 1985, 2nd ed.

[145] D. V. Lindley. The philosophy of statistics. *Journal of the Royal Statistical Society. Series D (The Statistician)*, 49:293–337, 2000.

[146] D. V. Lindley. Seeing and doing: the concept of causation. *International Statistical Review*, 70:191–214, 2002.

[147] D. V. Lindley. *Understanding Uncertainty*. John Wiley, New York, 2006.

[148] H. A. Linstone and M. Turoff. *Delphi Method: Techniques and Applications*. Addison-Wesley Educational Publishers Inc., Glenview, IL, 1975.

[149] J. H. C. Lisman and M. C. A. van Zuylen. Note on the generation of most probable frequency distributions. *Statistica Neerlandica*, 26:19–23, 1972.

[150] M. J. Machina. Comparative statistics and non-expected utility preferences. *Journal of Economic Theory*, 47:393–405, 1989.

[151] J. S. Malkus and R. H. Simpson. Modification experiments on tropical cumulus clouds. *Science*, 145:541–548, 1964.

[152] H. Markowitz. The utility of wealth. *The Journal of Political Economy*, 60:151–158, 1952.

[153] J. Marschak. Rational behavior, uncertain prospects, and measurable utility. *Econometrica*, 18:111–141, 1950.

[154] J. Marschak. Expected utility analysis without the independence axiom. *Econometrica*, 50:277–323, 1982.

[155] K. O. May. Intransitivity, utility, and the aggregation of preference patterns. *Econometrica*, 22:1–13, 1954.

[156] M. McCord and R. De Neufville. Lottery equivalents: reduction of the certainty effect problem in utility assessment. *Management Science*, 32:56–60, 1986.

[157] J. S. Mill. *A System of Logic. Ratiocinative and Inductive (Collected Works of John Stuart Mill)*. University of Toronto Press, Toronto, 1973 (first edition 1843).

[158] A. M. Mood, F. A. Graybill, and D.C. Boes. *Introduction to the Theory of Statistics*. MacGraw-Hill, New York, 1974.

[159] J. Neyman. On the application of probability theory to agricultural experiments. Essay on principles. *Statistical Science*, 5:465–472, 1923.

[160] A. O'Hagan. Bayesian statistics: principles and benefits. *Frontis*, 3:31–45, 2004.

[161] J. D. Olive. *Statistical Theory and Inference*. Springer, Switzerland, 2014.

[162] G. Parmigiani and L. Y. T. Inoue. *Decision Theory: Principles and Approaches*. John Wiley and Sons, New York, 2009.

[163] J. Pearl. Causal diagrams for empirical research. *Biometrika*, 82:669–710, 1995.

[164] J. Pearl. Causal inference in statistics: An overview. *Statistical Surveys*, 3:96–146, 2009.

[165] J. Pearl. The causal foundations of structural equation modeling. In R. H. Hoyle, editor, *Handbook of Structural Equation Modeling*, pages 68–91. The Guilford Press, New York, 2012.

[166] J. Pearl. Trygve Haavelmo and the emergence of causal calculus. *Econometric Theory*, 31:152–179, 2015.

[167] J. Pearl and D. Mackenzie. *The Book of Why: The New Science of Cause and Effect*. Basic Book, New York, 2018.

[168] K. Pearson. *The Grammar of Science*. Cambridge University Press, Cambridge, UK, 2015 (first edition 1892).

[169] P. Perny, P. Viappiani, and A Boukhatem. Incremental preference elicitation for decision making under risk with the rank-dependent utility model. In *UAI' 16 Proceedings of the Thirty-Second Conference on Uncertainty in Artificial Intelligence, June 25-29, 2016*, pages 597–606. Jersey City, New Jersey, 2016.

[170] M. Peterson. *An Introduction to Decision Theory*. Cambridge University Press, Cambridge, UK, 2009.

[171] L. Piccinato. *Metodi per le decisioni statistiche*. Springer-Verlag, Berlin, 1996.

[172] G. Pompilj. Logica della conformità. *Archimede*, 4:22–28, 1952.

[173] G. Pompilj. *Teoria dei Campioni*. Veschi, Roma, IT, 1961.

[174] A. Power, R. W. Kates, R. A. Howard, J. E. Matheson, and D. W. North. Seeding hurricanes. *Science*, 179:744–747, 1973.

[175] J. W. Pratt. Risk aversion in the small and in the large. *Econometrica*, 32:122–136, 1964.

[176] J. W. Pratt, H. Raiffa, and R. Schlaifer. *Introduction to Statistical Decision Theory*. The MIT Press, Cambridge, MA, 1995.

[177] J. Quiggin. A theory of anticipated utility. *Journal of Economic Behavior & Organization*, 3:323–343, 1982.

[178] H. Raiffa and R. Schlaifer. *Applied Statistical Decision Theory*. Clinton Press, Boston, MA, 1961.

[179] F. P. Ramsey. Truth and probability. In R. B. Braithwaite, editor, *Foundations of Mathematics and Other Logical Essays*, pages 156–198. Kegan, Paul, Trench, Trubner & Co. Ltd., New York, 1931.

[180] C. P. Robert. *The Bayesian Choice: From Decision-Theoretic Foundations to Computational Implementation*. Springer Science+Business Media. Inc., New York, 2007.

[181] C.P. Robert, N. Chopin, and J. Rousseau. Harold Jeffreys's theory of probability revisited. *Statistical Science*, 24:142–172, 2009.

[182] J. M. Robins. Latent variable modeling and applications to causality. In M. Berkane, editor, *Complex Stochastic Systems*, pages 69–117. Springer Verlag, New York, 1997.

[183] J. M. Robins. Association, causation, and marginal structural models. *Synthese*, 121:151–179, 1999.

[184] J.M. Robins and S. Greenland. The role of model selection in causal inference from nonexperimental data. *American Journal of Epidemiology*, 123:392–402, 1986.

[185] V. K. Rohatgi and E. Saleh. *An Introduction to Probability and Statistics*. John Wiley & Sons, New York, 2011.

[186] P. R. Rosenbaum and D. B. Rubin. The central role of the propensity score in observational studies for causal effects. *Biometrika*, 70:41–55, 1983.

[187] H. L. Royden and P. M. Fitzpatrick. *Real Analysis*. Pearson, 2010.

[188] D. B. Rubin. Estimating causal effects of treatments in randomized and nonrandomized studies. *Journal of Educational Psychology*, 66:688–701, 1974.

[189] D. B. Rubin. Direct and indirect causal effects via potential outcomes. *Scandinavian Journal of Statistics*, 31:161–170, 2004.

[190] L. J. Savage. The theory of statistical decision. *Journal of the American Statistical Association*, 46:55–67, 1951.

[191] L. J. Savage. *The Foundations of Statistics*. John Wiley and Sons, New York, 1954.

[192] U. Schmidt. Alternatives to expected utility: formal theories. In S. Barberà, P. J. Hammond, and C. Seidl, editors, *Handbook of Utility Theory*, pages 757–837. Springer, Boston, MA, 2004.

[193] T. Seidenfeld. Why I am not an objective Bayesian; some reflections prompted by Rosenkrantz. *Theory and Decision*, 11:413–440, 1979.

[194] G. Shafer. *A Mathematical Theory of Evidence*. Princeton University Press, Princeton, NJ, 1976.

[195] B. Shipley. *Cause and Correlation in Biology*. Cambridge University Press, Cambridge, UK, 2000.

[196] B. Skyrms. *Causal Necessity*. Yale University Press, New Haven, CT, 1980.

[197] B. Skyrms. Causal decision theory. *The Journal of Philosophy*, 79:695–711, 1982.

[198] P. Slovic, B. Fischhoff, and S. Lichtenstein. Response mode, framing, and information-processing effects in risk assessment. In D. E. Bell, H. Raiffa, and A. Tversky, editors, *Decision Making*, pages 152–166. Cambridge University Press, Cambridge, UK, 1988.

[199] C. A. B. Smith. Consistency in statistical inference and decision. *Journal of the Royal Statistical Society. Series B (Methodological)*, 23:1–37, 1961.

[200] M. E. Sobel. An introduction to causal inference. *Sociological Methods & Research*, 24:353–379, 1996.

[201] M. E. Sobel. Causal inference in the social sciences. *Journal of the American Statistical Association*, 95:647–651, 2000.

[202] M. E. Sobel. The scientific model of causality. *Sociological Methodology*, 35:99–133, 2005.

[203] C. Spearman. "General intelligence," objectively determined and measured. *American Journal of Psychology*, 15:201–293, 1904.

[204] T. P. Speed. Complexity, calibration and causality in influence diagrams. In R.M. Oliver and J.Q. Smith, editors, *Influence Diagrams, Belief Nets and Decision Analysis*, pages 43–54. John Wiley, New York, 1990.

[205] P. Spirtes, C. Glymour, and R. Scheines. *Causation, Prediction, and Search*. The MIT Press, Cambridge, MA, 2000.

[206] C. Starmer. Developments in non-expected utility theory: the hunt for a descriptive theory of choice under risk. *Journal of Economic Literature*, 38:332–382, 2000.

[207] H. Sundqvist, R. A. Howard, J. E. Matheson, and D. W. North. Hurricane seeding analysis. *Science*, 181:1072–1073, 1973.

[208] G. Szpiro. Measuring risk aversion: an alternative approach. *The Review of Economics and Statistics*, 68:156–159, 1986.

[209] A. Tversky and D. Kahneman. Rational choice and the framing of decisions. *The Journal of Business*, 59:S251–S278, 1986.

[210] A. Tversky and D. Kahneman. Advances in prospect theory: Cumulative representation of uncertainty. *Journal of Risk and Uncertainty*, 5:297–323, 1992.

[211] T. J. VanderWeele. *Explanation in Causal Inference: Methods for Mediation and Interaction*. Oxford University Press, New York, 2015.

[212] G. Vitali. *Sul problema della misura dei gruppi di punti di una retta*. Gamberini e Parmeggiani, Bologna, IT, 1905.

[213] J. von Neumann and O. Morgenstern. *Theory of Games and Economic Behavior*. John Wiley and Sons, New York, 1944.

[214] P. Wakker and D. Deneffe. Eliciting Von Neumann-Morgenstern utilities when probabilities are distorted or unknown. *Management Science*, 42:1131–1150, 1996.

[215] P. P. Wakker. Uncertainty aversion: a discussion of critical issues in health economics. *Health Economics*, 9:261–263, 2000.

[216] P. P. Wakker, I. Erev, and E. U. Weber. Comonotonic independence: the critical test between classical and rank-dependent utility theories. *Journal of Risk and Uncertainty*, 9:195–230, 1994.

[217] A. Wald. Statistical decision functions. *The Annals of Mathematical Statistics*, 20:165–205, 1949.

[218] M. C. Weinstein, H. Feinberg, A. S. Elstein, H. S. Frazier, D. Neuhauser, R. R. Neutra, and B. J. McNeil. *Clinical Decision Analysis*. Saunders, Philapdelphia, PA, 1980.

[219] H. White and K. Chalak. A unified framework for defining and identifying causal effects. Technical Report, Internal Report, Italian Radiotelevision RAI, University of California San Diego (CA), 2006.

[220] J. Woodward. *Making Things Happen. A Theory of Causal Explanation*. Oxford University Press, Oxford, UK, 2003.

[221] S. Wright. Correlation and causation. *Journal of Agricultural Research*, 20:557–585, 1921.

[222] G. U. Yule. An investigation into the causes of changes in pauperism in England, chiefly during the last two intercensal decades (part i). *Journal of the Royal Statistical Society*, 62:249–295, 1899.

[223] A. Zellner. *An Introduction to Bayesian Inference in Econometrics*. John Wiley and Sons, New York, 1971.

Index

Printed in the United States
by Baker & Taylor Publisher Services